Guidelines for
Safe Automation
of Chemical Processes

Publications Available from the

CENTER FOR CHEMICAL PROCESS SAFETY
of the
AMERICAN INSTITUTE OF CHEMICAL ENGINEERS

Guidelines for Safe Automation of Chemical Processes
Guidelines for Engineering Design for Process Safety
Guidelines for Auditing Process Safety Management Systems
Guidelines for Investigating Chemical Process Incidents
Guidelines for Hazard Evaluation Procedures, Second Edition with Worked
    Examples
Plant Guidelines for Technical Management of Chemical Process Safety
Guidelines for Technical Management of Chemical Process Safety
Guidelines for Chemical Process Quantitative Risk Analysis
Guidelines for Process Equipment Reliability Data, with Data Tables
Guidelines for Vapor Release Mitigation
Guidelines for Safe Storage and Handling of High Toxic Hazard Materials
Guidelines for Use of Vapor Cloud Dispersion Models
Safety, Health, and Loss prevention in Chemical Processes: Problems for
    Undergraduate Engineering Curricula
Safety, Health, and Loss prevention in Chemical Processes: Problems for
    Undergraduate Engineering Curricula—Instructor's Guide
Workbook of Test Cases for Vapor Cloud Source Dispersion Models
Proceedings of the International Process Safety Management Conference
    and Workshop, 1993
Proceedings of the International Conference on Hazard Identification and
    Risk Analysis, Human Factors, and Human Reliability in Process
    Safety, 1992
Proceedings of the International Conference/Workshop on Modeling and
    Mitigating the Consequences of Accidental Releases of Hazardous
    Materials, 1991.
Proceedings of the International Symposium on Runaway Reactions, 1989
Proceedings of the International Conference on Vapor Cloud Modeling,
    1987
Proceedings of the International Symposium on Preventing Major Chemical
    Accidents, 1987
1991 CCPS / AIChE Directory of Chemical Process Safety Services
Audiotapes and Materials from Workshops at the International Conference
    on Chemical Process Safety Management, 1991
Electronic Chemical Process Quantitative Risk Analysis Bibliography

Guidelines for Safe Automation of Chemical Processes

CENTER FOR CHEMICAL PROCESS SAFETY

of the

AMERICAN INSTITUTE OF CHEMICAL ENGINEERS

345 East 47th Street, New York, New York 10017

TP
155.7
G85
1993

Library of Congress Cataloging-in Publication Data
Guidelines for safe automation of chemical processes.
     p.   cm.
  Includes bibliographical references and index.
  ISBN 0–8169–0554–1
    1. Chemical process industries—Automation—Safety measures.
I. American Institute of Chemical Engineers. Center for Chemical
Process Safety.
TP155.7.G85 1992
660′ .2815′0289—dc20   91–46144
                           CIP

CONTENTS

LIST OF FIGURES

LIST OF TABLES

PREFACE

The American Institute of Chemical Engineers (AIChE) has a 30-year history of involvement with process safety for chemical processing plants. Through its strong ties with process designers, builders, operators, safety professionals and academia, the AIChE has enhanced communication and fostered improvement in the high safety standards of the industry. AIChE publications and symposia have become an information resource for the chemical engineering profession on the causes of accidents and means of prevention.

The Center for Chemical Process Safety (CCPS) was established in 1985 by the AIChE to develop and disseminate technical information for use in the prevention of major chemical accidents. The CCPS is supported by a diverse group of industrial sponsors in the chemical process industry and related industries who provide the necessary funding and professional guidance for its projects. The CCPS Technical Steering Committee and the individual technical subcommittees overseeing individual projects are staffed by representatives of sponsoring companies. The first CCPS/AIChE project was the preparation of *Guidelines for Hazard Evaluation Procedures*. Since that time, a number of other *Guidelines* have been produced. One of the projects initiated in 1987 has led to the publication of this book, *Guidelines for Safe Automation of Chemical Processes*.

The chemical industry today is becoming increasingly automated with the advent of programmable electronic systems for measurement and control. While increased automation, including alarm systems, controls and interlocks, reduces the potential for operator error, new types of faults may occur in the automated control system. The complexity of computer-based instrumentation systems increases the potential for design and maintenance errors. To ensure a high level of safety, plant management, experienced operations personnel, process specialists, safety professionals and electrical/instrumentation/control design specialists must all collaborate intelligently and with a general knowledge of the key safety philosophies and issues. The goal of this book is to provide the designers and operators of chemical process facilities with a general philosophy and approach to the safe automation of chemical processes.

While these *Guidelines for Safe Automation of Chemical Processes* are intended for use by control system specialists, they also include important information on how control system design and operation is integrated with process design, safety evaluations and plant management. Further, we expect that those primarily concerned with process design or with safety assessments will also gain from this book a better understanding of the strengths and pitfalls associated with chemical process automation. The engineer who needs a deeper understanding of the details of the discipline can consult the literature cited in the chapter references.

These *Guidelines* are a technical document intended for use by engineers and other persons familiar with the manufacture and use of chemicals. They are not a standard and make no attempt to cover the detailed legal requirements that apply to the construction and operation of chemical processing facilities. Meeting such requirements is a minimum basis for design and operation of all facilities and for satisfying general good practices reflected in industry and corporate standards. Over and above minimal compliance with safety regulations, each organization ultimately has the responsibility for developing its own safety philosophy and practices; thus, there are many alternate routes to achieving a high level of safety.

The CCPS Process Control Safety Subcommittee has co-authored these *Guidelines*, attempting to represent the spectrum of practices used by leading companies in the industry. The final draft of the book has been reviewed by CCPS sponsor organizations.

ACKNOWLEDGMENTS

The American Institute of Chemical Engineers (AIChE) wishes to thank the Center for Chemical Process Safety (CCPS) and those involved in its operation, including its many sponsors whose funding and technical support made this project possible. Particular thanks are due to the members of the CCPS Process Control Subcommittee who collaborated to write this book. Their dedicated efforts, technical contributions and guidance in the preparation of this book are sincerely appreciated.

The members of the Process Control Safety Subcommittee are:

Thomas G. Lagana, CHAIR, *Hercules Incorporated*
Suzanne M. Kleinhans, VICE-CHAIR, *Merck and Company*
Elio A. Comello, *Esso Chemical Canada/Exxon Chemical Company*
Thomas G. Fisher, *The Lubrizol Corporation*
Dallas L. Green, *Rohm and Haas Company*
Charles D. Hardin, *Hoechst Celanese Chemical Group*
Victor J. Maggioli, *E. I. du Pont de Nemours and Company (retired)*
Richard F. Palas, *Amoco Corporation*
Clark Thurston, *Union Carbide Chemicals and Plastics Company, Inc.*
Stanley Weiner, *Monsanto Chemical Company (retired)*

Former members were:
 Lynn W. Craig, *Rohm and Haas Company*
 Richard J. Lasher, *Exxon Chemicals USA*
 Ronnie L. Powell, *Union Carbide Corporation*
 Chris Smith, PAST CHAIR, *Rhône Poulenc Company*

Thomas W. Carmody, Edward L. Mullen and Elisabeth M. Drake (who also participated in the writing) of the AIChE/CCPS staff were responsible for the overall administration and coordination of this project.

The members of the CCPS Process Control Safety Subcommittee wish to thank their employers for providing time and support to participate in this project. Also, we thank many colleagues for their assistance as well as all those CCPS sponsors and members of the Technical Steering Committee who reviewed and critiqued this book prior to publication. Special thanks are due to the Du Pont Board for Process Control, to the Du Pont Technical editing team of R. J. Gardner, R. P. Grehofsky, W. H. Johnson Jr., C. G. Langford, and to Phyllis J. Drupieski who prepared the corrected final manuscript.

Russell H. Till managed production and design of the book.

GLOSSARY

ALPHA-TEST The first functional test of a device or system.

AUDIT(PROCESS SAFETY AUDIT) An inspection of plant/process unit, drawings, procedures, emergency plan, and/or management systems to verify compliance with safety standards.

AVAILABILITY The probability that a system will be able to perform its designated function when required for use (e.g, that an IPL performs its protection function when a demand is placed on the IPL). Another term frequently used is *Probability of Failure on Demand* (PFD). Availability = 1 – PFD.

BASIC PROCESS CONTROL SYSTEM (BPCS) The control equipment and system which is installed to regulate normal production functions.

BPCS CONTROLLER A controller dedicated to performing BPCS functions.

BETA-TEST A device or system test comprising production components that have satisfactorily passed an alpha-test. The scale of the test is such that the equipment is installed in an operating environment and its operation is monitored to determine performance against expectations.

CHEMICAL PROCESS QUANTITATIVE RISK ANALYSIS (CPQRA) The process of hazard identification followed by numerical evaluation of incident consequences and frequencies, and their combination into an overall measure of risk when applied to the chemical process industry.

COMMON MODE FAILURE A single failure that directly affects multiple systems in ways that are not consequences of each other.

COMMON CAUSE FAILURE A single condition (e.g., corrosive environment) that slowly degrades multiple systems in ways that may eventually lead to failures.

CONFIGURATION (1) Interconnected equipment forming a system. (2) Application programming using vendor-embedded software.

DEMAND A condition or event that requires a protective system or device to take appropriate action to prevent or mitigate a hazard.

DIAGNOSTIC PROGRAM A troubleshooting program for identifyir:g hardware or software malfunctions.

DIVERSITY A number of independent and different means to perform the same overall protective function.

EMERGENCY SHUTDOWN SYSTEM A term that is being superseded by *Safety Interlock System.*

FAIL-SAFE A concept that defines the failure direction of a component/system as a result of specific malfunctions. The failure direction is toward a safer or less hazardous condition.

FAIL-TO-DANGER An equipment fault which inhibits or delays actions to achieve a safe operational state should a demand occur. The fail-to-danger fault has a direct and detrimental effect on safety.

FAULT TOLERANCE That property of a system which permits it to carry out its assigned function in the presence of one or more faults in the hardware or software.

FINAL CONTROL ELEMENT A device that manipulates a process variable to achieve control.

HAZARD A chemical or physical condition that has the potential for causing damage to people, property, or the environment.

HUMAN ERROR Mistakes by people, such as designers, engineers, operators, maintenance personnel, or managers that may contribute to or result in undesired events.

HUMAN/MACHINE INTERFACE (HMI) The operators' windows to monitoring and keys, knobs, switches, etc. for making adjustments in the process.

INDEPENDENT PROTECTION LAYER (IPL) A system or subsystem specifically designed to reduce the likelihood or severity of the impact of an identified hazardous event by a large factor, i.e. at least by a 100 fold reduction in likelihood. An IPL must be independent of other protection layers associated with the identified hazardous event, as well as dependable, and auditable. (See Section 2.2.2.)

INTEGRITY LEVEL An indicator of SIS performance as measured by PFD (i.e., 1 – availability). (See Section 2.3.4.2.)

INTERLOCK SYSTEM A system that, in response to a predetermined condition, initiates or prevents a predefined action.

LOGIC SOLVER The portion of the SIS that performs the logic function (electromechanical relays, the CPU of a PES, etc.).

MAN/MACHINE INTERFACE (MMI) See Human/Machine Interface.

MITIGATION SYSTEM A system designed to respond to a hazardous event sequence by reducing its impact consequences.

OPERATOR'S WORKSTATION A device that allows the operator to communicate with the PES. It can be used to enter information, to request and display stored data, to actuate various preprogrammed command routines, etc. (See HMI).

PROBABILITY OF FAILURE ON DEMAND (PFD) See *Integrity Level.*

PROCESS HAZARDS ANALYSIS TEAM (PHA team) The group of operational, process, instrument/electrical/control, and safety specialists who are responsible for the safety and integrity evaluation of the process from its inception through its implementation and transfer to plant operations, to meet corporate safety guidelines.

PROGRAMMABLE ELECTRONIC SYSTEM (PES) A computer-based system connected to sensors and final control elements for the purpose of control, protection, or monitoring.

PROGRAMMABLE ELECTRONIC SYSTEM (PES) CONTROLLER A distributed control system controller, PLC system controller, single loop controller, etc.

PROGRAMMABLE LOGIC CONTROLLER (PLC) A computer, hardened for an industrial environment, for implementing specific functions such as logic, sequencing, timing, counting, and control. Although these are more commonly called *Programmable Controllers*, the acronym PLC is used in this book because PC is more commonly used in referring to a personal computer.

PROTECTION LAYER (PL) Protection layers typically involve special process designs, process equipment, administrative procedures, the basic process control system (BPCS) and/or planned responses to imminent adverse process conditions; and these responses may be either automated or initiated by human actions.

QUALITATIVE METHODS Methods of design and evaluation developed through experience.

QUANTITATIVE METHODS Methods of design and evaluation based on theory and mathematical analysis.

REDUNDANCY The employment of two or more devices, each performing the same function, in order to improve reliability.

RELIABILITY Probability that a component or system will function correctly under stated conditions for a stated period of time (for an IPL, it is reflected in the impact of spurious shutdowns; for a BPCS, it is reflected in the mean successful operating period).

RISK A measure of potential human injury, environmental damage, or economic loss in terms of both incident likelihood and the magnitude of the consequent potential injury, damage, or loss.

RISK ANALYSIS The development of a risk estimate based on engineering evaluation, and mathematical techniques for combining estimates of incident consequences and frequencies.

RISK ASSESSMENT The process of making risk estimates and using the results to make decisions.

SAFETY INTERLOCK SYSTEM (SIS) An instrumented IPL whose purpose is to take the process to a safe state when predetermined conditions are violated.

SAFETY INTERLOCK SYSTEM CONTROLLER A controller dedicated to performing SIS functions.

SAFETY LAYER See *Independent Protection Layer*.

SAFETY REQUIREMENT SPECIFICATION The SRS is a compilation of information found in the PHA report, logic diagram, process technology documents, P&ID, etc., including both functional requirements and safety integrity requirements.

SEPARATION   The physical and functional isolation of all hardware and software elements. Physical separation is defined as the requirement that the basic process control function (regulatory control—BPCS) and the safety interlock function (SIS) be performed in different logic solvers. Functional separation is achieved through the elimination of common-mode failures in execution of the BPCS and the SIS functions. This may require the separation of BPCS and SIS sensors, final elements, I/O components, the logic solvers, the software operating systems, and the application programs. Some communication may be allowed between separate components and systems as long as no common mode failures can occur. (See Appendix B)

SOFTWARE LIFE CYCLE   The activities occurring during a period of time that starts when the software is conceived and ends when the software is no longer supportable for use. The software life cycle typically includes a requirements phase, development phase, test phase, integration phase, installation phase, and a maintenance phase. (See IEC/TC 65A: Software for computers in the application of industrial safety-related systems.)

SUPERVISORY CONTROL COMPUTER   A computer, with HMI display and input devices, designed for accepting BPCS status inputs and either reporting data to the operator for process manipulation or indirectly regulating the process by manipulating setpoints in the BPCS. Process computer applications realize advanced levels of control, such as economic optimization, constraint control, model-based control, override control, multivariable control, statistical process control, etc.

USER-APPROVED (UA)   Refers to PES equipment status. (See Section 3.3.2.1).

USER-APPROVED SAFETY (UAS)   Refers to PES equipment status. (See Section 3.3.2.2).

VIDEO DISPLAY UNIT (VDU)   Any of several types of HMI that use video technology.

WATCH DOG TIMER (WDT)   A timer implemented to prevent the system from looping endlessly, providing inaccurate communications, or becoming idle because of program errors or equipment faults.

ACRONYMS

AIChE	American Institute of Chemical Engineers
API	American Petroleum Institute
ASME	American Society of Mechanical Engineers
BPCS	Basic Process Control System
CSA	Canadian Standards Association
CCPS	Center for Chemical Process Safety
CMA	Chemical Manufacturers Association (U.S.)
CPI	Chemical Process Industry
CPU	Central Processing Unit
CPQRA	Chemical Process Quantitive Risk Analysis
DCS	Distributed Control System
DCSC	Distributed Control System Controller
DDC	Direct Digital Control
DIN	Deutsches Institut für Normung e.V. (The Standards Institution of Germany)
EMI	Electromagnetic Interference
ESD	Emergency Shutdown (*see* SIS)
ETA	Event Tree Analysis
FAT	Factory Acceptance Test
FDT	Fractional Dead Time (*see* PFD)
FM	Factory Mutual, Inc. (U.S.)
FMEA	Failure Modes and Effects Analysis
FMECA	Failure Modes Effect and Criticality Analysis
FSSL	Fail-Safe Solid State Logic
FTA	Fault Tree Analysis
HAZOP	Hazard and Operability Study
HMI	Human/Machine Interface
HSE	Health & Safety Executive (U.K.)
IEC	International Electrotechnical Commission (Switzerland)
IEEE	Institute of Electrical and Electronic Engineers (U.S.)
I/O	Input/Output
IPL	Independent Protection Layer
ISA	Instrument Society of America
ISO	International Organization for Standardization (Switzerland)
LAN	Local Area Network
LEL	Lower Explosion Limit
MMI	Man/Machine Interface (*see* HMI)

MTBF	Mean Time between Failures
MTDF	Mean Time to Detect Failure
MTTF	Mean Time to Failure
MTTR	Mean Time to Repair
NDFIT	Nondestructive Fault Insertion Testing
NEC	National Electric Code (U.S.)
NEMA	National Electrical Manufacturers' Association (U.S.)
NFPA	National Fire Protection Association, Inc. (U.S.)
OSHA	Occupational Safety and Health Administration (U.S.)
PES	Programmable Electronic System
PFD	Probability of Failure on Demand
PHA	Process Hazard Analysis
PI	Proportional–Integral
PID	Proportional–Integral–Derivative
P&ID	Piping and Instrumentation Diagram
PLC	Programmable Logic Controller
QA	Quality Assurance
QRA	Quantitative Risk Assessment
RBD	Reliability Block Diagram
RFI	Radiofrequency Interference
RTD	Resistance-Temperature-Detectors
SIS	Safety Interlock System
SRS	Safety Requirements Specification
SSDC	Single Station Digital Controller
SSSC	Solid State Sequence Controller
SLC	Single Loop Controller
TÜV	Technischer Überwachungs-Verein e.V. (Technical Inspection Association of Germany)
UA	User Approved
UAS	User Approved Safety
UL	Underwriters Laboratory, Inc. (U.S.)
UPS	Uninterruptible Power Supply
VDU	Video Display Unit
WDT	Watchdog Timer

1

INTRODUCTION

1.0 OBJECTIVE

The preparation of this document was sponsored by the American Institute of Chemical Engineers (AIChE) Center for Chemical Process Safety (CCPS) as a part of their continuing effort to improve the safety performance of the chemical processing industry (CPI) through education of engineers and others who design, start-up, operate, maintain, and manage chemical processing plants. In particular, this book provides guidelines for the safe application of automation systems to the control of chemical/petrochemical processes.

1.1 SCOPE

This book is directed not only toward those responsible for the design, installation, use, and maintenance of process control systems, but also to the broader community of engineers who are responsible for the safe design, operation, and management of chemical processes. In the past, instrumentation and process control systems were designed by specialists who often were not full participants in process design development. As control systems become increasingly automated and more complex, it is even more important to safe design and operation that process design engineers and instrumentation and control specialists understand each other's disciplines and work together to provide facilities where instrumentation and control systems are fully integrated with process design to provide inherently safe facilities.

These guidelines provide current information to improve safety in the application of both the Basic Process Control System (BPCS) and the Safety Interlock System (SIS). The primary emphasis is on application of Programmable Electronic Systems (PESs), but the principles may be applied to all types of control system hardware. The term "PES" applies to all types of digital control systems: Distributed Control Systems (DCSs), Programmable Logic Controllers (PLCs), Single-Station Digital Controllers (SSDCs), and other microprocessor-based equipment that may be used for control applications.

The complete control system is covered, from the field-mounted process sensors through the control modules, the Human–Machine Interface (HMI), and the final control elements. The guidelines are applicable to both new and existing process control systems.

1.2 LIMITATIONS

The discussion of safety issues in this book is limited to the direct or indirect applications of instrumentation and control devices that can prevent and/or mitigate identified unacceptable process conditions. The focus of this book is on human and environmental safety and does not stress the issue of property loss. These guidelines are not intended for the nuclear industry, nor the military. The special safety concerns related to the discrete parts manufacturing industry, materials handling industry, or the packaging industry are not addressed in this book even though they may have some applicability in the chemical/petrochemical industry. Neither do these Guidelines cover fire protection systems. This book does not provide guidelines for the identification of potentially hazardous conditions nor does it address application of prevention and/or mitigation techniques that do not involve automation. This book suggests guidelines for how to determine whether or not you need an SIS and provides guidelines for how to design one if it is needed. The reader is referred to other CCPS publications as well as other texts and standards for information covering these other topics, namely, *Guidelines for Technical Management of Chemical Process Safety, Guidelines for Chemical Process Quantitative Risk Analysis, Guidelines for Hazard Evaluation Procedures*, and *Guidelines for Safe Storage and Handling of High Toxic Hazard Materials*.

These Guidelines were written by a group of experts who are leaders in the safe automation of chemical processes. The book also has been reviewed by more than a dozen sponsor companies and organizations who support CCPS. Thus, the Guidelines in this book represent a spectrum of the current practices of industry leaders in this area.

1.3 OVERVIEW OF THE CONTENTS

Each of the seven chapters following this introduction addresses a phase of the automation process. While some elements of sound process control and automation are presented as a starting point, primary emphasis is on specific issues that impact safety, rather than operability and reliability.

There are many good references addressing basic considerations in the selection of instrumentation and application to the control of chemical processes. References are listed at the end of each chapter. The reader is encouraged to use additional sources in applying sound engineering practices to the application of automation systems.

A brief description of each chapter follows.

1.3.1 Chapter 2—The Place of Automation in Chemical Plant Safety . . . A Design Philosophy

The chemical processing industry is in transition due to worldwide competition, increasing governmental regulations, and customer demands for product consistency and purity. These changing conditions require the use of more automation and less dependence on the human element in day-to-day operations. Rapid technological changes in control systems are also introducing additional pressures and opportunities that become important considerations. The impact of these factors on the application of automation to chemical processes is discussed.

Proper application of control systems improves safety in the operation of chemical processes. The use of modern technology offers additional enhancements if properly applied. Chapter 2 offers guidelines to accomplish this for both the Basic Process Control Systems (BPCSs) and the Safety Interlock Systems (SISs). The need to integrate these systems into an overall automation system is discussed.

A model is presented that steps through the different phases that should be included in the design of automation systems. It emphasizes the need for conducting hazard analysis, performing risk assessments, and identifying the various means that will be used to prevent and/or mitigate any hazards identified. Since both the BPCS and the SIS play important roles in this effort, the factors that are recommended for consideration in their selection, design, and application are discussed. Specific details covered in succeeding chapters of the text are referenced.

The concepts of "protection layer" and "Independent Protection Layer" are introduced. Guidelines are presented for identifying and evaluating if protection layers qualify as "Independent Protection Layers" (IPLs) using a set of specific criteria. Once the protection layers are defined and the need for additional protection in the form of automated protection layers is identified, a means for determining the design criteria for the required SIS IPL is presented. The need for each company to develop specific criteria in this area is emphasized, since decisions involve judgments of acceptable risk.

Finally, this chapter presents a summary of a safety design philosophy. The guidelines contained in the following chapters are based on this philosophy.

The approach presented in this chapter illustrates how a qualitative technique can be utilized if more rigorous quantitative techniques are not appropriate.

Readers are cautioned to satisfy their own guidelines or criteria when classifying systems (i.e., establishing SIS design requirements) and to use these guidelines in making distinctions among classifications.

1.3.2 Chapter 3—Techniques for Evaluating Integrity of Process Control Systems

This chapter discusses the importance of analyzing failure modes that may exist in control system components. It also covers the various techniques that can be utilized in evaluating the effects of these failures on the overall safety of the process. While both qualitative and quantitative techniques are presented, the reader is referred to other references for specific details.

The role of a Process Hazards Analysis (PHA) team is described and a framework for risk assessment and control throughout the life cycle of a process is presented. Particular emphasis is placed on the integration of process control system safety evaluations with overall risk review and control activities.

The types of safety reviews conducted during the evolution of a process design are described with emphasis on how automated process control system safety is integrated with overall process risk mitigation.

A technique for the certification or approval of equipment for particular applications within the automation process is suggested. A procedure to accomplish this is presented and both the benefits and shortcomings are discussed.

An extensive list of references is provided for the user.

1.3.3 Chapter 4—Safety Considerations in the Selection and Design of Basic Process Control Systems

Chapter 4 gives guidance in the application of control system technology, field instrumentation (process sensors and final control elements), operator/control system interface considerations, and process controllers in the Basic Process Control System (BPCS).

The safety impact of process sensors is discussed. Safety considerations in applying single-loop (pneumatic, analog, and digital) and distributed control system controllers are discussed. The application of varying types of final-control elements (e.g., control valves) is also presented. Emphasis is on the safety aspects rather than on general application and selection guidelines, since these can be found in other texts and references.

Operator/control system interface considerations are covered from the viewpoint of information overload or adequacy of information available to the operator. Guidelines are presented for selecting and supporting various types of hardware in the BPCS.

Information is also provided relating to safety concerns in power supply, grounding and distribution systems, installation of specific components, communication considerations between systems, and the use of advanced control techniques.

1.3.4 Chapter 5—Safety Considerations in the Selection and Design of Safety Interlock Systems

Chapter 5 addresses the specific issues related to the safety interlock systems, SISs, (commonly referred to as Emergency Shutdown systems, or ESDs, by many) that may be required to improve process safety. A method of selecting the most appropriate hardware for a given system is presented, along with criteria to follow in the system design. Special requirements for SIS software are also discussed.

The concept of fail-safe design is discussed. Communication considerations that may be required to maintain SIS security are covered. The concepts of separation, redundancy and diversity are presented with discussions of their impact on the overall system integrity. Methods for integrating the SIS reliability and availability requirements to obtain acceptable system performance are discussed. Guidance is provided to determine the appropriate separation of the BPCS and the SIS in terms of hardware, software, personnel, and function.

The model developed in Chapter 2 serves as the basis for the details provided in this chapter.

1.3.5 Chapter 6—Administrative Actions to Ensure Control System Integrity

This chapter addresses both the need for and the types of administrative procedures and actions that may be required to maintain any control system in a safe operating mode for the long term. It describes the content of procedures related to documentation, maintenance, operation, security, testing, bypassing, and other areas that apply to both the BPCSs and SISs.

Special emphasis is given to the management of changes to control system design and functional logic in the BPCS and the SIS.

Suggestions are presented for minimum levels of administrative control procedures. Also, the issue of built-in "hard" controls versus "soft" procedures is addressed.

There is an emphasis on the need for written procedures rather than depending on verbal instructions, ensuring the ability to audit.

The use of simulation techniques in control system applications is briefly discussed in this chapter. Also covered is a discussion of the types of personnel and the skills required for performing work on BPCSs and SISs. The need is emphasized to have different individuals perform design work for the BPCS and the SIS in order to avoid common faults on the same process.

1.3.6 Chapter 7—An Example: These Guidelines Applied to Safe Automation of a Batch Polymerization Reactor

To demonstrate the approach and application of these guidelines to a practical design and implementation situation, we have selected instrumentation of a typical reactor for the batch polymerization of Vinyl Chloride Monomer (VCM). This example follows the model developed in Chapter 2 and works through each step to highlight the need for different approaches at different stages in the application, design, process hazard analyses, and installation of current technology for the improvement of safety in chemical processing. The design and implementation follow guidelines presented in Chapters 4, 5 and 6. The reader is encouraged to use the model as a guide in applying the necessary steps to achieve the safe design of their control and safety systems.

1.3.7 Chapter 8—The Path Forward to More Automated, Safe Chemical Plants

There are many new and promising developments on the horizon in the area of process control safety. Many of these techniques are still in the developmental stage and their use is very limited at present. This chapter offers some insight into some of the techniques that are seen as potentially beneficial in improving process safety and suggests possible ways they might be used. It also points out some areas where additional research work would be beneficial in enhancing process control safety.

1.3.8 Other Information

In addition to the information already described, this book contains a glossary, a list of standards organizations, and references at the end of each chapter pertinent to the design and application of control systems in the chemical/petrochemical process industries.

An extensive index is also included for quick reference to specific topics within the book.

Appendices are included with information on several subjects that expand upon the material in the specific chapter. These provide additional reference materials for the user in applying the principles outlined in this book.

2

THE PLACE OF AUTOMATION IN CHEMICAL PLANT SAFETY. . . A DESIGN PHILOSOPHY

2.0 CHEMICAL PLANT OPERATIONS IN TRANSITION

The chemical processing industry is in transition. Business competition from a worldwide manufacturing community, increasing government regulation of the workplace, and customers who demand documented product consistency and ever-increasing purities cannot be ignored. Changes are occurring in operating methods and means to minimize costs while variability in the manufacturing process is being reduced. In addition, health and safety concerns have led to plant designs with significantly more process measurements and automatic controllers in an attempt to minimize personnel exposure to chemicals under all conditions.

Technological Advances in Instrumentation

Fortunately, the need for change in manufacturing practices has been coupled with significant technical advances in measurement and control equipment on all fronts. Microprocessor-based process sensors and controllers plus precision-throttling control valves now make practical the implementation and maintenance of complex process control strategies. Modern control centers include video displays showing pictorial images of plant equipment and process conditions at the operator's workstation. These graphic displays displace the older process instrument panels, augmenting the operator's concepts of what is happening inside process equipment and providing assistance to the operator when infrequent responses to changing plant conditions are necessary. Powerful control algorithms can be implemented at the lowest level of these modular, digital-instrument systems; process measurements are logged, monitored for alarm conditions, and made available for trend displays as standard instrument system functions; and product quality information is collected and archived by instrument system data storage units.

7

Changing Roles for Plant Operators

The chemical plant operator is responsible for the hour-to-hour operation of the manufacturing facility. In the past, this required continuous monitoring of plant variables and frequently, the repetitive adjustment of a large number of valves. Although automatic controllers were provided to improve production and safety, each controller came with an auto/manual switch; and, frequently many of the panel-mounted controllers would be in the "manual" mode. Taking process samples, performing simple analytical techniques, and monitoring process conditions by visual inspection of tank levels, fluid color, line temperatures, etc. were once done manually. The operator was literally "in touch" with the chemical process.

In modern chemical plants, control centers, usually located external to the processing area, are being expanded to contain more and more complex instrumentation. Plant status information is provided on video displays, which are situated at individual workstations, to provide maximum information transfer between the instrumentation and the responsible plant operator. Modern plants typically have several control rooms where a small number of operators monitor large segments of the manufacturing facility. The control room operator(s) remains in communication with outside workers by radio and occasional direct contact. Although batch-process plants may still require an operator to work both in the control room and the manufacturing area, the time away from the operator's workstation is minimal, and control room coverage is usually provided by others when the operator must be absent.

The operator was the first line of defense in older plants when processing equipment failed to perform. Typical operating activities included restarting pumps when they stopped, evaluating the consequence of equipment malfunctions, and manually adjusting operating conditions when the plant parameters wandered from a known safe-operating envelope. The operator remains the primary monitor of plant operating events in the modern production facility, but the past direct operator involvement with the process is becoming less and less practical.

Changing Issues in Safe Process Control

The heavy reliance on the operator in older types of process control activities resulted in occasional incidents resulting from human error. Increasing process automation, including alarm systems and automatic control actions, reduces the potential for operator errors initiating chemical plant incidents; however, faults may occur in automated control systems as well. The complexity of computer-based instrument systems increases the likelihood of human errors in design and maintenance which prevent the proper functioning of these systems. The challenge then becomes to design integrated human–machine systems that result in improved process safety by making effective use of automated process control technology while guarding against potential new sources of control system failure.

2.1 PLANT AUTOMATION

Most process plants employ automatic control systems to achieve consistent product quality, to minimize the manual labor of the production staff, to reduce human error in doing repetitive tasks, to improve equipment availability and production efficiency, and to enhance operational safety. Control systems in a modern chemical manufacturing facility can be separated into two groups: those systems that perform basic process control functions; and those installed for process safety purposes. Typically both applications make use of the same process control technology.

2.1.1 Basic Process Control

Instrumentation for process measurement, display, and regulation is found at all chemical processing sites. These instruments and controls must be designed, used, and maintained in a prescribed manner if an acceptable level of plant operational safety is to be achieved. The control equipment that is installed to support normal production functions is referred to in this book as the Basic Process Control System or BPCS.

Although the BPCS may consist of pneumatic, hydraulic, or electronic instruments with panelboard-mounted display devices, the basic controls in most new facilities will be composed of control and display modules built using programmable, computer-based components. These programmable control system modules may be located in several different places within the plant. State-of-the-art distributed digital controllers and video-display-based operator workstations contribute to the safe operation of the process through better communication of plant status information to the operating staff and by providing more powerful and reliable control algorithms for automatic regulation of processing operations. Installation of a supervisory control computer, which functions as part of the BPCS, is frequently found in large processing facilities. These computers collect data for management reporting purposes and may perform advanced control functions, for example, unit optimization, batch reactor recipe management, or statistical process control. Often the supervisory control computer provides powerful process data analysis tools to assist the operator when responding to process disturbances; however, these computers can not achieve the integrity required of process safety systems and should not be installed as primary risk control systems.

The advantages of programmable electronic BPCSs are widely recognized, but the implementation of these controls can result in the near-total dependence on the automation system for even minimum levels of plant operation. A failure in the BPCS can be the prime cause of a serious incident. The consequence of a BPCS failure may be (1) the total loss of information display for the operator during normal plant operations; (2) inadequate and slow access to plant status data during a plant upset; (3) disrupted communication among control modules, which causes unpredictable signals to be sent to control valves; or (4) loss of signals from and to the field instrumentation and valves.

BPCSs using programmable electronic technology have proven to be reliable where both redundant components are used throughout the system and sufficient effort is spent during the design phase to eliminate common-mode failures. A simple fault in a basic control system module will quickly become apparent since each module is equipped with extensive self-diagnostics and is directly involved in regulating a dynamic process. Where redundant or fault-tolerant control module configurations are used for closed-loop control functions, a fault can often be repaired without interrupting plant operations. However, the modular component structure of the digital control system allows many different configurations (system architectures) to achieve the same functional end. Attention to operational safety considerations when designing the BPCS is essential to minimize the number and consequence of plant upsets originating when BPCS failures occur.

2.1.2 Process Safety Systems

Frequently, instrumentation and control devices are installed to perform a protective function in addition to the process regulation provided by the BPCS. Instrument systems designed for protective functions differ from those designed for process regulation. These process risk-reducing systems monitor the status of key process variables, solve process state logic equations to identify acceptable operating domains of the plant, warn the operator when abnormal conditions are imminent, and, in some cases, automatically interrupt manufacturing operations to avoid the occurrence of potentially hazardous events.

In the past, protective controls were constructed from a large number of signal trip devices and electromechanical relays, and may now include computer-based modules identified as Programmable Electronic Systems (PESs). Although PESs provide the same logic functions that have been achieved using signal trip devices and electromechanical relays, these programmable units permit implementation of more complex monitoring functions.

Protective controls are normally separated from the Basic Process Control System and are identified as the plant Safety Interlock System or SIS. The SIS monitors the plant for conditions approaching an unsafe operating state. When unsafe operating boundaries are approached, alarms are sounded and automatic actions are taken. SIS actions often reduce operating rates, pressures, temperatures, etc., since an abrupt production stoppage can contribute to equipment damage and chemical releases. In some plants, SIS action will automatically initiate a complete shutdown of processing activities. The SIS is required to take action when (1) a critical variable exceeds a specified limit, (2) an operator enters a command, or (3) a failure within the SIS system itself removes the required level of automatic protection.

The design of protective control systems has traditionally been based on the principle of fail-safe component selection. That is, each component in a SIS would be selected so that all likely control equipment failures would result in placing the associated production facility in a safe state. Instruments containing solid-state components make fail-safe design more difficult since failures at the semiconductor chip level are equally likely to be in the On or Off state. Furthermore, identification of failure states for complex PES modules containing multiple microprocessors is much more difficult than for the electromechanical devices used in the past. Failure of an instrument system to perform a BPCS function is observable in most cases, and often is easily corrected soon after it occurs; but, failure of a system installed to perform a protective task may not be apparent until a hazardous plant condition develops. Consequently, experience with process safety systems has shown that acceptable performance of SISs that include PES modules depends on the

- selection of equipment with proven performance and powerful self-diagnostic capabilities;
- use of redundancy for selective signal processing functions;
- design of a system architecture that facilitates a maintenance program of periodic proof-testing.

SISs must be protection systems with high availability, that is, they must function reliably when an abnormal condition (a demand) occurs. However, faults within the SIS system (components or logic) can interrupt normal operations, causing spurious trips. These spurious trips can cause:

- unnecessary shutdowns and start-ups;
- dangerous shutdowns and start-ups;
- loss of pre-alarm information;
- confusion resulting in improper, dangerous responses.

Consequently, selection of highly reliable components and the extensive use of redundant circuits are necessary to reduce the frequency of these SIS-caused process interruptions. The goal of this text is to document design and implementation methodologies that result in SISs characterized by both high availability and operating reliability.

2.1.3 Total Process Control System

The total process control system, shown in Figure 2.1, will frequently consist of both BPCS and SIS components integrated at the communication level to provide information to the operating staff in a central control room. The modular structure of modern distributed control systems permits wide variations in control system architectures. The Figure 2.1 arrangement emphasizes the separation of the SIS controller from the regulatory control provided by

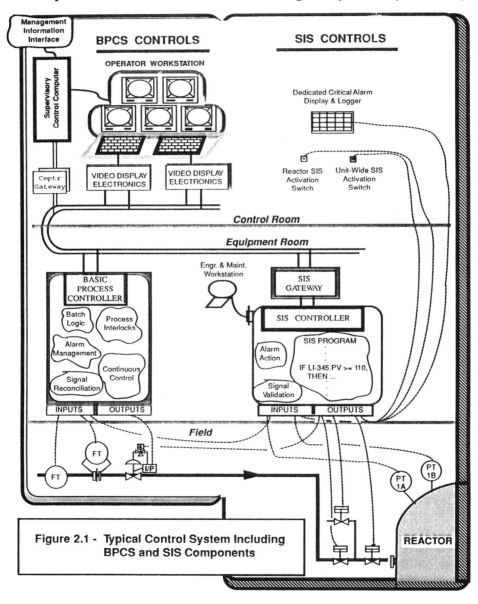

Figure 2.1 - Typical Control System Including BPCS and SIS Components

the BPCS. The design guidelines contained in Chapter 4 address the safety considerations applicable in the design of the BPCS; and Chapter 5 provides design information that focuses on the SIS. Both BPCS and SIS controls make important contributions to chemical plant safety.

2.2 A FRAMEWORK FOR CHEMICAL PROCESS SAFETY

Chemical plant development typically starts with the definition of production requirements. Next competing production technologies are identified and evaluated with respect to controlling criteria (economics, reliability, regulatory and corporate requirements, acceptable risks, etc.). Selecting a preferred process technology includes identifying hazards and making some qualitative and quantitative judgments on process risk control methods.

Many chemical processes contain concentrated energy sources and process materials that under abnormal conditions can be harmful to the environment, people, and process equipment. The AIChE/CCPS charter focuses on chemical plant risks related to the potential release of hazardous materials that could affect public health or safety. These Safe Automation Guidelines, which provide guidance in the design and operation of automated process control systems, are one important element in achieving safe operations. Plants with the potential to cause harm should be constructed with multiple, independent protection layers which are designed to avoid and to mitigate the harmful effects of each potential hazardous event. Instrumentation both in the form of the BPCS regulatory controls and the SIS interlocks play important roles in many protection strategies.

2.2.1 Protection Layers

Within each process, risk reduction should begin with the most fundamental elements of process design: selection of the process itself, the choice of the site, and decisions about hazardous inventories and plant layout. Maintaining minimum inventories of hazardous chemicals; installing piping and heat exchange systems that physically prevent the inadvertent mixing of reactive chemicals; selecting heavy-walled vessels that can withstand the maximum possible process pressures; and selecting a heating medium with maximum temperature less than the decomposition temperatures of process chemicals are all process design decisions that reduce operational risks. This focus on risk reduction by careful selection of the basic process operating parameters is a key step in the design of a safe process. A further search for ways to eliminate hazards and to apply inherently safe design practices in the process development activity is recommended. Unfortunately, even after this design philosophy has been applied to the fullest extent, potential hazards still remain in many chemical processing facilities; and additional protective measures must be applied to control risk.

The application of multiple protection layers in an operating facility is illustrated in Figure 2.2. In this figure, each layer of protection consists of a grouping of equipment and/or administrative controls that function in concert with other layers to control process risk. After a basic process is chosen, the detailed process design provides the first level of protection. Next comes the BPCS, supplemented by operator supervision, with a further layer provided by the alarm system and operator-initiated corrective actions. An automatic safety interlock system (SIS) is installed to take corrective action when

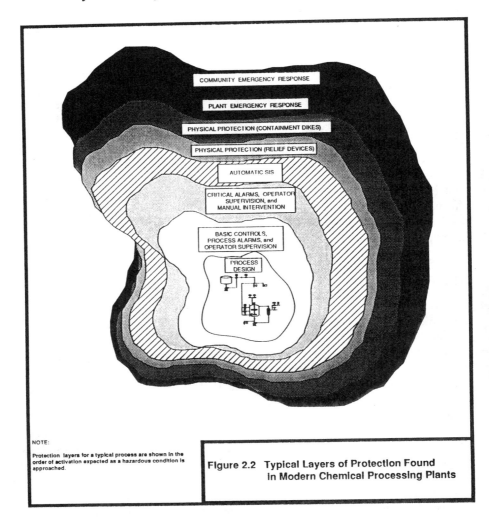

NOTE:

Protection layers for a typical process are shown in the order of activation expected as a hazardous condition is approached.

Figure 2.2   Typical Layers of Protection Found in Modern Chemical Processing Plants

failures occur in the process and BPCS protection layers. Physical protection is incorporated in the form of venting devices to prevent equipment failure from over pressure. Should all of these protection layers fail to function and a release occurs, a dike is present to contain liquid spills and emergency response plans are in place at the plant and in the community to further address the hazardous event.

The protection layers shown are in the order of activation during an escalating incident. The number and type of protection layers provided to address a specific hazard are determined by the application of company criteria and guidelines along with professional judgment. A **passive** protection layer mitigates potential hazards by virtue of design decisions (equipment selection, plant layout, etc.). An **active** protection layer initiates specific action when a hazardous event is likely. (Instrumentation is often part of an active protection layer.) In most cases, multiple protection layers are implemented for each identified process hazard.

Protection layers may involve special process designs, process equipment, administrative procedures, and planned responses to imminent adverse process conditions; and these responses may be either automated or initiated by human actions. Administrative procedures directing staff activities (which often include reliance on process instruments and alarms) can serve as a protection layer. Examples of these include standard operating procedures, process unit emergency procedures, preventive maintenance programs, plant emergency response procedures, and community emergency response procedures. Administrative controls may be applied both to reduce the frequency and the severity of a potentially hazardous event. Although these controls can be effective, they do depend on the timely and correct implementation of procedures by people. Safety-related procedures require a separate audit activity to verify that the procedures are effective. Often the protective strategy selected for a processing facility will include process surveillance and automatic corrective action provided by a SIS. In other instances, however, protective measures may depend both on instruments and operator action (e.g., manual intervention following an alarm system warning). The integrity of each instrumented protection layer is determined by the type of instrumentation selected, the number of measurements and critical alarms, the degree of dependence on the operator, the structure of the SIS installed, the maintainability of the instrument system, and the effectiveness of that maintenance.

Most failures in well-designed and operated chemical processes are contained by the first one or two protection layers. The middle levels guard against major releases and the outermost layers provide mitigation response to very unlikely major events. For major hazard potential, even more layers than those shown in Figure 2.2 may be necessary.

In summary, a protection layer is a distinct part of the process and plant design that is intended to avoid the occurrence or to reduce the effect of a specific hazardous event. When significant hazards cannot be avoided by inherently safe process and process equipment selection, instrumented protective functions assume greater importance. Assurance becomes necessary that these protection layers function so that a failure of one of the inner layers does not disrupt the effectiveness of an outer, backup layer. We use the term, "Independent Protection Layer," to indicate protective systems that are designed to prevent or to mitigate identified events and that meet tests of specificity, independence, dependability, and auditability.

2.2.2 Independent Protection Layers

Most chemical processes are designed with multiple protection layers, strategically placed to address each major process hazard; and many of these protection layers include instrumentation. When process-related threats to people and the environment are the source of substantial risks, the integrity required of these layers increases. Protective controls capable of addressing these higher risk levels are defined in this book as Independent Protection Layers, IPLs. A protection layer or combination of protection layers qualifies as an IPL when

1. the protection provided reduces the identified risk by a large amount, that is, at least by a 100-fold reduction.
2. the protective function is provided with a high degree of availability—.99 or greater.
3. it has the following important characteristics:
    —*Specificity:* An IPL is designed solely to prevent or to mitigate the consequences of one potentially hazardous event (e.g., a runaway reaction, release of toxic material, a loss of containment, or a fire). Multiple causes may lead to the same hazardous event, and therefore multiple event scenarios may initiate action of one IPL.
    —*Independence:* An IPL is independent of the other protection layers associated with the identified danger.
    —*Dependability:* It can be counted on to do what it was designed to do. Both random and systematic failures modes are addressed in the design.
    —*Auditability:* It is are designed to facilitate regular validation of the protective functions. Functional testing and maintenance of the safety system is necessary.

Only those protection layers that meet the tests of availability, specificity, independence, dependability, and auditability are classified as **Independent Protection Layers.**[1]

1. The BPCS often functions as a protective layer, but this multipurpose instrument system fails the test of specificity as the BPCS's first purpose is to regulate the process in day-to-day operation. Modifications are routinely made to the BPCS; in many facilities, this is done without extensive after-change validation of control system functional integrity. Therefore the BPCS should not generally be considered an IPL.

2.2.3 Safety Interlock Protection Layers

Automation equipment is an integral part of many different protection layers. When instrumented protection layers are advanced to the level of Independent Protection Layers, these control systems are identified as the SIS (illustrated in Figure 2.1) and are given special design and maintenance attention to achieve the protective performance characteristics outlined in Section 2.2.2. SISs are designed using instruments that employ a wide range of measurement and control technologies. Mechanical and computer-based field-mounted sensors may be used; PES devices, direct-wired relays, and solid state logic components are used alone or in combination as the SIS logic solver; and output signals are sent both to throttling control devices and block valves. Normally, the logic solver(s) are separated from similar components in the BPCS. Furthermore, SIS input sensors and final control elements are generally separate from similar components in the BPCS, but some sharing of process sensors and final control elements is acceptable to many chemical companies if failure of the sensor or final control element does not cause an initiating event. (See Section 2.3.3.)

The installed effectiveness (i.e., the integrity) of an instrumented IPL is determined by SIS architecture; by equipment selection and other design decisions of the control engineer; by installation procedures; and later, by the attention of the plant staff in periodic testing and maintenance of the layer. SIS interlock integrity can be described in qualitative or quantitative terms, and both methodologies can be used in the design of these systems. Comparison of a proposed or existing SIS with a set of design standards, known to provide an acceptable level of safety performance based on existing operating experience, provides a qualitative measure of protection layers integrity. Quantitative analysis of a proposed or existing SIS can be made to identify the level of availability and the frequency of spurious process interruptions that can be expected. Then this expected performance of the SIS can be compared with the quantified risk level required for a specific application.

The level of protection provided by a protection layer can be regulated through strict adherence to design and maintenance guidelines. Acceptable performance of SISs that include PES components has been achieved by many chemical processing companies, and some have published company-specific guidelines that are used to direct the design of these safety systems (e.g., Du Pont, 1986; and Monsanto, 1986). These documents have two common characteristics: (1) grouping of chemical plant operating risks into a few categories listed in order of severity and (2) definition of SIS designs (equipment selection, system architecture, and maintenance practices) acceptable for each risk category. A similar methodology is developed in this text to illustrate the application of good engineering practice in the design of SISs.

2.3 CHEMICAL PLANT SAFETY SYSTEM DEVELOPMENT

This section contains a brief description of the tasks associated with the identification of process hazards, the definition of Basic Process Controls, the assessment of process risks, and a methodology for the design of protection layers that include instrumentation. Figure 2.3 introduces a model of the chemical plant safety systems development activity in which the major tasks and work sequence recommended for the complete design of an instrumented safety system are identified.

Activities described by this model apply to new chemical process designs, to modifications of existing processes, and to facility modernizations in which control system changes are made. The entry point will vary depending on job requirements. Management of Change procedures (described in Chapter 6, Section 6.1.7) apply to all modifications of the BPCS and SIS after the original control system has been installed and the safety performance of the SIS has been validated to be in compliance with the Safety Requirements Specification. Following is a more detailed look at the sequence of activities recommended for the design of an instrumented safety system.

2.3.1 Process Hazards Analysis Team

Early in the development of a chemical process, a group should be identified to guide the process development and subsequent production facility design from a risk identification and control viewpoint. This Process Hazards Analysis (PHA) Team (or equivalent group) requires participants knowledgeable in process design, hazard assessment, chemical plant instrumentation, plant operations, and maintenance. The PHA Team needs to be given initial guidelines by the project and plant management concerning overall safety standards and goals as well as any other special requirements. Although good communication among the PHA reviewers and facility designers is very important, some members of the team should not be closely involved in the detailed design of the new facility to assure an independent assessment of risks created by the new process.

2.3.2 Process Hazards Definition

First the chemistry must be defined and the inherent hazards characteristic of a process must be identified in order to assess and control process risks effectively. Changes to an existing process should be subject to the same hazard identification procedure that is followed for new processes. Activities required in the first phase of the design process that may result in the installation of an SIS are shown in Figure 2.3–1.

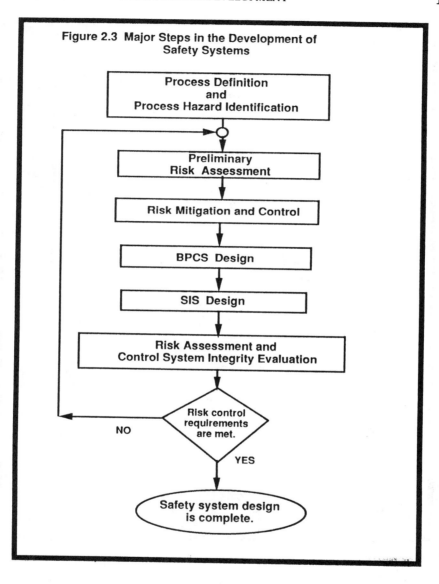

Figure 2.3 Major Steps in the Development of Safety Systems

Detailed information is required in order to identify process hazards [Ref. 2.4]. Typically needed for the analysis are: the chemical, physical, and toxicological properties of all chemicals handled; reactive chemical data and flammability properties for all chemicals including explosive characteristics for energetic materials; an approximate heat and material balance for the process; a preliminary definition of process equipment; and a site layout.

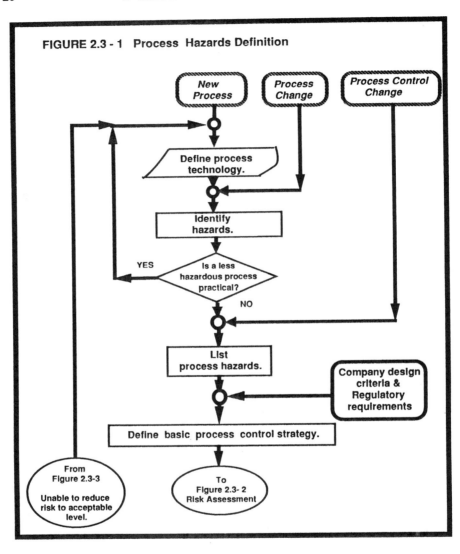

FIGURE 2.3 - 1   Process Hazards Definition

After the hazards of the specific process are identified, process modifications that would reduce operational risk are explored. Reduction in storage of potentially hazardous chemicals, unit operations with small in-process inventories, reduced operating temperatures and pressures, and change of solvents and feedstocks to less hazardous materials are all examples of modifications that can increase the inherent safety of a process. Some hazards may remain after all practical alternative process designs have been considered.

Next a basic process control strategy is developed (or reviewed and evaluated if an existing process is under study) as part of operability and process definition reviews. This basic strategy consists of both flow sheet control concepts and a control equipment layout plan. Basic control concepts include definition of control loops for regulation of material, energy, and momentum balances; process and product inventories; and product quality parameters. These automatic controllers not only maintain process conditions at the desired target conditions but also prevent hazardous process conditions from occurring. Chapter 4 describes in detail the safety considerations that should guide the selection of the BPCS and the preparation of a control equipment layout plan.

The process hazard definition and risk control work process includes several recycle paths (Figures 2.3–1, 2.3–3) in which the basic process design, process and control equipment selection, and control strategy may be modified to minimize inherent process hazards. Process and equipment changes are made first to minimize reliance on additional layers of risk control measures which tend to be complex and more dependent on user maintenance procedures.

2.3.3 Risk Assessment

Process risks can be evaluated after the process hazards have been identified and the basic process control strategy has been developed. Process risks are normally assessed by considering both hazardous event consequence (severity) and probability (likelihood). This work process is shown in Figure 2.3–2.

Methods to identify hazardous events (such as "What-If," Hazard and Operability studies (HAZOP), Failure Modes and Effects Analysis (FMEA), and Preliminary Hazards Assessment) are described in the CCPS Guidelines series [Ref. 2.1, 2.4].

Identifying the consequence of hazardous events requires assessment of site-specific conditions (population density, in-plant traffic patterns, meteorological data, etc.). Several methods are available for estimating hazardous event impact areas. Vapor cloud dispersion modeling techniques and other simulation studies may be required [Ref. 2.5].

Hazardous event likelihood can be established either by qualitative or quantitative techniques; and in some cases event likelihood can be determined from historical data. Fault Tree Analyses (FTA), which requires failure rate data, can be used to calculate event likelihood. Equipment manufacturers, government-sponsored research projects, and many professional groups have collected reliability data. In addition, the CCPS-published *Guidelines for Process Equipment Reliability Data* and *Guidelines for Human Reliability Data* provide selected reliability data that describe failure rates for process and control equipment and for human performance.

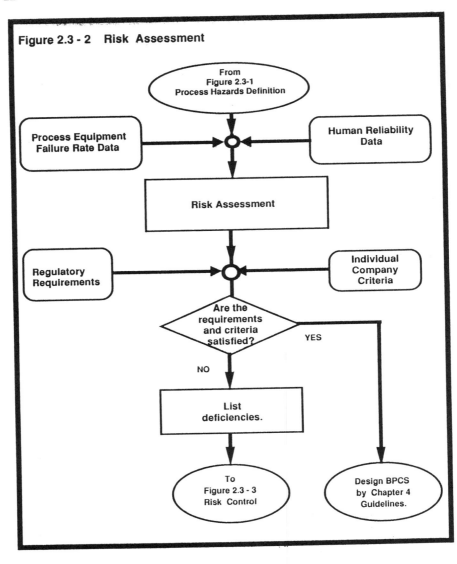

Figure 2.3 - 2   Risk Assessment

Risks associated with process operations can be ranked by considering numerical parameters linked to event severity and likelihood. The hazard and risk assessment work process shown in Figure 2.3–2 results in a summary of process risks (risks remaining after all process modifications are made and the BPCS is defined) that require further risk reduction. This work directly leads to the specification of risk reduction measures (including SISs) that are able to reduce the likelihood that an identified hazardous event will occur.

2.3.4 Risk Control

The prevention and mitigation measures that are appropriate for each risk can be specified after the hazardous event likelihood and consequence are understood. A qualitative methodology for linking hazardous event severity and likelihood with the integrity requirements for an instrumented protection layer is introduced in this section. An interlock design classification system is proposed for use in communicating SIS performance requirements. Detailed design guidelines that support this risk-based SIS interlock specification methodology are presented in Chapter 5. As an alternative, SIS performance specifications may be stated in quantitative terms which are described in Chapter 3.

The risk control work process which includes the integrity validation of the resultant SIS is described in Figures 2.3–3 and 2.3–4. Risk reduction measures specified by the PHA Team that rely on process control systems are documented in a Safety Requirements Specification, SRS. The SRS is a compilation of information found in the PHA report, logic diagram, process technology documents, P&ID, etc. The SRS should contain both a list of SIS interlocks and critical alarms as well as specific installation considerations and functional test requirements for the safety-related controls. In addition, reliability and availability expectations of the SIS are included in the specification. This specification, which is used as a key reference document during the SIS design, is the standard against which the SIS is first validated and then maintained.

2.3.4.1. Risk Reduction Process

Most company's safety procedures include a methodology for classifying potential hazards according to relative risk. Typically a qualitative risk-ranking methodology utilizes a company-specific ranking matrix that encompasses the two dimensions of risk: (1) hazardous event likelihood (probability) and (2) event severity (consequence). An example of a process risk-ranking model is summarized in Figure 2.4 [Ref. 2.6.3]. This model includes descriptors of hazardous event likelihood and severity that are used in this section and in Chapter 3 to develop a risk-based methodology for the design of SISs. The SIS design methodology developed here may require adjustment to be compatible with an individual company's guidelines and methods.

The PHA team (a multidisciplinary group) has the task of identifying the potential hazards and then rating each hazard against the two risk dimensions of the matrix. Event severity is based on potential release quantities and duration, and makes use of accepted consequence estimation techniques for hazardous vapor dispersion, fire, explosion, etc. Event likelihood is based on estimated frequencies and rough probability estimates tied to the number of independent elements or operations that must fail for an incident to occur.

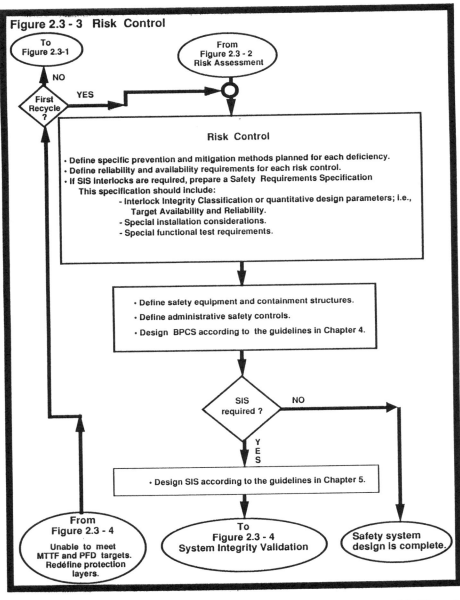

Figure 2.3 - 3   Risk Control

Estimates of these failures can be combined to arrive at an event likelihood. For example, the likelihood of a human error under low stress conditions is in the order of 1 error per 100 demands. An operation requiring ten correct human operations per year results in a potential demand on the protection system of 0.1 potential events per year. If the process (assume that an over

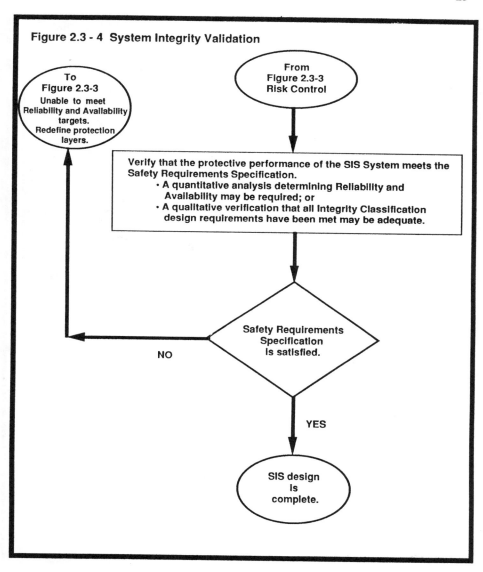

Figure 2.3 - 4 System Integrity Validation

From
Figure 2.3-3
Risk Control

To
Figure 2.3-3
Unable to meet
Reliability and Availability
targets.
Redefine protection
layers.

Verify that the protective performance of the SIS System meets the
Safety Requirements Specification.
• A quantitative analysis determining Reliability and
  Availability may be required; or
• A qualitative verification that all Integrity Classification
  design requirements have been met may be adequate.

Safety Requirements
Specification
is satisfied.

NO

YES

SIS design
is
complete.

pressure and equipment rupture is the hazard) includes a specific pressure
relief valve which has an availability of 0.999, then the likelihood of an
overpressure hazardous event is 0.0001 per year:

0.1 potential events x (1 − 0.999) probability of a protection failure.

SIS interlocks often are identified as risk reduction measures in the preliminary risk control work of the PHA Team. Consequently estimates of SIS

EVENT SEVERITY

Minor Incident = Impact initially limited to local area of the event with potential for broader consequence if corrective action is not taken.

Serious Incident = One that could cause:
- Any serious injury or fatality onsite or offsite.
- Property damage of $1 million offsite or $5 million onsite.

Extensive Incident = One that is five or more times worse than a **Serious** Incident.

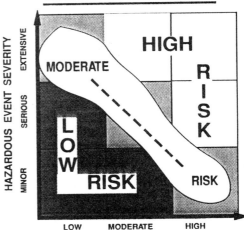

QUALITATIVE RANKING OF RISKS

HAZARDOUS EVENT SEVERITY — MINOR, SERIOUS, EXTENSIVE

MODERATE
HIGH RISK
LOW RISK
RISK

LOW   MODERATE   HIGH
UNMITIGATED
HAZARDOUS EVENT LIKELIHOOD

EVENT LIKELIHOOD

Low = A failure or series of failures with a very low probability of occurrence within the expected lifetime of the plant.
Examples: • Three or more simultaneous instrument, valve, or human failures.
• Spontaneous failure of single tanks or process vessels.

$$f < 10^{-4}, /yr$$

Moderate = A failure or series of failures with a low probability of occurrence within the expected lifetime of the plant.
Examples: • Dual instrument or valve failures.
• Combination of instrument failures and operator errors.
• Single failures of small process lines or fittings.

$$10^{-4} < f < 10^{-2}, /yr$$

High = A failure can reasonably be expected to occur within the expected lifetime of the plant.
Examples: • Process leaks.
• Single instrument or valve failures.
• Human errors that could result in material releases.

$$10^{-2} < f, /yr$$

f = Frequency of hazardous event, events per year.

Figure 2.4   Process Risk Ranking Model
(Stickles, Ozog, and Long; Arthur D. Little, Inc., Facility Major Risk Survey, AIChE Health and Safety Symposium, March 1990)

interlock fail-to-danger characteristics enter into the determination of hazardous event likelihood. Single signal-path SIS interlocks (built using standard industrial components and designed to be fail-safe) are assumed as the standard interlock design when estimating hazardous event frequencies. These single signal-path SIS interlocks typically have an availability of 0.99 [Ref. 2.8]. Other SIS interlock designs make use of redundant components and signal paths; and, consequently have higher availability than the standard single path design. *Identification of (1) the need for a protective interlock and (2) the performance requirements of an SIS is part of the risk reduction work process.*

Risk reduction requirements, which can be defined by specifying availability indices for the total protection system, often exceed that achievable using a single independent protection layer. A protection system that includes the combination of two IPLs may be expected to achieve an availability of no less than 0.9999. This level of safety system performance can only be achieved by managed design and maintenance practices that help avoid the inadvertent introduction of common failure modes. Since common mode failures are frequently of human origin with system maintenance, testing, and design being prime common failure sources [Ref. 2.7], the elimination of these failure modes is difficult to achieve in instrumented safety systems.

Potential hazards with relatively high risk rankings are the first candidates for risk-reducing actions, and frequently PHA Team recommendations include the addition of SIS protection layers. However, the simple specification of an interlock to prevent the occurrence of a potentially hazardous event is insufficient. More information is required for the control system designer and plant support staff if the requested SIS is to provide the intended level of risk reduction.

2.3.4.2. SIS Integrity Levels

Three distinct levels of safety performance can be identified and are used in this publication to structure qualitative design guidelines for instrumented protection layers. Higher interlock integrity levels will be found in some standards (Level 4, avionics and nuclear industries) but these SIS designs are generally not used for chemical plant applications. Redundant components and signal paths plus the extensive use of active diagnostics are used in the design of these interlocks to achieve varying degrees of fault tolerance to SIS component failures; and, thereby to achieve varying degrees of protection system integrity. High-integrity SIS interlock architectures may also be needed for lower process risk protection functions when assurance of process uptime (operating reliability) is a major concern. Relationships between Interlock Integrity Level assignment and the set of SIS architectures that are used in this book to illustrate the risk-linked interlock design methodology are outlined in Table 2.1 and are illustrated in Figure 2.5.

Several petrochemical companies have been using an interlock integrity identification system for internal design procedures for several years. These

Table 2.1 SIS Interlock Design Structure for Each Integrity Level

Integrity Level	Minimum Interlock Design Structure
1	*Nonredundant.* Best single path design.
2	*Partially redundant.* Redundant independent paths for elements with lower availability.
3	*Totally redundant.* Redundant, independent paths for total interlock system. Diversity should be considered and used where appropriate. A single fault of a SIS system component is highly unlikely to result in a loss of process protection.

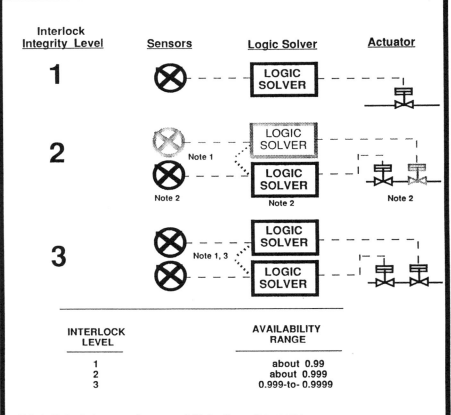

INTERLOCK LEVEL	AVAILABILITY RANGE
1	about 0.99
2	about 0.999
3	0.999-to- 0.9999

Note 1 - Redundant sensor values are available for diagnostic purposes.

Note 2 - Sensor, logic solver, and/or final element may be redundant as availability needs dictate.

Note 3 - The performance of two identical PES integrity Level 1 interlocks will not equal that of one PES integrity Level 3 SIS, because of common mode faults.

Figure 2.5 - Examples of SIS Structures

identification systems differ from company-to-company as would be expected: some use Integrity Level 0 for process control interlocks which this text assigns to the BPCS; many companies number their highest integrity level interlocks as an Integrity Level 1 or an Integrity Level A and their lowest as an Integrity Level 4 or 5 (D or E). This book uses an interlock numbering system which increases with interlock integrity because this structure is consistent with other standard development activities in the United States (ISA-SP84) and in Europe (IEC SC65A Secretariat 122 and 123).

Although this use of three SIS interlock integrity levels is somewhat arbitrary, the three integrity-level grouping permits a clear separation of system architectures, component design features, and system support requirements.

The risk reduction provided by an interlock increases with the Integrity Level (IL) number, but each interlock (IL 1, 2, and 3) is considered to be only one Independent Protection Layer in this book. An IL 3 interlock as shown in Figure 2.5 appears to be the application of redundant IL 1 SISs; however, an IL 3 interlock requires design and maintenance practices which far exceed those specified for an IL 1 interlock. The performance of two identical PES IL 1 interlocks will not equal that of one PES IL 3 SIS. Integrity Level 1 interlocks share more common mode failure possibilities than IL 3; do not have the level of active diagnostics required of IL 3 interlocks; and require a lower level of maintenance than do IL 3 interlocks. Specific design requirements (equipment technology, system architecture, fail-safe design considerations, application programming, diagnostics, HMI requirements, etc.) for each of the three SIS Integrity Levels are presented in Chapter 5.

SIS interlocks in each integrity level should be designed with "fail-safe" performance characteristics. This means that SISs include design features and an output state that signals the malfunction of an essential (SIS) component or of an energy source. "Essential components and energy sources" are those required for performing the interlock function.

When a system failure is detected, the "fail-safe" SIS may:

- Initiate the interlock action that automatically takes the process to a safe state.
- Signal the failure without being be able to continue monitoring the process until the fault is corrected.
- Initiate changes in process conditions to lower operating risks while the SIS is in a failed state.
- Initiate action that automatically signals the failure, replaces the failed component, and continues monitoring the process.

2.3.4.3. Risk Linkage to SIS Integrity Levels

A methodology that permits linking corporate management risk standards to the three SIS Integrity Levels defined in this text is shown in Figure 2.6. Each

SIS interlock, from field sensor to final control element, is designed as a function of (1) the consequence of the hazardous event, (2) the likelihood of the event, and (3) the number of other Independent Protection Layers address-ing the same cause of a hazardous event. *The minimum number of Independent Protection Layers required to address a process risk can be derived from the user company's safety policy.*

This qualitative SIS classification methodology provides the PHA team with a practical way to communicate the number and integrity of instru-mented safety layers judged appropriate for a specific process hazard. A simple application of the methodology is described in Figure 2.7. A more complete example of one method for selecting and documenting SIS interlock

Figure 2.6 Linkage of process risk to SIS integrity classifications.

APPLICATION OF THE SIS INTEGRITY CLASSIFICATION METHODOLOGY

A company plans to use the risk-linked SIS design approach defined in this book to manage the application of instrumented safety systems. The company safety policy states:

> A minimum of two Independent Protection Layers shall be provided for all identified potential hazardous events which are considered of Serious or Extensive Severity regardless of the likelihood of the occurrence.

This policy resulted from recognition that common-mode failure possibilities are inherent in all technology and that the potential for human errors cannot be eliminated from the SIS design and maintenance activity.

The following steps are taken:

Form a PHA Team

A multi-discipline team is formed to identify the risks associated with the new process and to define appropriated risk mitigation measures.

Identify Process Risks

The PHA Team reviewed the new process and identified an overpressure scenario in which a single instrument failure and an operator error would likely result in a fatality. The Risk Ranking Model of Figure 2.4 was used to categorize the event severity (**Extensive**) and likelihood (**Moderate**).

Identify Risk Control Measures

The PHA Team selects an approved pressure relief valve to protect against the occurrence of this hazardous event. The relief valve is an independent protection layer (i.e., reduces the likelihood of the event by at least 100 fold) and is required, in this case, by a regulatory agency.

Since the safety policy requires two IPLs to protect against an Extensive severity event, the PHA team decided that a high process pressure SIS should also be applied to mitigate the identified risk.

Determine the Interlock Integrity Classification

The pressure interlock design requirements are determined by the use of Figure 2.6 which requires the following data for entry into the appropriate interlock matrix:

• Event Severity	Extensive
• Event Likelihood without either IPL in service	Moderate
• Total Number IPLs (SIS I/L + Non-SIS IPLs)	2

The need for an integrity level, 2 SIS interlock is indicated from Figure 2.6. Therefore, each SIS element is analyzed and appropriate redundancy is provided to achieve 0.999 availability.

Note that if the company's safety policy had permitted the use of one IPL to mitigate this hazardous event (and if the regulatory agency had not required a relief valve), Figure 2.6 would recommend additional risk review before specifying only one safety interlock to mitigate an Extensive severity event.

Prepare Safety Requirements Specification

The PHA Team then prepares a Safety Requirements Specification that specifies

    (1) a pressure relief valve and

    (2) an integrity level 2 pressure SIS

to mitigate the identified process risk.

Figure 2.7 Application of the SIS integrity classification methodology.

integrity levels is presented in Chapter 3, Section 3.1.4. In addition the SIS-process risk linkage of Figure 2.6 is used in the control system design example contained in Chapter 7.

Design guidelines for each of these three SIS Integrity Levels are presented in Chapter 5. These guidelines address the types of logic, selections of SIS instrumentation, and implementation of alarms for each interlock integrity level. Finally, in order to achieve the needed SIS performance, site dedication is required to ensure implementation of and compliance to well-defined plant procedures that are described in Chapter 6. Alarm and interlock test procedures, management of change to the SIS, and complete documentation of these changes are all key elements in the maintenance of instrumented safety systems.

2.3.4.4. *Quantitative Design Methodologies*

Quantitative reliability assessment of Safety Interlock Systems has been described in recent publications [Ref. 2.1, 2.2, and 2.3]. Reliability block diagrams can be used to model safety systems, and methods to determine the probability that a Safety Interlock System will function when and only when needed. Signal flow diagrams are used to help identify critical paths of safety system components that must be functioning in order for the SIS to carry out a specific task. The analysis makes use of approximation methods based on Kinetic Tree Theory [Appendix E of Ref. 2.1] to simplify calculation of total safety system availability. A quantitative availability measure can readily be calculated for these automation systems. Availability is a parameter that is frequently used in defining the likelihood that a SIS will provide the needed protection when a plant initiating event occurs. Verification of the protective performance of a SIS system designed by the guidelines in Chapter 5 can be done using the techniques outlined here and further discussed in Chapter 3.

In order to utilize a quantitative methodology, target availability and reliability measures must be established for the SIS as part of the total plant risk control strategy, refer to Figures 2.3–3 and 2.3–4. An SIS architecture is selected by the control engineer, equipment vendors are identified, and a preventive maintenance program is defined. The availability of the proposed SIS system is then calculated using instrumentation reliability parameters. Once the calculated availability and reliability of the proposed SIS approximately matches the target availability, the SIS design is complete.

If the calculated availability falls significantly short of the specified value, additional modifications must be made to the proposed SIS to increase the calculated availability. When the SIS cannot be practically modified to make the calculated availability equal to the target value, the total set of protection measures originally proposed to control the risk must be reevaluated. In most cases, company criteria for risk control can be met by increasing the number of administrative, process equipment, and instrumented protective measures. Conversely, if the calculated availability for the SIS is considerably higher than the target value, an opportunity to achieve cost savings is identified. Redesign of the SIS with less redundancy would likely result in the calculated availability more closely matching the target value and would result in a cost savings. Often the higher availability may be judged worth the incremental cost because of the safety improvement.

A final quantitative check of the SIS should be made to determine if SIS-originated trips or poor SIS reliability will create potential hazards during operation. If so, modifications to the SIS are necessary. If not, the SIS design is complete.

Although quantitative methods for calculation of protective system PFD are well documented, the complexity of analysis and the inherent difficulties in establishing meaningful values of needed reliability parameters make this methodology useful only for the most rigorous safety system design efforts. A specialist trained and experienced in safety system reliability analysis is needed when a quantitative analysis is to be done.

2.4 SAFETY SYSTEM DESIGN PHILOSOPHY

The popularity, increased functionality, and wide use of PES equipment for BPCS regulatory and process interlock applications ensure the application of this equipment in plant safety systems. SISs that achieve levels of operating reliability and plant security equaling or surpassing those of protection layers built using electromechanical relay devices can be implemented using programmable electronic devices. Compared with electromechanical Safety Interlock Systems, PES SISs:

- Offer expanded operator awareness of plant conditions.
- Provide self-documentation of the installed safety logic.
- Make possible more extensive automatic avoidance of unsafe plant events.
- Provide increased diagnostics capability.

Although electromechanical and pneumatic relays will continue to be applied for small SISs and where special considerations favor the use of this proven safety system technology, the advantages of integrating the SIS into the total plant control system make PESs the technology of choice for the majority of instrumented safety systems. In many cases, SISs using PES equipment are competitive in cost with designs that use traditional SIS technologies.

PES technology offers a multitude of options to the SIS designer. Effective use of this technology in safety applications is aided by the establishment of individual company guidelines that identify the level of risk that is to be controlled by an SIS. Design criteria and guidelines for three integrity levels of SIS systems are presented. The risk-linked design methodology presented here is applicable to all technologies (both electromechanical and PES) used for instrumented safety systems.

Protection layers built with PES modules employ complex digital subsystems, which will continue to evolve as computer technology advances. Consequently, careful design and a detailed plan of plant administrative controls are essential for maintaining the integrity of these SISs. The following design and maintenance principles should be considered for the long-term, successful use of SISs containing programmable modules:

1. SIS logic solvers should be separated from and independent of BPCS logic solvers performing regulatory functions required in the normal operation of the plant.
2. Instruments and control modules selected for SISs should have proven reliability records in similar service and should be subjected to formal company approval procedures.
3. Instruments and control modules used in SISs should be applied to achieve fail-safe behavior.
4. Self-diagnostics and testing should be incorporated in the SIS design and contingency measures identified when faults are detected.
5. A master shutdown switch (or switches) independent of programmable logic should be provided. (In applications where sequential shutdown action is required, use of programmable logic to accomplish the shutdown sequence may be acceptable, subject to PHA approval.)
6. When bypasses of individual shutdown initiators are required, the following is recommended:
   —Bypass action should be alarmed and annunciated at an attended location.
   —Installation of a master bypass (a switch that bypasses multiple trip initiators simultaneously) should be avoided.
   —Output bypasses are not allowed.
7. Communication among the SIS logic solver, the BPCS, and other network devices is valuable, but the SIS should be designed with internal security to ensure that corruption of SIS programs by BPCS modules is prevented.
8. Unattended engineering workstations that permit access to the SIS application program should never be left connected to the SIS logic solver.
9. Master records of safety system logic should be maintained. Diagnostics capable of comparing and identifying differences between the running software and the master record should be provided.
10. Continuing integrity of a SIS protection layer requires the long-term commitment of plant management. Periodic testing of the protective functions, verification of SIS logic, limited access to and responsible management of the control program, and procedures for maintaining the SIS modules at appropriately tested revision levels all require special attention.

2.5 REFERENCES

2.1 CCPS/AIChE, *Guidelines for Chemical Process Quantitative Risk Analysis*, New York (1989).

2.2 Hogstad, P. and Bodsberg, L., *Reliability of Safety Systems, Models and Data.*, Foundation for Scientific and Industrial Research at the Norwegian Institute of Technology, Trondheim, Norway (1988)

2.3 *Programmable Electronic Systems in Safety Related Applications,* U. K. Health and Safety Executive, Her Majesty's Stationery Office, London (1987).

2.4 CCPS/AIChE, *Guidelines for Hazard Evaluation Procedures*, 2nd Edition with Worked Examples, New York (1992).

2.5 CCPS/AIChE, *Guidelines for Use of Vapor Cloud Dispersion Models,* New York (1989).

2.6 Stickles, Ozog, and Long, Arthur D. Little, Inc., *Facility Major Risk Survey,* AIChE Health and Safety Symposium, Orlando, Florida, March 1990

2.7 CCPS/AIChE, *Guidelines for Improved Human Performance in Process Safety,* New York (1992).

2.8 Gruhn, P., *Matching Safety Control System Performance to Process Risk Levels,* AIChE's Process Plant Safety Symposium, Houston, TX, February, 1992.

3

TECHNIQUES FOR
EVALUATING INTEGRITY OF
PROCESS CONTROL SYSTEMS

3.0 INTRODUCTION

Building on the safe design philosophy outlined in Chapter 2, this chapter presents information on techniques for assessing and verifying the safety and integrity of processes with automated Basic Process Control Systems (BPCSs) and Safety Interlock Systems (SISs). Methods used by industry to certify or approve components or systems are presented. Some of these techniques primarily focus on process safety with appropriate consideration of control systems; others principally focus on the evaluation of control system safety and integrity with appropriate consideration of relevant process and ambient conditions. An overview of some of the common techniques available for evaluating total system integrity, with emphasis on the BPCS and SIS, is given, with references to more detailed sources of information on each technique. The integrity evaluation philosophy provides background for discussions of safety considerations in the selection and design of BPCSs (Chapter 4), of SISs (Chapter 5), and administrative actions to minimize the potential for incidents due to human error (Chapter 6).

The best way to achieve safe plant operation is to have inherently safe processes, operating in correctly designed and maintained equipment. Total inherent safety, however, is not always achievable in practice. Typical plants involve a spectrum of risks ranging from minor operational problems to those that have the potential to impact the health and safety of workers or the public. Therefore, protection layers should be provided to prevent or mitigate each potentially hazardous event. Events with more serious potential hazards may require more layers of protection than events with lesser potential impact. A properly designed BPCS, operated by trained and alert operators, is one of the first lines of defense, beyond sound process design, in preventing incident-initiating events from ever occurring. Protection layers beyond the BPCS include the relief systems, dump systems, as well as SIS and accident mitigation systems (including emergency response procedures). Typical protection layers in modern chemical process plants are shown in Figure 2.2.

Taking action at the earliest step in the general sequence of events leading to a potential incident, as discussed in Chapter 2 and illustrated in Figure 3.1, minimizes the impact of an abnormal event. This requires that the protection

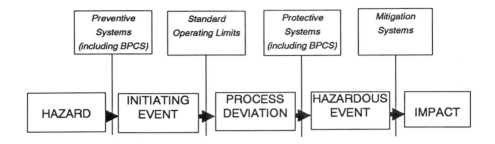

Figure 3.1 Anatomy of an incident.

layers associated with an event scenario be effective, and that they possess a sufficient degree of integrity (availability). Therefore, as indicated in Figure 2.3, risk assessment and mitigation are integral parts of the chemical process design procedure. The safety evaluation techniques described in this chapter can be applied, with varying levels of detail, as the design progresses and can aid in the determination of the need for IPL requirements for various portions of the system. Some techniques are more appropriate for early design stages, when only a preliminary design is available. At later stages, others are useful for detailed design and system integrity verification. Figure 3.2 shows the life cycle of the process and indicates where different types of safety analysis may be useful. The method and training for safety analysis meetings is discussed later (Figure 3.3).

Shortly after a project decision has been made and a tentative site selected, the key management personnel *responsible* for project implementation should meet to establish general safety requirements and guidelines for the project. This meeting provides an opportunity to establish the criteria that will be used by the design and process hazard reviewers in making risk-related decisions. Typically, this group would include the plant manager, the project manager, the person who will be responsible for leading process hazard analysis activities for the project and/or a safety–health–environmental specialist, and others, as appropriate. At this meeting, basic design goals should be reviewed or established, with due consideration to corporate, OSHA, EPA, local, or other applicable requirements. While this book focuses on safety, the environmental requirements may have a significant influence on safety—for example, if relief valve discharges must be collected and treated in a closed system, the hazards of the downstream systems will also require consideration. These environmental constraints and similar issues need to be identified prior to the start of design.

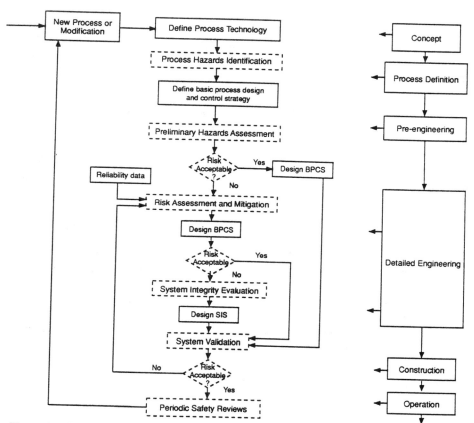

Figure 3.2. Process hazards analysis team activities during the process life cycle. The PHA team activities are shown in the dashed boxes.

Commitment to the implementation of management systems necessary to ensure levels of human performance, maintenance, auditing, and documentation, etc., to achieve the required overall system safety goals should also be an output of this meeting.

The results of this initial management meeting will provide input to the design teams, where groups of specialists work together in developing the design. The process engineers focus mainly on the process functions and equipment design; the instrumentation and control specialists are concerned with the measurement and control system design and integrity. However, safe operation requires an integration of process and control system functions. Thus, in the earliest stages of developing a project involving hazardous materials, a formal method for process hazards analysis should be established. In this book, we will identify the safety review group as the Process Hazards Analysis (PHA) Team. While different organizations may call the safety review group by different names, the role of some review group is essential. For many toxic or reactive chemicals, some form of PHA is required in the United States under OSHA regulations entitled *Process Safety Management of Highly Hazardous Chemicals* (29CFR 1910.119). More discussion of this regulation is provided in Chapter 8. The PHA team, or its equivalent, should include qualified personnel with operations–maintenance experience, with process expertise, with electrical–instrumentation–control expertise, as well as risk specialists who can perform various safety evaluations, using techniques described in this chapter. The PHA team operates in an evaluation mode. This team should be qualified to make effective and unbiased risk related decisions. Results of the PHA team's analyses are incorporated into the process and control system design.

To achieve the most effective implementation of a safe design, the risk experts must communicate clearly with the process and instrumentation design specialists. The diversity of the team will facilitate the understanding of interactions between different portions of the total system and process and in the identification of sources of common mode or common cause failures. This chapter is intended to provide instrumentation–control system designers with some understanding of process hazards assessment and of the more common techniques used by risk analysts; further, we hope that risk specialists will find the design portions of this book helpful in increasing their understanding of the important safety issues in control system design.

At each of the stages shown in Figure 3.2, the PHA team should communicate with the engineering and the control system design specialists to review the status of the present design and decide upon actions required to ensure a safer design, including sound procedures for operations and maintenance. In the process definition stage, options for inherently safer design are evaluated and hazards are identified.

Preliminary hazard assessment methods are applied to pre-engineering designs. More detailed analyses are used to evaluate the detailed engineering designs for the process and BPCS. These analyses are used to identify the need for an SIS and to establish its performance requirements. The PHA team should work with the SIS design team to assure that performance requirements are met. Throughout the construction phases, quality assurance programs focus on correct implementation of the system design. Verification testing is an important component of qualifying the system for start-up. The safety, process engineering, and control specialists also cooperate closely during startup, especially if any modifications to the design or procedures are needed. Periodic safety reviews by specialist teams are continued throughout the operational life cycle. Modifications involving any potential impacts on safety also require an appropriate level of safety evaluation.

Figure 3.3 shows when in the life cycle of a process the various types of hazard analysis techniques discussed in this Chapter are generally most

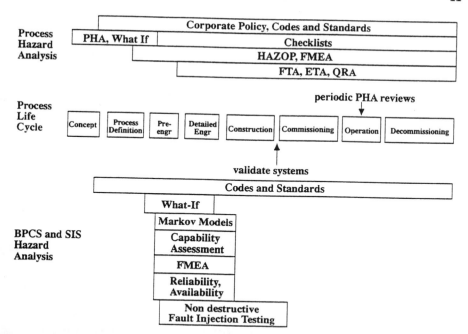

Figure 3.3 Relationship of hazard analysis techniques to process life cycle.

useful. Early in the life cycle, some of these techniques may be used in a preliminary manner to aid in process definition and pre-engineering design. When the detailed engineering is complete, the same techniques may be applied more rigorously for system validation. Those techniques primarily used for process hazards analysis are shown at the top of the figure; those used primarily for BPCS and SIS evaluations are shown at the bottom.

What is new in today's chemical plants is the role Programmable Electronic Systems (PESs) play in BPCSs and SISs. PESs utilize vendor-designed, vendor-supported, and vendor-modified communication, hardware, and software systems. The user must determine if these systems meet the integrity requirements for their use as protection layers. This chapter addresses some techniques for evaluating BPCS and SIS effectiveness in achieving and maintaining the integrity level required.

The use of PESs in BPCSs and SISs introduces failure modes more difficult to identify and respond to than those of analog controllers and direct-wired relay logic. However, many causes of control system failures (e.g., sensor, final control element, human error) are the same in both types of plants. Additionally, PESs, placed in existing environments (e.g., utility power, grounding, electrical noise, HVAC, surge protection) with temperature, humidity, dust and contaminant levels suitable for older control technologies, may not provide the reliability required without modifications to the existing environments.

Table 3.1 Failure Contributors to 17 Incidents Observed during the 4 Years of Operation in PES-Controlled Plants in the Netherlands (Ref 3.1)

Failure Category	Detailed Failure	Factor Involved in	
		Number of Failure Contributors	Percent of Failure Contributors
Hardware	Computer hardware	3	18
	Connection Hardware		
	—Electronic	0	0
	—Pneumatic	0	0
	—eEectrical	1	6
	SIS system hardware	0	0
	Equipment hardware	5	29
Software	System software (manufacturer's shell)	1	6
	Application software implementation (i.e., software written for the process plant and installed during and after implementation)	2	12

Characteristics of operating incidents due to the use of PES-controls were identified in a study where 17 incidents were analyzed (Ref. 3.1). Data were collected for a four-year period from PES-controlled plants in the Netherlands. Table 3.1 shows a breakdown of the causes of the incidents, analyzed from a human-factors perspective, using a mechanistic model of human behavior.

Many of the incidents involved more than one failure contributor and thus, the total of the failure contributors exceeds the number of incidents. A more detailed breakdown of the failure contributors and their relationship to the incident are given in Table 3.2.

Human error contributed significantly to the 17 incidents. Accordingly, much of the content of this book addresses procedural, maintenance, testing, training, documentation, auditing, and organizational contributions to safe process operation when using PES controls as discussed in:

- Chapter 6, "Administrative Actions to Ensure Control System Integrity."
- Chapter 4, Section 4.6, "Operator–Control System Interfaces" (Human error due to insufficient and/or incorrect information supplied to the personnel involved).
- Chapter 5, Sections 5.1.2.8, 5.6.4, and 5.6.10.1, Human machine interfaces and operational description guidelines for the SIS.

Table 3.1 also shows that some of the failures were not dependent on human error: However, PES hardware failures and software design errors and failures contributed to many of the 17 incidents assessed.

Table 3.2 Breakdown of Causes of the 17 Incidents (Ref. 3.1)	
HUMAN AND SOFTWARE ERRORS	INCIDENT NUMBER
Interface does not display actual plant status.	1, 6, 8, 16
Installation error leads to incorrect information.	3, 8
Alarm set incorrectly	4
No alarm (maintenance error)	4, 5
No alarm (design)	4, 5
Operator misses information due to overload.	2, 13, 15, 17
No independent means of cross-checking provided	1, 3, 6, 16
Operator fails to cross-check.	8
Trip disabled/manual override	1, 8, 11
Overreliance on computer	9, 11, 14, (15)
Inadequate knowledge	(3), 11
Failure to Update Operator's Information	12, 17
Incorrect control signal (maintenance error)	10
Design error: plant	4, 5, 8, 17
Design error: computer control system	7
HARDWARE FAILURES	
Equipment hardware	2, 5, 13, 14, 15
Computer hardware	7, (11), 16
Connection hardware	6
(Note: Numbers represent coded incidents. Incident numbers in parenthesis indicate that the error category may have contributed to the incident, but this is uncertain.)	

Accordingly, this chapter addresses some techniques that may be used for evaluating control system integrity. Some are based on documented demonstration of satisfactory performance; others employ analytical techniques. Both qualitative and quantitative analysis techniques are discussed, but qualitative evaluation is always the starting point. Often qualitative evaluations are based on the knowledge of experienced engineers with an implicit quantification, reflecting years of familiarity with similar real equipment and systems. For systems where experienced-based performance data are inadequate, where significant safety implications require high-integrity design, or where complex interactions make qualitative evaluation difficult, additional quantitative analyses may be used.

Section 3.1 provides information on some safety and integrity evaluation techniques that are often used in making process safety evaluations at various stages in the evolution of process and control system design. Section 3.2 describes appropriate techniques and the typical outputs of process hazard assessment activities during the different stages of review. Section 3.3 presents a description of equipment approval and certification methods for BPCS and SIS systems and components. Finally, Section 3.4 lists some additional guidelines and standards, that relate to process control safety.

3.1 SAFETY AND INTEGRITY EVALUATION TECHNIQUES

The chief purpose of project PHA reviews is to ensure that the total process and control system design meets the safety criteria required by all relevant codes, standards, and regulations as well as meeting corporate safety goals. Each organization has the responsibility for developing safety goals, whether qualitative or quantitative. Figure 2.4 is an example of one approach to establish some general semi-quantitative design safety criteria. Figure 2.6 is an example of one approach to select appropriate integrity levels for SIS design.

Ensuring that the required integrity of the process safety systems is achieved is likely to be an iterative process involving refinements of the control system by the designers in response to periodic evaluations provided by the PHA team. Techniques for accomplishing this are discussed in Chapters 4 and 5.

The Center for Chemical Process Safety (CCPS) publication *Guidelines for Hazard Evaluation Procedures* (Ref. 3.2) presents qualitative and semiquantitative techniques for assessing the potential levels of hazards present in a process. Another CCPS book, *Guidelines for Chemical Process Quantitative Risk Analysis* (CPQRA) (Ref. 3.3) describes methods to quantitatively assess the integrity of the chemical process system.

For analysis of systems requiring detailed quantitative risk assessment, utilize a knowledgeable risk analyst on the evaluation team, since these techniques require specialized knowledge. Some companies may have corporate experts in risk assessment available for this purpose; otherwise, it may be necessary to retain an experienced consultant for assistance in applying fault tree analysis and the other analytical techniques described briefly in this chapter.

All the techniques described in the following sections are subject to some limitations and need to be used with judgment and understanding of the uncertainties. Uncertainties may be present due to several different factors, including

1. *Database limitations.* The quality of failure-rate data for system components varies widely because of the size of the database and the possible inclusion of inapplicable or biased data. Failure-rate data on widely used control system components may be quite extensive, whereas data on human failure rates under specific conditions may be limited and subject

to wide uncertainty. Sometimes "educated guesses" are made when data are lacking.

2. *Failure-rate data limitations.* Often, a mean or expected failure rate is used for component failures that are statistically distributed as a function of time in most quantitative reliability and failure analysis techniques. Usually the "typical" value given in tabulations is the median. Uncertainty analysis can account for the variance in the failure rates and propagate these variances through reliability and availability models to obtain a probability distribution for the system reliability or availability.

3. *Inadequacies in models.* Analysts may overlook some potential failure sequences or system dependencies that can lead to errors in analysis. Models may be based only on steady-state effects and be unsuitable for identifying problems relating to system dynamics. Of course, the analysis is always limited by the adequacy and quality of the failure rate data and model estimates should be treated accordingly.

It is important to keep in mind the quality of information being used in integrity evaluation studies, so that potential uncertainties are understood when decisions are being made about risk reduction.

If deficiencies in a BPCS or SIS are identified early in the project, corrective measures can usually be incorporated with minimal impact on cost and timing. However, if deficiencies are not identified until the late stages of a project or when the equipment is operational, additional components or significant modifications to the system may be necessary, causing cost and timing problems.

Note that the techniques described in this chapter focus on equipment and operating systems without specific attention to the application software. Verifying the integrity of application software is covered in Chapters 4, 5, and 6. While software itself does not pose an inherent hazard, deficiencies in software may lead to misoperation of other system elements that can lead to hazardous events. Thus, software must be tested thoroughly and systematically to establish the necessary degree of confidence that the total system behaves in an acceptable manner under a wide variety of upset–failure conditions. The response of the software to abnormal process events is often tested by extensive simulation testing techniques; the response to related hardware component failures is tested by the simulation of erroneous input or output data. Also important is the evaluation of potential failures caused by inappropriate or erroneous operator input and the related ergonomics of the operator interfaces.

Errors during maintenance and provision for future upgrades or modifications in the software system must also be addressed.

The software testing procedures need to be carefully developed, reviewed, and documented and then reviewed as part of the periodic PHA team review activity.

3.1.1 Qualitative Techniques

Sound, experienced, engineering judgment is essential to safety and integrity evaluation. Even when codes of practice, standards, and design guidelines are available, the skill and experience of the individuals performing the PHA directly affect the value of the assessment. A multidisciplinary expert PHA team is always needed. Although the team approach is expensive and time-consuming, a single-discipline approach may lead to "blindspots" in the design.

Safety standards and codes of practice represent the collective judgment of groups of experts and these documents are often used as design criteria. Some companies use benchmarking, guidelines, vendor practices, or comparison with similar operational technologies, as an evaluation tool.

For new technology applications, where past experience is limited, safety analysts may "brainstorm" potential failures, using techniques like "What-if" analysis. These techniques are generally part of a preliminary hazard analysis.

Once a process design progresses into the pre-engineering stage, more thorough evaluation techniques are usually required. These include Failure Mode & Effect Analysis (FMEA), Hazard and Operability (HAZOP) studies, and similar techniques used to review process equipment and control systems. Many applications require a combination of techniques.

3.1.1.1. Operating Experience

If a new process or control system is a carbon copy of an existing system, experience with currently operating systems may be the basis for assessment. Caution should be exercised with this method, however, since evaluators may be lulled into a false sense of security (all the faults in an existing system may not have been discovered) and other items for assessment may not be evident.

Some typical BPCS/SIS operating experience questions are

- Do field performance documents (i.e., operating and maintenance records) confirm system acceptability?
- Has this equipment operated in an equivalent environment (e.g., electrical, particulate, maintenance, service, chemical operations, powerline conditioning, operating hours)?
- Is the installed database adequate to obtain operating experience? Well-documented operating experience on your own plant site may be the best indicator of a system's ultimate performance.

3.1.1.2 Standards and Codes of Practice

Standards and codes of practice may be used to assess the safety integrity of a design or an installation. The standards and codes of practice are the result of the experience and efforts of committees of experts in the field. These experts have combined their experience to form composite documents that enable designers to follow a predetermined set of rules. The safety integrity achieved by this process is usually not quantified. Standards and codes of practice often provide minimum guidelines on what should be done to reduce hazards. However, the minimum safe guidelines for an office building may not be adequate for a chemical process control room in a hazardous area. Therefore, ensure that the standards and codes of practice used are both current and applicable to the actual situation. CCPS's *Guidelines for Safe Storage and Handling of High Toxic Hazard Materials* (Ref. 3.4) contains useful information on general safe practices for some types of chemical processes.

3.1.1.3. Design Guidelines

Design guidelines are often used to achieve a safe equipment design or to assess such a design for safety purposes. They are most useful when the risks are well known from a long background of experience. As with codes and standards of practice, the design guideline is based on the experience and judgment of a committee of experts. The limitation of this approach is apparent when new ideas are proposed by designers. New designs are accompanied by new risks and a requirement for new design guidelines; but without the background of experience in the new field, these guidelines cannot be formulated. Therefore, design guidelines can undergo a continual process of modification and discussion and lag the introduction of new technology.

Design guidelines include safety principles that have evolved over many years, in every industry, and at every site. Safety principles that have evolved with reference to SIS systems include the following concepts:

- Power failure should move the protective system toward a safer condition.
- High risk fault conditions should be monitored by at least two independent, measurements and may require measurement diversity.
- Common-mode failures (e.g., power loss) should be evaluated.
- No single equipment fault should invalidate the protection provided.

These principles generally require the use of redundancy, diversity, and fail-safe design concepts. These are now accepted design factors for SISs. Similar principles and techniques may also apply to BPCSs.

3.1.1.4. "What-if" Analysis

"What-if" analysis is often used in the earliest stages of process design development to identify potentially serious hazards. In essence, it is a subjective review of the total system based on hypothesizing a wide range of potentially serious failures without regard to likelihood. It is performed by a multidisciplinary expert team (e.g., the PHA team) that conducts a methodical review of the process by formulating questions concerning potential hazards in the operation and maintenance of the equipment. The team may include design, safety, operations, and maintenance personnel. Questions asked are relative to a specific initiating event. Responses are examined and recommendations where appropriate are proposed.

"What-if" analysis can be used later in the design also. BPCSs or SISs can be analyzed, starting with the output devices and working back through the logic system to the input devices, by asking "What-if" questions to see which problems could cause an unsafe condition (Ref. 3.5). For example, if a programmable logic controller (PLC) uses plug-in subassemblies, some typical "What-if" questions might be (Ref. 3.6):

- What happens if a subassembly is removed under power?
- Will subassembly removal cause spurious data results?
- Does the internal watchdog timer (WDT) monitor the subassembly removal?
- Can the subassembly be reinserted without spurious data being generated?
- What happens if a subassembly is effectively removed - by dirt, corrosion, or vibration - under power?
- What happens if an interconnecting cable between modules or sub assemblies is removed with the system ON?
- What happens if an input wire is broken, shorted, or grounded?
- What if a plug-in module fails, creating an output signal on one or more channels?
- Does any individual or combination of "What-if" situations create an unsafe condition?

Using "What-if" analysis in conjunction with appropriate checklists is a systematic approach. The checklists impart structure to the analysis by acting as a reminder of the types of questions that should be asked.

3.1.1.5. Checklists

Checklists can be particularly useful when designs contain equipment that is proven and well understood. This technique requires knowledge of the pro posed equipment and system configuration planned for a given application. Checklists are often a summary of the applicable codes and standards, supplemented by "What if" questions and by the failure modes observed from earlier FMEA/HAZOP analyses during previous design reviews for systems using the same or similar PES equipment. Checklists provide a way to verify that a BPCS or SIS conforms to existing codes and standards.

Extensive checklists are provided in the Health and Safety Executive (HSE) document on PESs (Ref. 3.5). Two types of failures identified in the HSE guidelines are random hardware failures and systematic failures. Systematic failures always occur under the same set of conditions; random failures do not. Random failure impacts may be reduced by redundancy; systematic ones may not. The HSE guideline contains two, 16-section, structured checklist questionnaires. One set of checklists is for qualitative assessment and the other for common-cause failures in redundant configurations. Only a multidisciplinary expert team familiar with all aspects of hardware and software specifications, design, testing, installation, operation, and maintenance is able to respond to all items on the checklists.

The disadvantage of checklists is that any items not included might be overlooked. As a result, the consequences of all the potential equipment failures may not be fully examined. Checklists may be satisfactory if there is little or no innovation and all the deficiencies have been encountered before. They are not sufficient when the design is new.

3.1.1.6. Hazard and Operability (HAZOP) Studies

The HAZOP study was developed not only to identify hazards in a process plant but also to identify operability problems that, though not hazardous, could compromise the plant's ability to achieve design productivity. HAZOP studies cover process segments that include equipment, controls (unfortunately, this is not always done well if control system experts are not involved) and procedures. It is an alternative to FMEA, and is often used in evaluating process system safety for new installations or for modifications.

This approach also requires the formation of a multidisciplinary team of experts (e.g., a PHA team) in various aspects of process design and operations. The team leader follows an established procedure in which the piping and instrumentation diagrams (P&IDs) are systematically analyzed component by component. For each component, the team leader uses a set of guidewords to help identify any process deviations that could lead to abnormal operation or a hazard. For example, for the variable "flow," the guidewords might be: "more," "less," "no," "reverse," or "other than." For a batch process, the specific process step has to be identified. A spreadsheet format allows the consequence of the deviation to be noted along with any provisions in the design to mitigate the effects of the deviation (e.g., a high-flow signal regulating an upstream control valve or actuating a shutdown valve). Excursions in temperature, pressure, composition, and other process variables are analyzed for each component. The team leader helps establish the level of detail for the analysis by choosing what is included in each component. The HAZOP technique looks for system changes around the selected component (for PESs, this might be flows of information or current, voltage levels, etc.), and then looks for resultant changes in the system. HAZOP is difficult to apply in detail to highly interconnected systems and software. While the actions of the control system are important to a process HAZOP, the PHA team may not want to use the HAZOP as its primary method of validating the PES controls and interlocks. Some teams might be tempted to use simplified rules such as assuming that only one level of the control system will fail at a time. However, this is a dangerous assumption in systems which are subject to common mode failures and covert faults, unless the reliability and availability of the control system are validated by an independent method.

In HAZOPs, the magnitude of the hazardous impact is considered along with the likelihood of the hazard. This can be done qualitatively, or relative rankings can be done, along the lines of the categories indicated in Figure 2.4 or with some similar criteria. For example, hazards might be ranked in terms of high, medium, or low consequences and likelihood, in terms of high, medium or low frequency of occurrence.

The HAZOP team considers the protection layers available for each significant deviation, considering the potential hazardous impact and the likelihood. If there is potential for a very major hazard, additional quantitative analysis may be required to make a judgment on the level of risk involved. Usually, a HAZOP study results in a number of recommendations for design, equipment, or operating philosophy improvements. Any questions that are unresolved during the HAZOP analysis are flagged for followup and should be resolved before final acceptance of the design.

Because HAZOPs focus on process deviations, they can be used in a very qualitative way early in the process design - after a basic flow sheet and general types of equipment are identified. At this stage, the HAZOP may identify some areas of concern that need special attention in the detailed design. When the BPCS design is complete, the HAZOP format can be used to identify further improvements for the BPCS as well as additional needs for an SIS. A detailed HAZOP, when the design is complete, is a useful tool for process safety evaluation and validation. However, HAZOPs are limited in their ability to identify common mode failures, so they usually need to be used in conjunction with other techniques for PES control validation.

A more detailed description of the HAZOP technique is given in the CCPS book, *Guidelines for Hazard Evaluation Procedures* (Ref. 3.2). A typical HAZOP worksheet approach is described in Section 7.4.3 for the selected polyvinyl chloride (PVC) batch polymerization process example. As a result of a HAZOP following the design of the BPCS, the requirements for the SIS design are established along with the integrity levels (see Section 2.3.4.2) required for each interlock to meet the overall corporate system safety goals. This is discussed further in Section 3.1.4.

3.1.1.7. Failure Mode and Effect Analysis (FMEA)

FMEA is used to analyze all the failure modes of a given item of equipment for their effects on other components and the final effect on process safety. FMEA is concerned with failure and the effects generated by each failure mode, and may also estimate the probabilities of failure (Refs. 3.7 and 3.8). A multidisciplinary PHA Team that includes appropriate vendor experts may be used. Each piece of equipment is considered in turn. This method is equipment-oriented and is a bottom-up approach that can be very thorough if all failure modes can be identified. This approach may also be used in predicting the equipment reliability.

Data from the evaluation constitute the basis for determining where changes can be made to improve total system integrity. Failures are categorized and may have criticality rankings based on experience. The difference between FMEA and Failure Modes Effect and Criticality Analysis (FMECA) lies in the ranking of the results, which is part of the latter exercise but not of the former. The FMECA method is qualitative.

The typical steps involved in determining the consequences of malfunctions or failures of all components are

1. Identify battery limits and list all components in order to select additional PHA Team members for vendor-supplied equipment, if appropriate.
2. Identify all failure modes for each component, considering all possible operating modes.
3. Determine the effects of each failure on other components and the overall system.
4. Evaluate, or calculate (optional), the probability of each failure mode.
5. Evaluate the consequences of each failure mode.

This process can be extremely time-consuming, but it provides a systematic approach for identifying the consequences for all identifiable failures.

FMEA is component-failure-oriented (i.e., as opposed to process-deviation-oriented) and embodies the concept of system failure resulting from component failures. It is difficult to model the system configuration if it has parallel or redundant components. It also is not designed to treat combinations of failures, although this can be done to some minimal extent on simple systems.

While FMEA has often been used to analyze process and mechanical systems, it can be adapted to the review of details of the control system hardware configuration and can be used as a framework for identifying component failures that might disable large groups of outputs or control functions simultaneously. This type of review cannot be done until there is a firm understanding of the control system design. Preferably, the control system FMEA would be done after completion of a process HAZOP or FMEA, so that the PHA team will have a good understanding of the process itself. Vendors may be able to provide listings of failure modes and frequencies for purchased systems.

The control system FMEA is very likely to uncover the need for changes in system hardware, so it is most effectively done when the design is nearly complete, but before field work has begun on its installation. When used at the end of the design phase, changes can be made more quickly and at less cost than for later changes that require a field modification.

The PHA Team can use a schematic of the control system, including data highways, field devices, PESs, other logic devices, etc. Information required for the evaluation includes vendor information describing the operation and the potential failure modes for each component of the control system, details of I/O assignments to and from the control system components, location of equipment, and descriptions of process control computer software. Descriptions of user-configurable options for each component need to be known, especially if they relate to how the system responds to various component failures. Details on power supply, emergency power, and grounding are also essential.

Of particular importance is the identification of components, usually down to the functional level, such as communication cards, hard disk drives with system configuration, data highways, etc. Failure mode and frequency information come from many sources, including vendor information, other users, experience with similar systems, and information in control system maintenance procedures. Checklists are another resource for identifying less obvious failure modes.

A description of any active or passive diagnostics is needed, including the response to each detected failure and the subsequent functioning of redundant devices, spares, or back-up systems.

The PHA Team also will consider the software and control philosophy for the process. The PHA Team needs access to the control programs, and a description of operator access to the various software systems. It is important to know who will be authorized to change various levels of control (e.g., set points, sequence logic, interlock set points) and how such changes will be monitored and documented. Safeguards to prevent unauthorized access should be described.

It is also important to understand how process data will be communicated to the operator. This includes HMIs to convey the normal process operating information, control system status, and all alarms and interlocks.

The FMEA proceeds by selecting each hardware device, describing its function and listing all the potential failure modes. For each failure mode, the chain of events caused by the failure is followed far enough to judge whether or not a potentially hazardous impact might result. The PHA team will consider both the magnitude of the potential impact and its likelihood (estimated approximately from the chain of events), leading to a judgment that the safety level is adequate or that further modifications are needed. Some of the modifications may, for example:

- Require redesign.
- Require addition of another component.
- Result in reconfiguring multiple PESs to fail in different ways.
- Note that certain pairs of wires must be run in separate channels.

- Add diagnostics, alarms, or interlocks.
- Result in modifications to maintenance or testing procedures.

Of course, good follow-up on action items and documentation of the FMEA is essential.

3.1.2 Quantitative Techniques

For systems with the potential for major hazards, it may be necessary to perform some quantitative analyses, in addition to the qualitative analyses described in the preceding section. These may be used for new designs, modifications to existing systems, or evaluating existing installations. Some of these techniques may also be used as qualitative evaluation tools, employing the logical evaluation framework without quantification.

Quantitative techniques include Capability Assessment, Reliability Data base Analysis, Fault Tree Analysis (FTA), Event Tree Analysis (ETA), Reliability Block Diagrams (RBDs), Incident Analysis, Quantitative Risk Analysis (QRA), Markov Models, Monte Carlo (MC) simulation, and Nondestructive Fault Insertion Testing (NDFIT).

3.1.2.1. Capability Assessment

Once a protective system design is developed, a capability assessment should be made (i.e., an evaluation of the system's ability to meet safety requirements, taking into account the accuracy and the dynamics of the equipment used). This is of great importance where safety is a major consideration. Figure 3.4 shows an example where a cumulative effect of errors and delays (all within the manufacturer's specifications for equipment) result in an inability to shut down the plant in time to prevent a major accident, even with multiple protection layers. A capability assessment will identify problems of this type so that design modifications can be made to correct identified deficiencies. Usually, a full and detailed capability assessment of any design should be made before the reliability assessment of a control system is done.

3.1.2.2. Reliability and Availability Analysis

Reliability analysis is used to estimate the probability that a unit or a system will function correctly under specified conditions for a given time period. Reliability is important for normally operating components to meet production requirements.

$$\text{Reliability} = \text{Uptime} / \text{Total (process operating) Time}$$

While reliability is an important consideration in the selection of basic process operating components to minimize hazards associated with spurious upsets, for components of a SIS, the major parameter of concern is *availability*. Availability is the probability that a system is actually capable of performing its

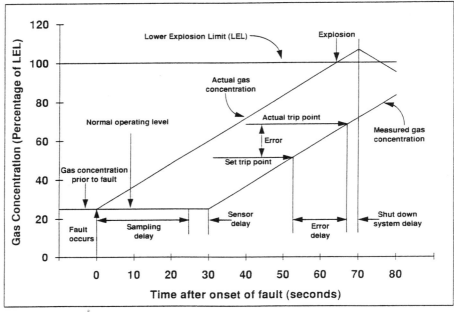

Figure 3.4 Capaility assessment example. Response of a plant and its proactive system (Ref 3.9).

designated function when required for use. Availability is important for stand-by components that must function on demand to perform a protective function. SIS components need to be "available" when abnormal process conditions demand their actuation. Note that if an SIS component has an undetected failure, it is unavailable for protection until the fault has been detected and repaired. For a single element in a repairable system:

$$Availability = Uptime/(Uptime + Downtime)$$

$$Uptime = MTTF = mean\ time\ to\ failure$$

$$Downtime = MTDF + MTTR$$

where
    MTDF = mean time to detect failure;
    MTTR = mean time to repair or replace;
    Availability = MTTF/(MTTF + MTDF + MTTR).
The mean time to detect failure often can be reduced by more frequent testing, whether manual or automated.

Availability analysis techniques are similar to those used in reliability analysis, except that the component/system (if repairable) is allowed to be repaired from a detected failed state when estimating the availability. For repairable systems, reliability analysis quantifies the probability of having a failure in a set period of time; availability quantifies the probability of the system operating at a set time, t, including its having failed and been repaired one or more times prior to time t.

The Integrity Levels used in SIS design, as described in Section 2.3.4.2, are distinguished by degrees of reduction in hazard event likelihood required, if any, after the practical means of risk reduction have been incorporated in the process and BPCS design. As an IPL, the SIS should have an availability of greater than 99% to reduce failure likelihood by a factor of more than 100. This availability is expected for Integrity Level 1. Integrity Levels 2 and 3 each provide an additional reduction in likelihood by about another factor of 10. Further information on SIS design to achieve these integrity levels is presented in Chapter 5.

Both reliability and availability are important considerations in safe automation of chemical processes. Reliability and availability of the BPCS reduce the frequency of unnecessary demands on the SIS systems while SIS availability is important to ensure that the SIS will function properly when a demand occurs.

There are a variety of sources available for finding reliability information (or failure rates) for different types of process equipment. The CCPS publication, *Guidelines for Process Equipment Reliability Data* (Ref. 3.10) discusses existing sources of data for a wide range of equipment and control system components. In addition, many of the larger companies have or are developing their own databases, based on specific maintenance and operating experience. Vendors also may have reliability data for their equipment as a result of reliability testing or from information obtained from the customers they service.

Data of these kinds can not only be used in evaluating the integrity of a new system or a system modification, but may also be reviewed to distinguish between equipment that operates reliably and equipment with chronic problems. Unusual operating problems or high maintenance demands for a particular type of equipment may suggest the need for modification or replacement. Designers tend to choose equipment with better field performance if they have access to relevant reliability databases.

An extension of reliability database analysis to identify problem areas is the development of company- or plant-wide incident analysis records. An "incident" is often defined as an abnormal process excursion that is part of a failure path leading to a safety problem. More serious incidents are sometimes called "near misses" and are often subjected to detailed analysis similar to that used in accident investigations. Useful lessons are often learned from accident and incident analysis that provide additional insight for quantitative integrity evaluation techniques.

3.1.2.3. Fault Tree Analysis (FTA)
Fault trees are logic diagrams that systematically display sequences of failures of either the process or the process control equipment leading to defined "top events." They aid the analyst in evaluating the likelihood (i.e., frequency) of major events, as a function of the magnitude of consequence (i.e., top event). Top events are usually selected by looking at the range of potential serious accidents. For example, top events might be selected from serious consequences identified in a "What-if" study, an FMEA, or a HAZOP analysis. Typical top events might include:

- Fire in a storage tank dike.
- Release of a volatile toxic vapor for 10 minutes from a major transfer line.
- Volatile, very toxic, vapor release from an instrument line failure.
- Explosion of a process vessel.
- Toxic or flammable gas vented from a relief valve.

The top event is defined by a quantity and rate/duration of release of a particular material of concern that can then be used to estimate a potential hazard zone based on hazard estimation models. For each selected top event, a fault tree is constructed that displays all the combinations of failures of more basic system components that might lead to the top event. Initiating failures are shown at the bottom of the diagram; simple logic gates, usually Boolean AND and OR gates, are used in developing each branch of the tree. Usually there are multiple independent pathways that can cause the top event. By assigning failure frequencies or conditional probabilities of failure to elements in the tree, an estimate of the likelihood of the top event can be made. Further, from the fault tree, one can see the major pathways that contribute to the likelihood of the top event and use this information to choose suitable mitigation methods to reduce the likelihood or consequences of the top event if the risk is not acceptable. An example of a fault tree is shown in Figure 3.5.

The fault tree shown does not include numeric values, but these can be computed by assigning failure rates or conditional failure probabilities to each item shown in the tree. Then, using Boolean algebra, the likelihood of failure of each of the highest level events on the tree can be estimated. If the risk needs further mitigation, then the branch of the tree that makes the largest contribution to risk is usually the best place for modification.

Great care must be taken in developing the trees for systems that have potential for common mode failures, covert faults, or components with long repair times. As with other techniques that have been designed for process safety evaluation, FTA needs to be supplemented with other validation techniques for PESs.

Further details and examples of fault tree analyses are given in CPQRA (Ref. 3.3), in the Nuclear Regulatory Commission *Fault Tree Handbook* (Ref. 3.11), and in IEC Standard 1025 (Ref. 3.12). Fault trees are fairly easy to review but

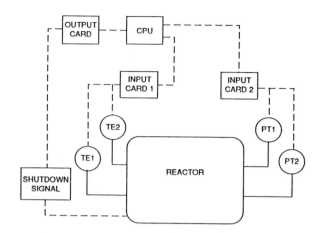

Example of Chemical Reactor Monitoring System (Ref. 3.3)

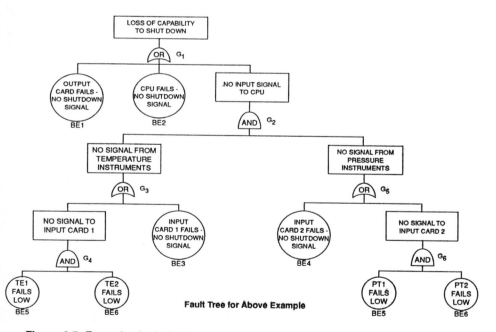

Fault Tree for Above Example

Figure 3.5 Example of a fault tree.

are difficult to construct properly because of constraints imbedded in the mathematical assumptions (e.g., independence of events, treatment of common-mode failures, elimination of cross terms, treatment of failure detection and repair times as instantaneous). They do not model degraded states of the system (e.g., valve partially closed) and do not represent a dynamic system (i.e., time and/or sequence of failure events are not modeled) because FTA describes systems at a certain instant of time, usually steady-state. It is therefore important that someone proficient in FTA be included as a member of the team constructing the fault trees. Results should also be used with care, as with any modeling approach, since many of the failure rate estimates are subject to much uncertainty. Nevertheless, fault trees are an increasingly common part of quantitative risk analysis studies for facilities that have the potential for major offsite hazards.

Fault trees can be used to verify that the design ensures that the layers of protection in the BPCS and SIS systems function as intended.

3.1.2.4 Event Tree Analysis (ETA)

Event trees are also logic diagrams, but they differ from fault trees in that they start with a single, initiating event and then branch outward into all possible sequences of subsequent failures that might lead to a hazard. The outcome events of an event tree include a list of events of varying severity from "no hazard" to some of those that might represent top events in fault tree analyses. The event tree provides systematic coverage of the time sequence of event propagation, through a series of protective systems actions, normal plant functions, and operator interventions, each having some probability of failure. Each path in the event tree may be quantified with a probability of occurrence. For example, in Figure 3.6, Event B, the coolant flow alarm working, has an annual probability, B, associated with the "YES" branch. Then the probability of the "NO" branch is (1-B). Working through the event tree, the probability of occurrence of each sequence is determined (e.g., ABCD for sequence 1). The risk of each sequence can be evaluated considering the likelihood and consequence of the sequence. However, multiple sequences may lead to the same consequence (top event in FTA), so a sequence likelihood may only be one portion of a particular top event–consequence likelihood.

Since event trees portray ordered sequences of failure events, they may be useful in understanding total delay times and identifying any response time inadequacies in a control system design, using capability assessment techniques. Further details on event tree analysis are given in CPQRA (Ref. 3.3).

Another application of event trees is their use in incident analysis, as part of an accident investigation. Special expertise in the details of event tree analysis is needed for their correct development and application. An example of an event tree is shown in Figure 3.6.

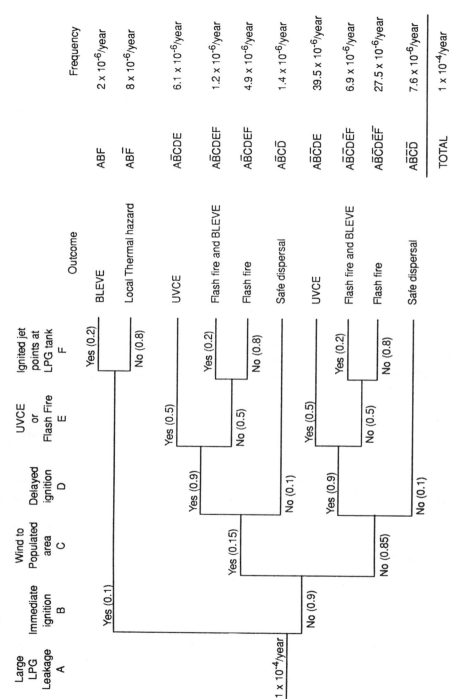

Figure 3.6 Example of an event tree

59

3.1.2.5 Reliability Block Diagrams (RBD)

One technique for evaluating system reliability is the use of reliability block diagrams (RBDs), diagrammatic representations of the reliability characteristics of a system, which show the subsystems that must function to avoid failure of the system. Thus it is necessary to represent the system as a number of functional blocks. Each block of the diagram represents the function of one separable subsystem (equipment/component); the blocks are joined by connecting lines to show the combinations of series items, standby redundancy, and active redundancy that make up the system. RBDs are good, simple techniques to approximate the reliability of a system, but they include a number of inherent assumptions, including: constant failure rates, instantaneous fault detection and repair, static operation, and simple common-cause modeling. RBDs cannot model degraded system states nor the time and/or sequences of failure events. IEC Standard 1078 (Ref. 3.13) provides additional information on this method.

A number of general rules should be noted when defining blocks.

1. Each block should represent the maximum number of components in order to simplify the diagram.
2. The function of each block should be easily identifiable.
3. Blocks should not contain any significant redundancy since the internal addition of component failure rates would not be valid.
4. Each replaceable unit should be a whole number of blocks.
5. Each block should contain one type of technology (e.g., electronic or electromechanical).
6. All systems within a block should be subjected to a common ambient environment.

Relability block diagrams are used in the prediction of overall system reliability or availability, and for identification of critical subsystems and potential "weak links" in the system.

Reliability prediction enables calculation and assessment of anticipated system reliability from the subsystem or component failure rates. It provides a quantitative measure of how close a design comes to meeting reliability objectives, and also permits comparisons between alternate design proposals. An RBD example is shown in Figure 3.7.

3.1.2.6. Quantitative Risk Analysis (QRA)

The purpose of a QRA is to gather and quantify all the potentially significant sources of risk for a particular activity and to provide quantitative estimates of the likelihood and consequences of the spectrum of concerned risks. The risk spectrum is often displayed as a risk profile showing estimates of the frequency of accident severity. Figure 3.8 shows a typical risk profile that is

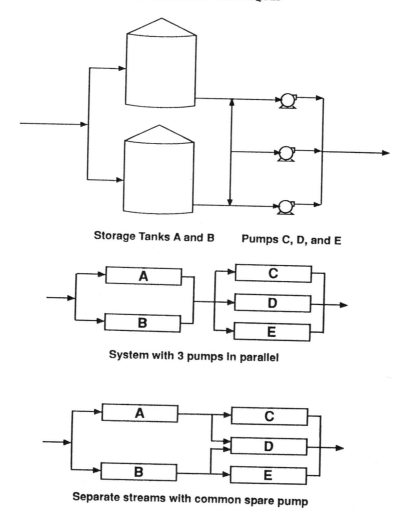

Storage Tanks A and B   Pumps C, D, and E

System with 3 pumps in parallel

Separate streams with common spare pump

Figure 3.7 Example of a reliability block diagram.

developed from aggregating all the fault trees for the facility after both the likelihood of each top event and the severity of the associated consequences have been estimated.

In developing fault trees, the performance of the BPCS, SIS, and operations and maintenance personnel are included along with process equipment failures in assessing failure paths and estimating the likelihood of failures.

For low-consequence accidents, it is usually unnecessary to conduct as detailed an analysis as would be needed for a process involving potential for

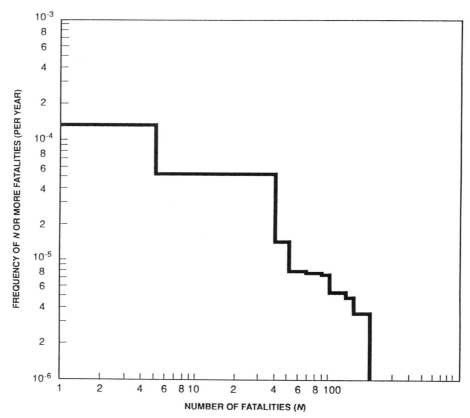

Figure 3.8 Example of a risk profile. Total societal risk for an example fractionating system. (Ref. 3.3)

major offsite impacts. CPQRA (Ref. 3.3) suggests criteria for selecting the appropriate level of analysis for the particular hazard levels involved. If there is a potential for significant offsite risk, it may be prudent to consider a QRA.

Some major companies use QRA results as guidance in developing general corporate risk-based safety criteria. These criteria tend to be in terms of broad levels of performance (e.g., Figure 2.4); it is not unusual to have inherent order of magnitude uncertainties in likelihood estimates of very rare major accidents.

Nevertheless, QRA, conducted at an appropriate level of detail, is a valuable method for evaluating system integrity relative to general safety criteria. It is useful in identifying the best opportunities for risk mitigation in complicated systems. For example, an unacceptable high frequency of a high-consequence risk may suggest a modification of the SIS, BPCS, or operating procedures.

3.1.2.7 Markov Models

Markov models are used widely in analyzing the reliability of complex systems, particularly when the sequence of failure is important or when repair is done on a continual basis. For these applications, a Markov model requires describing the system as a discrete-state, continuous-time process. The discrete-state is normally given as a unique, well-defined condition of the relevant system components (e.g., pump working, valve open, safety valve failed).

For Markov modeling to be applicable, two other conditions must be met:

1. The future state of the system must be independent of all past states except the immediately preceding state.
2. The transition rate, from one given state to another, must be the same at all times in the past and future. This requires using constant failure and repair rates for a given state. (Note: Monte Carlo simulation may also be applied to Markov models to perform uncertainty analysis and determine a probability distribution for system reliability and availability.)

Because of their flexibility in modeling common cause events, software, system dynamics and inspection/repair policies, Markov models are sometimes used to analyze the behavior of process control and instrumentation systems, but they require great expertise and care in their construction to ensure that possible dependent events are eliminated. Further, for complex systems, the models become cumbersome and may be impractical to solve using analytical techniques (i.e., closed form solution). Numerical techniques can then be used to analyze the Markov model. Additional information and references on Markov models are presented in CPQRA (Ref. 3.3 and in ISA S84). A Markov model example is shown in Figure 3.9.

3.1.2.8. Monte Carlo Simulation

In 1980, F. P. Lees (Ref. 3.14) noted that many practical problems in reliability and/or availability of complex systems cannot be solved by analytical methods. Such problems require numerical simulation, rather than analytical treatment of the effects of inputs described with probability distributions (e.g., equipment failure rates). Monte Carlo techniques represent the distribution as sequences of discrete random values. A sufficiently large number of these discrete values approximates the original continuous distribution.

For those applications where a base event is better described by a probability distribution than by a single value, it is often desirable to obtain the top event probability distribution. This may be difficult or impossible to do analytically. The Monte Carlo technique applies properly distributed random values for the base event likelihoods in order to determine the top event probability. By repeating this process many times, the probability distribution of the top event can be approximated.

Dual Programmable Logic Controller with Single I/O

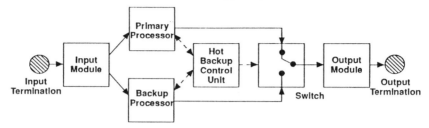

MARKOV MODEL

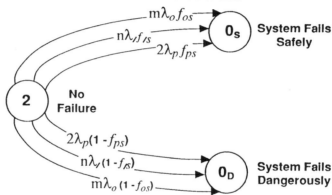

Safe & Dangerous Failure Rates & MTBFs

$$\lambda_s = 2\,\lambda_p\,f_{ps} + n\,\lambda_{,}f_{,s} + m\,\lambda_o f_{os}$$

$$MTBF_s = \frac{1}{\lambda_s}$$

$$\lambda_o = 2\,\lambda_p\,(1 - f_{ps}) + n\,\lambda_{,}(1 - f_{,s}) + m\,\lambda_o\,(1 - f_{os})$$

$$MTBF_o = \frac{1}{\lambda_o}$$

Where

λ_p is the failure rate of a processor.

n and $\lambda_{,}$ are the number and failure rate of input boards

m and λ_o are the number and failure rate of output boards.

f_{ps} is the fraction of processor failures that result in safe shutdown.

$,$ is the fraction of input board failures that result in safe shutdown.

f_{os} is the fraction of output board failures that result in safe shutdown.

Figure 3.9 Example of a Markov model (Ref 3.20).

In principle, the probability distributions of all base events can be studied simultaneously. However, this may require a large amount of computer time. The Monte Carlo method can be applied to other problems, including uncertainty analyses in fault and event tree analyses.

In 1981, Henley and Kumamoto (Ref. 3.15) developed the theory and procedure for Monte Carlo techniques in detail. Useful examples were provided by Shooman, 1968 (Ref. 3.16), and Lees, 1980 (Ref. 3.14). The Reactor Safety Study (Ref. 3.17) uses Monte Carlo sampling to generate failure distributions for fault tree components. The SAMPLE computer program, developed for that study, handles normal, log-normal, and log-uniform component failure distributions. Monte Carlo techniques have been used extensively to investigate CPQRA uncertainty. Further details on Monte Carlo analysis are presented in CPQRA (Ref. 3.3).

3.1.2.9 Nondestructive Fault Insertion Testing (NDFIT)

Since common-mode failures can affect multiple control loops in a PES, the integrity of the PES is crucial. When a new device becomes available that is attractive to a user, it is important that the user have some experience with the device before using it more extensively or in a critical application. One method to gain confidence in a new device is to perform nondestructive fault insertion testing. This testing is only part of the total evaluation of a PES. (See Section 3.1.) The nondestructive fault insertion testing should identify common-mode failures, safe and unsafe failure modes, and design deficiencies. It may also be used to identify the diagnostic coverage factor appropriate for the PES. The coverage factor is defined as the percentage of modules, units, wiring, etc., whose removal, absence or failure is determined by built-in test functions or by a suitable test program along with a properly functioning verification procedure.

All reactions to simulated fault injections are evaluated. The results are documented and the information is used to make corrections to the device or system (Ref. 3.18).

3.1.3 Human Factors

The operators, maintenance staff, supervisors, and managers all have roles in safe plant operation. However, humans occasionally make errors or are unable to perform a task, just as instruments and equipment are subject to malfunction or failure. Human performance is therefore a system design element. The human–machine interface (HMI) is particularly important in process control systems.

Human Reliability Analysis (HRA) is a discipline that provides values for human error for inclusion in the quantitative integrity evaluation methodologies (e.g., FTA and ETA). HRA identifies conditions that cause people to err and provides estimates of error rates based on past statistics and behavioral studies. Some examples of human error contributing to chemical process safety risk include:

- Undetected errors in design.
- Errors in operations (e.g., wrong set point).
- Improper maintenance (e.g., replacing a valve with one having the incorrect failure action).
- Errors in calibrating, testing, or interpreting output from control systems.
- Failure to respond properly to an emergency.

Humans are more likely to make an error if they are

- Required to make an important decision quickly under emergency conditions.
- Required to make multiple decisions in a short time span.
- Bored or complacent.
- Poorly trained in procedures.
- Physically or mentally incapable.
- Subjected to confusing or conflicting displays or data.
- Unqualified for their job.

The CCPS *Guidelines for Improved Human Performance in Process Safety* (Ref. 3.19) and CPQRA (Ref. 3.3) also discuss methods for estimating human reliability.

Separate evaluations of human task sequences in plant operation and maintenance are also useful inputs to the development of operating, maintenance, auditing, and emergency response procedures. Training programs and personnel qualifications are required to provide a high level of human reliability and are essential in critical areas.

Ergonomics indicates that, in order to minimize human error, the industries providing equipment, instrumentation, and maintenance systems should recognize and address the following considerations in their designs:

- Engineering console complexity.
- Access to information and information display.
- Number of operator consoles per operator.
- Lighting requirements.
- Seating requirements.

3.1.4 *An Integrated Safety Review Methodology for Identifying SIS Design Requirements*

In this section, a comprehensive framework is described for an overall evaluation of all risk reduction techniques employed in a process and BPCS design which leads to the identification of any need for additional SIS design requirements. Some companies are already using this general approach as a qualitative design strategy; the method described in this Section suggests one approach to formalization of the review procedure.

After the process and its BPCS designs are available, it is common to conduct a PHA team review on both the process itself and the process with the BPCS included using techniques like HAZOP, FMEA, FTA, etc., as discussed earlier in this Chapter. These techniques explore sequences of failures that may lead to specific process hazard events as a result of various initiating faults.

During the review process, potential "process hazard events" are identified (e.g., rapid release of x tons of toxic gas from reactor B, a continuous spill of y lb/sec from instrument line failure, spill of flammable liquid with ignition into sump with area of z ft^2) and are grouped into general categories related to type and extent of potential impact. The severity of each of these impact scenarios can be identified in general terms (extensive, severe, or minor) or quantitatively, as indicated in the example in Figure 2.4. Similar assessment categories may also be assigned to environmental or business risks, but these are beyond the scope of this book.

Acceptance of risk is a decision that could be considered inherent in some regulations and is the responsibility of company management. The relative risk criteria shown in Figure 2.4 are examples of numbers a corporation might choose to set as their acceptable risk levels. The example criteria suggested in Figure 2.4 indicate that an event with high ("extensive") severity potential, and a likelihood between 10^{-2} and 10^{-4} per year is judged to be high risk needing prompt control action. Below a likelihood of 10^{-4} per year, the system is judged to present moderate risk and to require plans for some further risk control. Although Figure 2.4 does not indicate the acceptability level for an "extensive accident," one might infer that a further reduction in likelihood by an order of magnitude or more is needed. An order of magnitude reduction to a level of 10^{-5} per year brings likelihood into the range of failure frequencies for storage tank rupture or for failures initiated by natural disasters. On the other hand, a minor severity accident is considered to be generally acceptable in the framework of Figure 2.4 if the likelihood of occurrence is less than 10^{-2} per year.

With this background in mind, the PHA team may wish to conduct further analysis. This could involve identification of major impacts of concern. The PHA team may group potential accidents into appropriate impact severity levels, and then generate lists of failure sequences that might lead to accidents in each of the selected impact categories. If they were conducting a FTA, the impact categories would constitute a set of "top events." For each "top event," a fault tree is developed and the highest level of the tree represents *independent* failure sequences that might result in the top event. The "primary causes" or "initiating events" that start the chain of events leading to the top of the tree, are shown at the lowest level in the tree and the succession of events involved is shown sequentially, working toward the top of the tree. These same "causes" could also be identified in the conduct of a HAZOP, FMEA or equivalent type of review procedure.

In the analysis of PES controls which may share data highways, have programming done by a single individual, have the same vendor, or be susceptible to other common mode failures, more sophisticated FTA techniques are needed. Thus, it is important to have a good understanding of the number of *independent* protection systems in the design. One of the essential features of the SIS is that it be able to function independently of the BPCS.

In Chapter 2, the concept of protection layers was introduced, as shown in Figure 2.2. Moving outward from the core, these layers often include:

- Process Design.
- BPCS and Operator Supervision.
- Critical Alarms–Operator Supervision–Manual Intervention.
- Automatic SIS.
- Physical Protection (relief devices).
- Physical Protection (containment dikes).
- Plant Emergency Response.
- Community Emergency Response.

Protection layers, or combinations of protection layers that prevent the hazardous conquences may also qualify as IPLs, if they meet the necessary criteria.

The design philosophy presented in this book recommends achieving as much risk reduction as practical in each of the inner layers before adding outer layers. This is based on the general philosophy that prevention always is preferable to mitigation of consequences. Thus, if the residual risk from a design is not acceptably low for a particular potential accident sequence, the search for additional risk control possibilities should start with the innermost layer (process design) and work outward.

If risk is not acceptably low after non-SIS risk reduction options are exhausted, the additional risk reduction required provides the basis for the SIS design.

Figure 3.10 presents the sequence of steps for one method that may be followed to establish SIS requirements. After PHA studies of the process and BPCS have been completed, the protection layers available to reduce the likelihood of each of the hazardous events identified can be described and documented. In the Chapter 7 example, two possible approaches for conducting the assessment are presented. The goal of the assessment is to identify any residual risks in the system which cannot be controlled satisfactorily by the

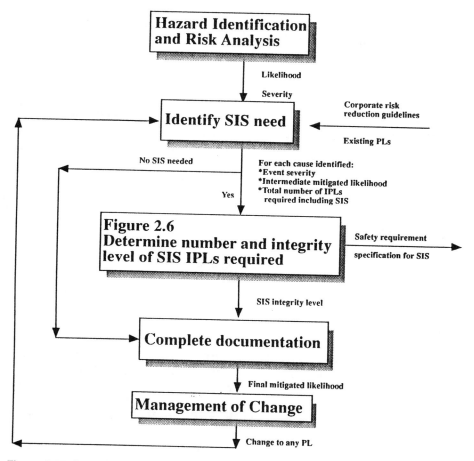

Figure 3.10 Sequence of steps in establishing SIS requirements.

non-SIS systems. Then the need for an SIS and the SIS Integrity Level can be found from Figure 2.6.

Subtle common mode failure possibilities are often present in SISs even when independent and diverse system elements are selected. Consequently, a separate review of all control-related protection layers is suggested after the SIS design is complete. The review is conducted systematically, starting with the highest integrity level interlocks. Usually, the design is based on providing the necessary level of availability required by the highest Integrity Level. Redundant and parallel elements as shown in Figure 2.5 are required in an SIS for a Level 3 system, with an availability target of 0.9999. The purpose of the review is to ensure that the designers have not overlooked any of the critical attributes of high integrity systems. The verification/validation step may be performed with the help of checklists, asking key questions about attributes such as availability, separation, diversity, fail-safe characteristics, test plans, etc. The checklists may contain some more subtle issues for example: Were separate elements of software programmed by different people? Can the component be tested with the system in operation?

There may be a number of alternative ways to validate total system safety. The method described here is given as an example to highlight the importance of designing independent protection layers into a risk control design strategy and of the limitations of certain types of analyses in assessing safety of PES-based control systems.

3.2 TYPICAL PROCESS HAZARD REVIEW ACTIVITIES AND OUTPUTS

Since a periodic process hazard review is an integral part of process and control system design, there are many ways to achieve the interactions required for safety. Each company or project team may choose different pathways to assure design integrity, but the important element is to create strong interaction between the process designers, the control system designers and the risk evaluators, while each contributes their specialist expertise.

In the example flow chart shown in Figure 3.2, process hazards need to be identified early in the selection of process technology. At the earliest stage, after potential process hazards are identified, alternative process routes are reviewed to see if a practical, inherently safer process can be found. This is particularly true for new technology; for existing technology, past experience provides a starting point for further improvements.

Once a process is selected, the types of potential hazards are reviewed in very qualitative terms to provide guidance in site selection and process component layout. After a location is chosen, a management policy meeting should be held to establish project design guidelines, with due consideration to applicable codes, standards, regulations (e.g., SARA Title III, OSHA), and local requirements. Typically, this management policy meeting would include the key personnel responsible for the project, such as the plant manager, the project manager, the person who will be responsible for leading process hazards assessment activities for the project, and/or a safety–health–environmental specialist, and others, as appropriate. An output of this meeting is guidance to the design and review teams on any special requirements and on criteria for risk acceptance.

The next step involves development of a preliminary process design and of a general control strategy. The instrumentation–electrical–control designers will consider the potential extent of hazards in the basic process design in selecting the control strategy.

At about this stage in the project, a PHA team is selected (including experts in process design, electrical–instrumentation–control system design, operations–maintenance, and risk assessment). Often, a preliminary hazard assessment is conducted using qualitative techniques ("What-if," checklists, or nondetailed analyses following a general HAZOP or similar format). Once a process design, including the BPCS design, is developed, a more detailed risk review is conducted. This review is a key part of safety documentation and a necessary guide to the design of the SIS.

Evaluation techniques, such as HAZOP and FMEA, were developed for process design review purposes and are good techniques if used properly. An incorrect analysis sometimes implemented in the past, focused on the process design and equipment by treating the BPCS as a "black box." It was assumed that the BPCS would malfunction and the SIS would function as a back-up. This approach should never be used because it could under estimate the potential for an escalating incident in the event of an SIS malfunction. Treating the BPCS as a black box does not provide a realistic understanding and verification of the control system design.

Some companies have now adapted HAZOP and FMEA methodologies to the systematic review of control system hardware and software. Key to these reviews is a good knowledge of the potential failure modes of control systems–subsystems hardware and software.

To establish the requirements for the design of the SIS, the PHA team should develop a comprehensive list of potentially significant hazardous events and identify potential initiating causes for each event. The hazardous event can be described in terms of potential impact (extensive, severe, or minor - or by some other measures) and design guidelines on risk provide a likelihood level above which the risk is deemed unacceptable. Again, these criteria may be qualitative or quantitative.

Significant events and potential causes are systematically reviewed to identify layers of protection provided in the basic design, equipment, BPCS, and procedures. If the risk level is low and the hazard is adequately controlled, no further action is needed. If the risk level does not meet desired criteria, the PHA team explores possibilities for incorporating additional layers of protection within the process design/BPCS framework. Only when such practical possibilities are exhausted, does the PHA team require further risk control by addition of a separate, diverse, instrumented safety interlock.

One possible format for evaluating the need for an SIS is discussed in Section 3.1.4. The PHA team not only identifies the process parameters that require safety interlocks, but also indicates the integrity level needed to reach levels of acceptable risk.

When the list of SIS parameters and integrity levels are given to the SIS designers, in a form similar to the Safety Requirements Specification discussed in Section 5.1, the electrical–instrument–control engineers will work with software specialists to ensure that the required system integrity is met or exceeded in the design. Sometimes lower integrity interlocks may be grouped into a higher level to avoid having two different levels of design and equipment. The PHA team may informally review selection of equipment/vendors, as well as plans for testing and validation of the SIS, as they are developed.

Some companies rely on the control system design team to establish the SIS design, based on the results of the HAZOP or equivalent evaluation. Still other companies complete both the BPCS and SIS designs before performing a detailed design review. If the technology is well established, or the design is based on familiar technology or similar systems, less scrutiny may be necessary in the early stages of process design. A final system integrity evaluation needs to be performed to satisfy that the completed system design meets all the necessary safety and operability criteria. In the pre-commissioning review, the PHA team will review and update earlier analyses and will carefully examine the documentation from testing of hardware and software. Management, operating, maintenance, and emergency procedures need to be considered in this pre-commissioning review to ensure that each component and the integrated system meet the overall corporate safety policies.

The PHA team activity continues beyond the design, construction, testing and start-up stages throughout the lifetime operation and maintenance, and ultimate decommissioning of the facility. Periodic, well-documented safety reviews are a continuing activity. Often changes are made to the process or the control system after original commissioning. These need to be evaluated and documented by a PHA team to assure that the integrity of the system is not diminished by modifications.

3.3 SYSTEM INTEGRITY EVALUATION CRITERIA AND CERTIFICATION METHODS

Chapter 2 (Section 2.3.4.2) outlines three integrity levels (i.e., 1, 2, and 3) that require varying levels of system and SIS fault response characteristics and availability. Of course, reliability is a very important element of BPCS design, particularly for processes that are hazardous when going through start-up and shutdown. BPCS and SIS designs that minimize unnecessary shutdowns also reduce unnecessary startups and thus improve the overall safety of the system.

However, for the SIS, availability to respond to a demand is the major concern because it determines what fraction of time the SIS will respond as intended. Target reliability and availability levels for the SIS are needed if validation of the SIS is to be performed. The control–instrumentation engineer identifies equipment vendors and defines the SIS system architecture and preventive maintenance program (testing). The reliability of the "proposed"

SIS system is then calculated using instrumentation failure rates. The calculated reliability of the SIS, as designed, is compared against the target reliability. If the calculated reliability is equal to or greater than targeted, then the availability of the SIS, as designed, is calculated and compared against the target availability. If the calculated availability is equal to or greater than the target availability, the SIS passes the validation. If either of these checks fails, the SIS must be redesigned.

With a qualitative analysis, the target reliability and availability values are based on experience or a relative comparison to a known standard. A frequently used technique is to design the SIS system and validate it by comparison to a time-proven set of standards based on operating experience. A qualitative methodology built around three levels of SIS system performance is described in Section 2.3.4.2. Although the number of SIS performance levels is somewhat arbitrary, three distinct levels of performance are identified. Several chemical companies have identified SIS system classification structures that contain anywhere from two to five distinct levels.

The Integrity Levels used in this book for SIS design are distinguished by degrees of availability, ranging from Integrity Level 1 design, where the system is about 99% available, to Integrity Level 3, where the system is up to 99.99% available. Further information on SIS design to achieve these integrity levels is presented in Chapter 5.

The user should consider the merit of grouping lower integrity level interlock into higher integrity level interlock systems to avoid multiple systems and reduce complexity. This may be an economic and operational decision only.

Another approach in evaluating SIS integrity involves a program where systems and components are precertified through some identified procedure for use in a specific integrity performance classification.

3.3.1 Certification of BPCS and SIS Systems

BPCS and SIS system certification methods may be specified by corporate, site, local, state, provincial, or national requirements. Designers and implementors of BPCS and SIS systems must be aware of these requirements, so their systems can be certified in a timely manner. Certification may be provided through

- Self-certification.
- Third-party certification.

The responsibility for certification of the installed system, however, remains with the user, no matter which method or methods are used.

3.3.1.1. Self-Certification

The user certifies that a system meets the integrity level as defined in Chapter 2. In self-certification, a user may conduct various integrity tests and evaluations to ascertain system suitability for the intended use in their plant. Caution should be observed by the user to ensure that there are available and capable specialists to accomplish self-certification. It is desirable that these specialists have not been involved in the original design team.

3.3.1.2. Third-Party Certification

Government agencies, insurance companies (e.g., Factory Mutual), and independent laboratories (e.g., Underwriters Laboratory) sometimes certify a system for use in a particular application. Some countries require a third-party certification for certain applications (e.g., burner management). Government regulations appear to be moving toward a more formal certification. Some companies may provide independent third-party certification that presents no conflicts of responsibility for project implementation.

3.3.2 Equipment Classification Methods

In addition to system certification, individual elements of hardware and software also require classification to assure that they meet the appropriate level for the required usage. This section provides a structured approach for equipment classification that enhances the manufacturers' and users' ability to work together to:

- Select suitable BPCS and SIS components.
- Improve BPCS and SIS performance.

Although this discussion may emphasize hardware, all elements of hardware and software, from the sensor to the final control element, should receive the same structured approach.

Industrial control equipment passes through a series of predictable stages from development to obsolescence. Product characteristics and performance specifications often change due to:

- Rapid development of microcomputer component technology.
- Software flexibility of intelligent instruments.
- Increased user demand.

Therefore, the user must have a methodology to maintain the control equipment as an approved product during these changes.

In the development phase of a PES-based product, the area most frequently affected by change is the software. Manufacturers may sell products

- With partial software (e.g., additional functions are planned and often described but not included in the first delivered devices) to remain competitive.
- Compatible with existing hardware but with new, different software to enter a new market area (e.g., a multiloop controller with software modified to become a sequence controller).

A close relationship between control system user and control system manufacturer is important during this phase of product development. A user may agree to perform *beta*-site testing (i.e., user testing of early production components) and thereby help in the development of a new control device. Significant design changes may be required to correct problems found during *beta* tests. Products at this stage of development should never be used in safety applications. Users who are eager to make use of new control technology may find themselves in an unplanned, unanticipated *beta*-site mode and may suffer from the problems of technology development.

Manufacturers change PES hardware and software as products are enhanced and when design faults are identified in established products. Since PES maintenance consists of removing modules where failures have been diagnosed and returning these units to the manufacturer for replacement, functional reliability of an installed PES depends on the owner's attention to: keeping the installed equipment updated with the supplier's latest product revision; and testing of the safety logic in concert with replacement of modified modules. This managed upgrade of installed digital systems is essential to maintaining the integrity and reliability of PESs. Therefore, the quantified reliability of the system is a dynamic parameter and has to be considered in continuing the process safety management.

To be useful, equipment classification relies on stable product characteristics. PESs may be classified by the

- Extent that performance is proven in chemical plant applications.
- Frequency that changes are made to the control product by the manufacturer.

The maturity levels of PES devices identified in this text are:

1. User-approved (for BPCS)
2. User-approved safety (for SIS)
3. User-obsolete

Criteria for each of these product maturity levels is contained in the following sections.

3.3.2.1. *User-Approved*

Before a product is classified as "user-approved," it must be *beta*-site tested. While the product is passing the *beta*-site test, it is checked against a list of criteria, such as

- Manufacturer's approval of hardware and/or software for the specified application.
- Timely user access to the review of all problems and engineering change orders (ECOs).
- User approval of hardware and/or software for the specified application.
- User approval of product documentation.
- User approval of the method used to document the application program.
- User approval of manufacturer's training and service for maintenance and operation.
- Analysis of diagnostics confirming that:
  —An internal watchdog timer (WDT) to monitor for system looping or hang-ups exists.
  —An external WDT, without susceptibility to common-mode faults, can be added if required.
- Manufacturer's compliance with quality control procedures and standards such as ISO 9000. Manufacturer's conformance to consensus standards as promoted by NFPA, ISA, IEEE, etc.

If the product testing and analysis result in acceptable compliance with established criteria, the user should then

- Consider the product for suitable applications.
- Train personnel in its programming and support.
- Install and start-up the controlled process.
- Gain operating experience.

For a product to receive the "user-approved" status, it should fulfill the following criteria:

- The product has operated properly for a minimum period of time in a single application. (Twelve calendar months may be a good "rule of thumb" to follow.)
- The user reviews the product performance, checks it against the latest manufacturer engineering change order (ECO) levels, and finds no changes that invalidate previous testing.
- The user audits and monitors the equipment for covert failures as well as overt failures.

Once the product has fulfilled the above criteria, the equipment may be classified as "user-approved" and may be used as a protection layer (i.e.,

BPCS). "User-approved" equipment should not normally be used in a safety layer (i.e., SIS). For details on BPCS design, see Chapter 4.

3.3.2.2 *User-Approved Safety*

For a product to receive the "user-approved safety" status, it must fulfill the following additional criteria:

- The product has operated properly.
- The plant wishes to use the product in separate SIS applications.
- The plant reviews the product performance, checks products against the latest producer ECO levels, and finds no changes that invalidate previous testing.
- The plant meets with the product manufacturer and jointly develops the following strategy:
  —All problems with the commercial product are jointly reviewed by the manufacturer and the plant for impact on SIS applications.
  —The manufacturer provides all the unsafe failure modes of the product.
  —The manufacturer identifies each unsafe failure mode and its frequency for the product.
  —The plant develops a program to qualify the product continuously.

When the above data are taken into account and the system requirements are met, the product may be classified "user-approved safety." This means that the product is suitable for use as a safety layer (i.e., in a SIS).

With automated control components, especially controllers, vendors are often able to supply different control capabilities to meet the user's specifications. Thus, it is important to assure that the vendor supplies information on failure modes that match the particular component, since components with the same hardware may have different failure modes, depending on the specific control capability provided.

3.3.2.3. *User-Obsolete*

A product becomes "user-obsolete" when support and maintenance of it no longer meet user requirements. "User-obsolete" equipment should not be used in new or retrofit applications. All safety equipment will eventually become obsolete. To maintain the original intended level of safety, it is important to initiate a timely replacement.

3.4 EMERGING DOCUMENTS

Following is a list of some of the known emerging standards and guidelines relating to process control safety:

a. IEC SC65A/WG9 "Software for Computers in the Application of Industrial Safety-Related Systems."

b. IEC SC65A/WG10 "Functional Safety of Programmable Electronic Systems: Generic Aspects."

c. ISA SP84 "Programmable Electronic Systems for Use in Safety Applications."

d. IEEE Project 1228 "A Standard for Software Safety Plans."

e. NATO Standardization Agreement (STANAG) 4404 draft entitled "Safety Design Requirements and Guidelines for Munition-Related Safety-Critical Computing Systems."

f. CCPS/AIChE *Guidelines for Improved Human Performance in Process Safety.*

3.5 REFERENCES

3.1 "Safety of Computer Controlled Process Systems," Technica, Inc., Public Report C970, prepared for Dienst Centraal Milieubeheer Rijnmond, Netherlands (1988).

3.2 *CCPS/AIChE Guidelines for Hazard Evaluation Procedures, 2nd edition with Worked Examples.* New York: American institute of Chemical Engineers (1992).

3.3 CCPS/AIChE *Guidelines for Chemical Process Quantitative Risk Analysis* New York: American institute of Chemical Engineers (1989).

3.4 *CCPS/AIChE Guidelines for the Safe Storage and Handling of High Toxic Hazard Materials.* New York: American institute of Chemical Engineers (1988).

3.5 U.K. Health & Safety Executive. *Programmable Electronic Systems in Safety Related Applications.* London: Her Majesty's Stationery Office (1987).

3.6 Bryant, J. A., "Is Your Plant's Control System Safe" (August 1979)

3.7 King, C. F., and D. F. Rudd. Design and maintenance of economically fault-tolerant processes. *AIChE Journal, 18,* 257 (1972).

3.8 IEC Standard 812: "Analysis techniques for system reliability—Procedure for failure mode and effects analysis (FMEA)," Geneva.

3.9 U.K. Atomic Energy Authority, Systems Reliability Service. *Risk Control and Instrument Protective Systems in the Process Industries.* Warrington, U.K. (1980).

3.10 CCPS/AIChE *Guidelines for Process Equipment Reliability Data.* New York: American institute of Chemical Engineers (1989).

3.11 *Fault Tree Handbook.* Washington, D.C.: U.S. Nuclear Regulatory Commission, NUREG-0492 (1981).

3.12 IEC Standard 1025: "Fault tree analysis," Geneva.

3.13 IEC Standard 1078: "Reliability block diagram method," Geneva.

3.14 Lees, F. P. *Loss Prevention in the Process Industries* (Vols. 1 and 2). Stoneham, MA: Butterworths (1980).

3.15 Henley, E. J., and H. Kumamoto. *Reliability Engineering and Risk Assessment.* Englewood Cliffs, NJ: Prentice-Hall (1981).

3.16 Shooman, M. L. *Probabilistic Reliability: An Engineering Approach.* New York: McGraw-Hill (1968).

3.17 Rasmussen, N. C., *Reactor Safety Study, WASH-1400.* Washington, D.C.: U.S. Nuclear Regulatory Commission, NUREG-70/014 (1975).

3.18 Lasher, R. J. Integrity testing of process control systems. *Control Engineering* (October 1989).

3.19 CCPS/AIChE *Guidelines for Improved Human Performance in Process Safety,* New York: American institute of Chemical Engineers (1994).

3.20 Frederickson, T. and L. V. Beckman, "Comparison of Fault-Tolerant Controllers Used in Safety Applications." ISA Paper #90-1315 (1990).

4

SAFETY CONSIDERATIONS IN THE SELECTION AND DESIGN OF BASIC PROCESS CONTROL SYSTEMS

4.0 INTRODUCTION

This chapter addresses the safety aspects of the basic process control system (BPCS). As discussed in Chapter 2, the BPCS provides a key layer of protection required to prevent or mitigate hazards in a chemical processing plant. Figure 4.1 shows the BPCS as part of the total process-control system. The BPCS includes:

- The field instrumentation required for the regulation of the process (for example, measuring elements, sensors, signal transmitters, final-control elements and actuators, as well as various local, i.e., field-mounted, process controllers, analyzers, recorders, and indicators).
- The BPCS process controller(s) located in the control center and equipment rooms (for example, digital, distributed process control systems and panel-mounted analog or digital instrumentation for regulatory control and for verifying, diagnosing, and mirroring the safety interlock system (SIS), and sequencing systems for continuous and batch process control).
- The human/machine interface located in the control room for operator control and for engineering access to the BPCS.
- The supervisory control computers located in the control room for advanced, supervisory-level control, data acquisition, and plant optimization, as well as for communication with plant-level and corporate–management information systems. (Note that general-purpose computers should not be relied on for critical regulatory control. Loss of general-purpose computing should not lead to unsafe operation.)
- All the signal interconnections among these various components (wiring, cabling, and tubing).
- All system software, databases, and application software required by the process controllers and general-purpose computers.

The BPCS excludes specialized safety instrumentation and the safety interlock system (SIS). SISs are covered in Chapter 5. The basic process control system is a protective layer, but usually does not qualify as an independent protection layer. Safety concerns play a critical role in the definition of the BPCS for a particular process.

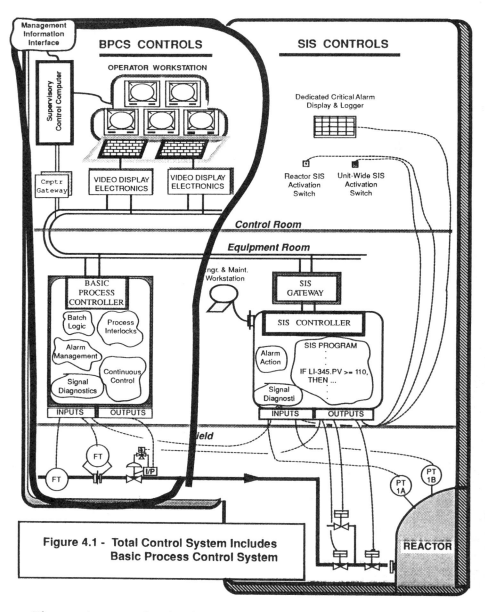

Figure 4.1 - Total Control System Includes
Basic Process Control System

The requirements for the design of the BPCS come from many sources. However, all design requirements fall into two categories: safety and function with maintainability a subset of each. Safety requirements come from the Process Hazards Analysis (PHA) process, while functional requirements come from business requirements.

Before designing the basic control strategy for a process, the control-system engineer must thoroughly understand the process. This fact *cannot* be over-emphasized. Because the chemical-process industry (CPI) is so diverse, this requisite, process-specific knowledge cannot be described in detail here. In general, however, the personnel responsible for BPCS control-system design must know the

- Purpose of the process and its economics.
- Basic nature and design of the process (continuous, batch, or semibatch).
- Fundamental process chemistry and physics.
- Potential hazards posed to employees, the community, the environment, and the operating facility.
- Equipment and process limitations.
- Steady-state and dynamic behavior of the process.
- Normal and abnormal operating states of the plant.
- Availability and quality of BPCS support services, utilities, etc.
- Duties and capabilities of the operation and maintenance groups.
- Sources and consequences of upsets to plant operation.

Once this basic understanding has been acquired, the next step for the BPCS designer is to determine what process variables have to be controlled. This takes into consideration the regulation of product quality, inventories, and the capacity (throughput) of the plant, equipment protection, economic viability (e.g., the minimization of raw material and utility usages), and safety and environmental concerns.

The third step in the strategy for the design of the BPCS is to determine what process variables or conditions should be directly measured on line or inferred from other on-line measurements. Reliability, accuracy, reproducibility, speed of measurement, ranging, and scaling must all be considered. See Section 4.3.

The fourth step is to determine what control variables or process conditions can be manipulated automatically.

The final and most involved step is to design the control algorithms (or laws) to achieve the desired regulation of the process. This accomplishes the closing of the loops by associating measured and manipulated variables in single-loop, negative-feedback controllers. The number of such loops must equal the number of degrees of freedom available for specifying the state of the process. Controller actions, initial tuning, and scaling must be determined. See Section 4.5. The means for implementing the recipes for the batch aspects of the process must be designed. See Section 4.10. In addition, advanced control (regulatory or supervisory) and/or optimization strategies may be appropriate. See Sections 4.5 and 4.12. The requirements for the operator/control-system interface, including alarms, must be determined. See Section 4.6.

All the above details must be thoroughly documented.

A number of references and guidelines are available to assist the control-system engineer with this task, for example, process-control textbooks, journal articles, codes, and standards. Previous experience also plays an important role. Computer simulation of the complete system is a useful and sometimes necessary way to acquire the requisite understanding of the process and to develop and test viable BPCS designs. This is especially true for new process designs, highly interactive processes, and processes exhibiting unusual dynamic behavior.

Designing, selecting, installing, operating, and maintaining the BPCS requires a team approach. The entire team of instrumentation and control specialists must address a number of safety-related issues, including those specified by the PHA team. Included among these are:

- The degree of risk reduction that is required of the BPCS.
- The regulation of the process in a safe and stable manner to minimize the frequency with which the SIS, if present, comes into operation.
- System reliability of the basic process control equipment, including the failure of equipment due to electrical overloads, power-line conditioning problems, etc., and to environmental effects, such as water, ice, corrosive vapors, signal interference, heat, and vibration. See Section 3.1.2.2.
- Failure modes of the various hardware and software components of the basic process control system.
- The display of the process condition to the operating personnel in a logical fashion so that whatever is occurring can be quickly, easily, completely understood, especially during a process upset.
- Personnel safety—protection of plant personnel from undue exposure to risk, such as that arising from faulty electrical (safety) grounding, high-voltage wiring, and hot surfaces.
- The selection of electrical equipment suitable for the area classification (i.e., an area that is likely to contain explosive vapors or dusts).
- The proper training of the personnel involved in operating and maintaining the basic process control system.

The detailed design of the BPCS follows the preliminary work of the PHA team as depicted in Figure 2.3 in Chapter 2. The detailed PHA is performed near the completion of the detailed BPCS design to verify that all safety requirements have been met and to allow time for implementing any changes that may be needed. A general design strategy for the BPCS, which takes into consideration the functional and safety requirements of the BPCS as outlined in this section, is shown in Figure 4.2.

Specifically, this chapter begins with a general discussion of instrumentation and signal types primarily from a safety viewpoint. This is followed by a discussion of field measurements and actuators. The operator interface to the control system is presented. The safety aspects of process control systems,

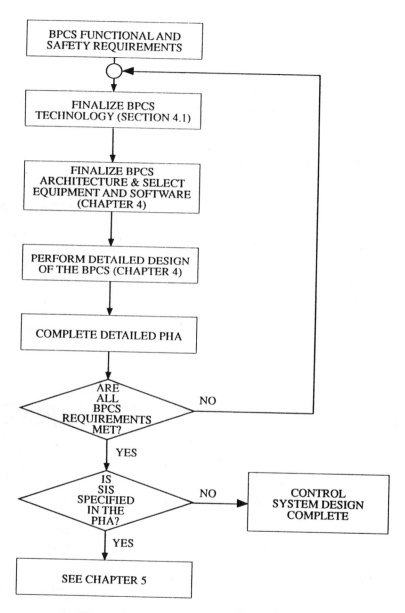

Figure 4.2 A general BPCS design strategy.

including such ancillary topics as power supplies and grounding, are discussed. Sequential process control, its connection with continuous process control, and its special safety concerns are presented. Safety issues involving the management of databases and the development of application software in digital control systems (for both distributed process control and supervisory control) are discussed. Finally, documentation of the BPCS is discussed.

4.1 BPCS TECHNOLOGY SELECTION

The PHA team should conduct their preliminary evaluation with a BPCS technology in mind. Considerations in selecting the BPCS technology are discussed in this section.

The latest developments in basic process control equipment are primarily digital in nature, but analog components for continuous processes and direct-wired systems for sequential processes are still quite common. Many basic process control systems are hybrid, employing a combination of analog and digital components. Furthermore, some chemical-processing facilities are instrumented exclusively with analog devices and direct-wired systems.

In what follows, the analog and digital approaches to the hardware implementation of the basic process control system are compared from the viewpoint of safety. The advent of large-scale integration (LSI) in electronics, and particularly the invention of the microprocessor, have rendered digital devices sufficiently powerful, reliable, and inexpensive to overcome virtually all the advantages offered by analog devices. However, it is still important to understand the basic differences between analog and digital implementations.

When control loops are implemented with analog devices, inputs are monitored continuously, control algorithms are solved continuously, and outputs are manipulated continuously. Conversely, when control loops are implemented with digital devices, inputs are sampled periodically, control algorithms are solved sequentially, and outputs are updated periodically. See, for example Ref. 4.1.

There are several safety implications of these facts. For one, a single component failure in an analog control system tends to affect only one control loop, whereas in a digital system, the failure of a single component (which may by design be shared among many loops) can affect multiple loops. However, highly reliable components and redundant configurations available in digital systems have surpassed the advantages of single-loop analog systems.

Another possible safety implication is that analog (PID) controllers are faster acting than standard digital ones, exhibiting no digital, sampling-time delays. This theoretically may allow higher controller gains and presumably tighter or more stable control with analog systems. But once again, LSI and high-speed, solid-state technologies allow digital devices to be dedicated to a single loop or to a very few loops, permitting sampling times to be sufficiently short to emulate analog controllers.

Digital devices offer superior flexibility, computational power, and ease of reconfiguration. For example, transport-delay (or pure-deadtime) calculations and PID control algorithms with very long integral (or reset) times can be readily implemented in digital controllers. Also, as just pointed out, digital systems exhibit wider bandwidth (i.e., produce more accurate signals, nearly free of noise) and are less susceptible to drift and general error than are analog devices. Furthermore, the use of sophisticated control algorithms for multi-variable and nonlinear control, the use of on-line optimization and process identification, the superior data-acquisition, display, and storage capabilities, the ability to validate sensor performance using a process model, and the superior communication capabilities of digital systems can all lead to better regulated and safer plant operations. Continuous automatic testing and diagnostics are also possible and can reduce the need for periodic, manual tests. However, there are potential negatives to digital systems. These include the possibility of information overload, the use of filtered and compressed data, failure modes in both hardware and software, larger areas of responsibility for the operators, and less familiarity on the part of the operators with details of the process.

Finally, the topic of the direct-wired-relay versus the programmable-electronic-systems (PES) approaches to sequential control parallels the earlier discussion concerning continuous process control. Thus, in general, the digital implementation of sequential control is more reliable, offers more computational power, and is more readily reconfigured than are separate relays and switches. Furthermore, communication is more easily handled digitally.

Important technical characteristics that pertain to specific types of control technology and that should be considered in selecting BPCS technology are discussed in the following sections.

4.1.1 Analog

Analog control systems come in two primary implementations: pneumatic and electronic. Both have been used for decades and are still encountered on a daily basis. But from the viewpoint of safety, there are a few differences that should be mentioned.

Electronic analog systems are inherently faster acting than pneumatic systems (as measured by standard frequency-response techniques), although deliberate damping can render an electronic system slower than a pneumatic one. Also, the amplifier gain in an electronic controller is somewhat higher than that in a pneumatic flapper–nozzle amplifier, making electronic controllers somewhat more accurate. Reliability is also an important safety concern, and modern electronic devices are generally considered as reliable as pneumatic devices under normal circumstances. However, pneumatic analog devices are safe (not an ignition source) in the presence of explosive vapors or dusts, whereas electronic devices must be made to comply with the requirements of the hazardous-area classification. Pneumatic analog devices are suitable for simple, local-control (field-mounted) applications. (See Section 4.2.1.2.)

Analog controllers often offer a manual station which allows the controller to be replaced while maintaining a signal to the final control element.

4.1.2 Digital

Most of the topics in this book involve modern, digital (PES-based) control systems, and the specific safety issues peculiar to them will be addressed. However, some general comments on digital systems are worth mentioning at this point.

There are a number of choices with digital systems: direct-digital control (DDC), distributed process control, and single-station digital controllers. Digital process control includes distributed control systems (DCSs) and programmable logic controllers (PLCs). The DDC and DCS approaches are illustrated in Figure 4.3. An issue arises with DDC and DCS digital systems that is not an issue with most analog systems: common-mode failures. DDC can lead to reliability problems because of the possibility of common-mode failures that could affect every loop in the system. DCSs, on the other hand, are distributed in nature (with common-mode failures affecting only a small number of loops), tend to be better hardened for the process environment, and offer greater software security. Thus, modern, multiloop, digital-based, basic process control systems are primarily implemented as DCSs. For a general discussion of DCSs, see Ref. 4.2. See also Section 4.5.

DCSs and PLCs are both capable of performing continuous and sequential control functions. However, PLCs are generally more appropriate for implementing interlock logic, and PLCs usually offer shorter sample times—i.e., faster scan rates—than do DCSs. See, for example Ref. 4.3.

In many cases, plants will have both types of programmable electronic systems (PESs). The choice of which type to use for the SIS is addressed in Chapter 5.

Redundancy is one way to improve the overall reliability of a system by increasing the mean time between system failures. In addition, redundancy may make repair easier and thereby improve system availability. In PESs, redundancy is a concern of both the hardware and software designers and affects the specification and selection of components by the user. (Software reliability is discussed in more detail in Section 4.11.)

Redundancy should be used in PES-based BPCSs when the availability of the control system is essential for either safety or business reasons, when

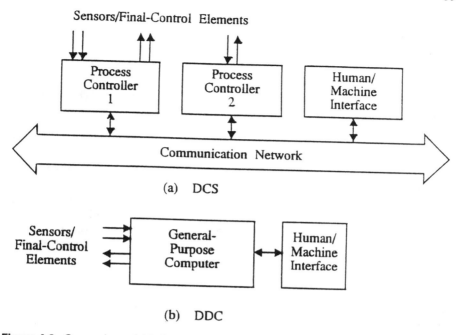

Figure 4.3 Comparison of (a) distributed control system (DCS) and (b) direct-digital-control system.

continuous production is required, or when on-line maintenance of the control system is required. A properly designed and installed redundant system should allow:

- The control system to continue to function in the event of a single failure.
- The redundant, backup unit to function automatically upon failure of the primary unit (in a manner that is bumpless and therefore transparent to the operators except for alarming the failure).
- Maintenance can be performed on one unit while a redundant unit continues to operate.

To accomplish this, each unit may monitor the other, or an independent, master computer may monitor all redundant units. Redundant systems should be able to identify component, software, or communications failures and transfer operation to the backup unit gracefully and safely.

In the following sections, redundancy in various modules of a typical PES-based BPCS is discussed. Sensors and final-control elements will also need to be made redundant in some digital, as well as analog, systems.

1. Operator Workstations. The operator workstation is made up of redundant visual-display units (VDUs), etc. Redundant operator display and data entry means are necessary in the BPCS. This includes redundancy in all the electronics; multiple functionality does not guarantee redundancy. Redundant operator stations are required to ensure access to the process and to alarm information (where annunciator panels do not exist).

2. Communication. Communications within the BPCS are critical. Redundant communication paths are necessary to ensure reliable communication. On-line diagnostics for automatically detecting hardware failures in both parallel and serial communication buses are essential, because such failures are often difficult to detect, diagnose, and repair. Communication controllers should also be redundant with automatic switch-over. Failure of a communication controller or path should be alarmed to the operator.

There are additional hierarchical structures for communication between the control and I/O processors. These must be examined for redundancy also.

3. Controller Processors. Redundancy is often necessary with automatic, bumpless switch-over depending on process considerations. This switch-over should be sufficiently fast to satisfy the dynamics of the process. Not all processors within a BPCS need be redundant. For example, special, higher-level, supervisory processors may not require redundancy. Such supervisory computers should not be used to implement critical regulatory-control functions.

4. Input/Output and Power. PES-based control systems may have some form of redundancy available for the I/O. If this built-in redundancy does not meet the needs for a particular application, the user may have to add redundancy.

A BPCS should also include redundant power supplies. (See also Section 4.8 for a discussion of power-supply and distribution systems.)

Back-up batteries necessary for volatile memory should be checked and replaced periodically, because their failure is not always apparent.

4.2 SIGNALS

Reliable and accurate signals are essential to the safe operation of a chemical-processing facility. The safety aspects of signals in the basic process control system are discussed in this section. The subject is organized according to the characteristics of signals, the media by which signals are transmitted, and the protection of signals from interference.

4.2.1 Signal Characteristics

In this section devices and signals are categorized by their energy levels for the purpose of safety.

4.2.1.1 Live Zero

Pneumatic signals in control systems are standardized. The most common range is 3 to 15 psig. Three to 15-psig signals are used for the linear operation of pneumatic controllers, transmitters, and computing devices, including valve positioners. Six to 30-psig signals are typically used to actuate spring- and diaphragm-operated control valves with high force requirements. Because the nonzero, low-range values of these signals represent the minimum value of the corresponding process variable—for example 0%—the signal is said to possess a "live zero."

The use of live zeros to represent process variables offers the advantage of more effective maintenance. For example, a pneumatic signal level of 0 psig would imply the complete absence of signal and perhaps indicate a disconnected signal line or a closed air-supply valve.

In signal measurement, live zero is not the same as zero suppression or elevation, which are forms of scaling. Scaling, in general, is a useful means of increasing the accuracy and/or precision of transmitted signals. By making the entire signal range useful, scaling has safety implications of its own, but does not affect reliability or maintainability.

The concept of live zero also applies to standardized electrical signals. One to 5-volt or 4 to 20 milliampere analog signals can allow the automatic detection of signal losses, such as broken wires or failed components, thus increasing reliability and safety. A live zero also improves signal accuracy by increasing the signal-to-noise ratio. In two-wire systems, a live zero has the additional advantage of allowing field-mounted devices to be powered with the same two wires used for signal transmission.

Today's PES-based basic process control systems may implement a form of live zero called "end-of-line detection." Here a signal is transmitted that checks the health of the signal transmission against opens, shorts, and grounds. The signal is nonintrusive so as to be detected not by the basic process control system, but by the diagnostics of the PES to ensure signal integrity.

4.2.1.2 Low-Energy, Intrinsically Safe and Nonincendive Electrical Signals

Pneumatic devices are safe, being incapable of igniting explosive vapors or dusts. Electrical devices may possess enough electrical or thermal energy to ignite hazardous atmospheres by arcing, sparking, or heat generation. Intrinsically safe and nonincendive circuits offer specialized engineering approaches to make complete systems of low energy electrical devices and wiring safe. The brief overview that follows is directed toward the user of electrical instrumentation, not designers of such equipment, and is not a definitive discussion. The Instrument Society of America (ISA) offers several good books on the subject, and the reader is encouraged to consult them. See, for example, Refs. 4.4–4.6. Several standards are also relevant, including Refs. 4.7–4.10.

Note: Many of these standards are updated on a scheduled basis, and the most recent revision should be consulted.

It is important that users of electrical control instrumentation understand the basic concepts of intrinsically safe and nonincendive circuits and how they affect safety. Equipment users should understand the relevant standards, the installation procedures provided by the equipment vendors, and the basic principles of this subject to properly use and maintain such equipment.

The areas of the plant in which intrinsically safe and nonincendive equipment may be used are categorized according to the types of ignitable vapors or dusts present and how likely these substances are present in the flammable range. The National Electrical Code (NEC) of the National Fire Protection Association, Inc. defines classes and groups of ignitable materials and divisions based on the probability that these materials are present. The basic tenets of nonincendive and intrinsic safety are that there should be no greater danger with the safe electrical device present than without it, that there should be at least two unrelated (independent) levels of protection afforded, and that the loss of either of these levels should be improbable. That is, no single fault should lie between a safe situation and a disaster. See, for example Refs. 4.11–4.13.

The major difference between nonincendive and intrinsically safe equipment is this: Nonincendive equipment will not cause ignition of a specified hazardous material in its most readily ignitable concentration thermally or by electrical means during normal operation. Intrinsically safe equipment, on the other hand, will not cause such ignition under either normal or abnormal conditions (e.g., involving faults such as reversed wiring, short circuits, broken light bulbs, and so forth). Therefore, intrinsically safe equipment can be used in areas of the plant where ignitable atmospheres are normally present (Division 1 locations), whereas nonincendive devices can be safely used only in areas where a process failure will release ignitable substances (Division 2 areas). This is an example of the two-fault tenet cited previously. In addition, in an intrinsically safe system, maintenance and calibration work can be performed on field devices without special precautions.

Formal certification of intrinsically safe and nonincendive devices by recognized third-party laboratories, such as the Factory Mutual Research Corporation (FM), Underwriters Laboratories, Inc. (UL), or the Canadian Standards Association (CSA), is often obtained by equipment manufacturers. However, such certification is based on normal atmospheric pressure, air composition, and temperatures. Therefore, special situations not covered by the standards of intrinsic and nonincendive safety can exist in some plants which make these techniques nonapplicable.

Intrinsic safety and nonincendive systems are concerned with the entire control-system loop, including those devices located in the control house that are connected to the field-mounted devices in hazardous areas of the plant. The design and installation of such equipment is governed by standards promulgated by ISA and the National Fire Protection Association (NFPA) referenced earlier. Care must be exercised in maintaining proper separation of safe from unsafe devices, in not exceeding the levels of the electrical variables specified (e.g., capacitance), and in protecting against power and grounding faults, which could render a system ignition-capable. Energy-limiting barriers are utilized to separate intrinsically safe devices from those that are not intrinsically safe. Continued safety requires careful inspection and maintenance procedures. Inspections should be performed periodically and should include checking for proper installation, the presence of unauthorized modifications, signs of damage or wear, and making sure the system is being used as intended. Safe devices should be clearly identified in the field so that special procedures will be observed.

Safe electrical devices are but one means of hazard reduction in the plant. For example, complete isolation of electrical devices from hazardous substances (such as by encapsulation or sealing) may be a suitable technique in some situations. The following section describes several other approaches that can be used with devices requiring high energy levels for their operation.

4.2.1.3 High-Energy Signals and Explosion-Proof Signal Housing

The techniques for risk reduction for low-energy electrical devices may not be practical for high-energy devices. Therefore, high-energy electrical devices *must* be made safe in other ways.

In addition to the techniques of encapsulation and sealing just cited, another preventive method is to prohibit contact of ignitable substances with hot or electrical devices by purging those devices with air. This prevents sufficient concentrations of ignitable substances from accumulating in the device. Inert gases should not be used in human-occupied areas for purging because of the risk of oxygen depletion. See, for example, Refs. 4.14 and 4.15.

Purged or pressurized enclosures may have requirements for deenergizing before opening for calibration or testing until power has been removed and enough time has elapsed to allow internal components to cool down sufficiently. Interlock switches for shutting off power to the cabinet, if present, should never be bypassed.

There are also techniques for actually managing an internal explosion and controlling its flame path to prevent it from spreading. The use of confining, so-called explosion-proof (or flame-proof) housing is an example of this. The housing containing the electrical device and its associated wiring is built so as to confine any explosion that might occur. The principle involves designing any gaps in the housing to quench the fire by conducting heat away from the flame front. Many devices are available in explosion-proof enclosures. Standards are set by Underwriters Laboratories, Inc., among others. It is important with explosion-proof housings that all enclosures be properly closed and that sealing surfaces not be damaged. And, as is the case with pressurized equipment, an explosion-proof housing should not be opened without powering down the device and allowing it to cool or gas-testing the area. These classes of devices also need to be inspected periodically for damage to the sealing surfaces.

Explosion-proof housings are designed for Division I locations. Other, simpler enclosures suitable for Division II applications also exist. An alternative to the approaches mentioned is to locate equipment in a nonhazardous area (for example, in the control house, as opposed to the field).

4.2.2 Transmission Media

The safety considerations in the selection of the media by which control-system signals are transmitted within a chemical-processing facility involve several general issues. In addition to the question of a potential ignition source, there are questions of accuracy and reliability. Common signal transmission media are discussed in the following sections. Also, see Section 4.2.3 on signal protection.

4.2.2.1 Tubing
Pneumatic signals suffer from transmission delay, especially over long distances, and very narrow or damaged tubing can lead to very poor dynamic transmission response. Similarly, restricting valves, especially if followed by relatively long tubing runs, and volume tanks should not be used where response time is critical.

The choice of material is influenced by the corrosive and thermal environment of the plant. Plastic tubing can fail at moderately high temperatures, although it should be noted that the deliberate use of plastic tubing may provide a means of helping to ensure the loss of signal in a fire (e.g., the signal going to a fail-closed valve). Failure of some plastic tubing is due to attack by ultraviolet light. In general, tubing needs to be supported and protected from damage.

4.2.2.2 Fiber Optics
From a safety standpoint, fiber-optic cable is an appropriate medium for plant digital-computer control, networking, and signal transmission. Fiber-optic cables are highly immune to all forms of electromagnetic interference, even without shielding or special precautions. Fiber-optic cable offers complete

electrical isolation and is resistant to most chemicals. In addition, fiber-optic cable can be safe (i.e. not an ignition source by keeping the energy level sufficiently low), but the transmitting and receiving equipment must comply with the hazardous-area classification. Fiber optics provide high-quality signals, with relatively low noise and low signal attenuation. However, the installation of fiber-optic cable requires special tools and procedures.

Numerous standards covering the testing of fiber-optic systems have been promulgated by the Electronic Industries Association (EIA). An introduction to the concepts of fiber optics is given in Ref. 4.1. The specialized application of fiber optics to sensors is discussed in Ref. 4.16.

4.2.2.3 Radio Waves

Telemetry may be used in industrial applications when it is more cost-effective than other media, but it is generally limited to line of sight. Radio waves are subject to a number of interference sources, such as radiofrequency interference (RFI) and electromagnetic interference (EMI), including lightning, although the degree of immunity to interference as well as the speed with which data can be transmitted (i.e., the baud rate for digital data) are dependent on the frequencies and modulation methods used. As is the case with fiber-optic signals, radio waves themselves can be low enough energy not to be an ignition source, but the transmitting and receiving equipment must be made safe in hazardous areas. See, for example, Refs. 4.1 and 4.17.

While telemetry is employed in supervisory control and data acquisition (SCADA) systems for remote applications, radio-wave communication requires careful consideration when used for critical applications in process control.

4.2.2.4 Wire

Wire includes coaxial and twisted-pair cables for digital signaling and communication, as well as twisted-pair conductors for electronic analog instrumentation. Chemical or atmospheric corrosion, mechanical protection, damage from fire, moisture, electrical isolation, and grounding problems are all relevant safety concerns to control and instrumentation specialists. Low-level signal wiring requires some type of protection (e.g., conduits, trays, etc.), proper grounding of the shielding, and segregation from other signals that could cause crosstalk problems.

Electrical signals are subject to various forms of noise and interference, with twisted-pair wiring generally more susceptible to interference than coaxial cable. Common-mode voltage interference can be generated by power circuits in electrical devices and is preventable with electrical isolation techniques. Normal-mode voltage interference is usually the result of improper grounding (i.e., ground loops). Coaxial cable can carry one signal (baseband) or multiple channels on different frequencies simultaneously (broadband).

Wiring is highly reliable and suitable when properly installed. However, leased wiring through outside utilities should not be used in critical control applications. Plant control and instrumentation specialists should be aware of the relevant ISA standards and the NEC of the National Fire Protection Association, Inc., for safe wiring practices.

4.2.3 Signal Protection

Signals in the basic process control system must be protected from interference. In areas of known fire risk, special precautions may be needed to ensure the integrity of control signals. In the following sections, the subject of signal protection is elaborated on.

4.2.3.1 Instrument-Air Quality
High-quality instrument air in chemical-processing facilities is essential. This is true even in modern facilities employing programmable electronic systems (PESs), where instrument air is used for valve actuation, instrument purging, etc. Instrument air should be free of particulates and corrosive substances and also dry and oil free, with a dewpoint of at least $-40°F$ at the maximum instrument air system operating pressure. If the coldest ambient temperature encountered is below $-40°F$, then the dewpoint must be at least $10°$ lower. Instrument air should come from a dedicated system and should never be employed for process uses, to operate pneumatic tools, or where check-valve failure could cause contamination or pressure loss in the system. Instrument air must also be maintained at proper pressures. A modern system should have a minimum of 80 psig pressure available to every user. There should be an operating plan to deal with the loss of instrument air. This should include alarming the loss and should include a backup system, reserve supplies, or other contingency plan. Nitrogen should only be used as a backup for instrument air in an emergency or when no other backup supply of air is practical. If nitrogen is used, a method of alarming this condition should be provided, and any areas where nitrogen might displace oxygen should have oxygen monitors with warning lights.

Several national standards and recommended practices address this general subject. See, for example, Refs. 4.18–4.20.

4.2.3.2 Electromagnetic Interference
Electromagnetic interference generally refers to those unwanted electromagnetic phenomena affecting electrical signals at lower than the radio-frequency band. This form of electrical noise commonly stems from motors, transformers, power conductors, and coils (e.g., solenoid-operated devices). Crosstalk, which can afflict alternating-current and pulsating signals, is a special type of EMI and originates from poor wiring practice. EMI is transmitted by

the wiring and can corrupt the transmitted signal. EMI can cause permanent shifts in signal zeros and spans and random output variations, and can trip alarms and shutdowns. Appropriate filters, along with the use of good wiring practice (twisted pairs, shielding, and separation may be required to attenuate EMI at the input terminals of a control device or data system. Alternating-(AC) and direct-current (DC) systems need to be separated from each other and not mixed in conduits or trays. Other types of EMI radiation, such as high-intensity light sources, can corrupt erasable, programmable electronics. Similarly, instrumentation should be protected from x-rays generated by nondestructive testing equipment. In particular, nuclear (gamma-ray) density detectors can be affected by the x-ray equipment used for testing welds and can lead to erroneous control action.

Electrical interference in general, and EMI in particular, are covered in Ref. 4.1.

4.2.3.3 Radiofrequency Interference

Radiofrequency interference arises from spurious signals in the radiofrequen-cy range (typically 0.5 to 500 MHz). RFI can be generated locally by ignition systems, two-way communications devices, welding equipment, control de-vices, computers, etc. It can also be picked up by one component in a circuit and then propagated to other components. RFI should be prevented rather than filtered out after it has corrupted the system.

Separation from radiofrequency sources, shielding, and protective housing design can be used, along with radiofrequency filters, to prevent RFI. Housing protection, though, can be compromised during periods of maintenance or calibration, where, for example, they become more vulnerable to hand-held communication devices.

RFI is covered by standards promulgated, for example, by the IEEE.

4.3 FIELD MEASUREMENTS

Accurate, reliable measurements of process and utility conditions are neces-sary inputs to any dependable BPCS and SIS. The control engineer should be aware that sensors can drift, give erroneous readings, and drive a feedback control loop into an unsafe condition. Therefore, sensors (and final-control elements) are directly connected to the process, field devices are often the weakest links in a control system. Where safety is an issue, the performance of the sensors and transmitter signals used by the control and information systems becomes a significant factor. Proper selection, installation, and main-tenance are all required to ensure that the information generated can and will be used in the intended manner. Sensor and transmitter selection should be based on proven reliability and the requirements of the application. Preferred equipment lists should be reevaluated periodically based on user experience. Installation and maintenance procedures should be based on the chemicals being handled, the type of process, the environment, the type of equipment selected, and its accuracy and reliability requirements, among other factors. Redundancy, including redundant process connections, may be needed to achieve these requirements.

4.3.1 Smart Transmitters

"Smart" transmitters are generally more accurate, more versatile, and more reliable than standard analog electronic transmitters. However, since these transmitters can be operated in different modes (digital, analog, etc.), care must be taken in their application. For example, a digital indicator, meant for use in the digital mode, can shift the unit's calibration if the transmitter configuration is changed to the analog mode. If a digital smart transmitter is used in an intrinsically safe system with barriers, the barriers must be different from those used with conventional instrumentation. Since digital transmitters can be easily reprogrammed, their use in safety related applications should be covered by special procedures or protected by electronic integrity, e.g., read-only communications (Appendix "D"). The sampling frequency of digital transmitters may be slower than analog units and may not be fast enough (signal update time) for some control applications. When a smart transmitter is used, its failure modes must be considered.

4.3.2 Selection Criteria

Field devices, including final elements, are the weakest link in most control system. Their performance has the largest effect on BPCS and SIS reliability and availability. Selection of field instrumentation involves the use of rigorous application criteria (both selection and installation) by experienced field-instrumentation specialists, with process designers providing special requirements for the normal operation of the process and also special operational characteristics for start-up, shutdown, and abnormal conditions.

Considerations in the selection criteria include:

- *Process Suitability:* fluid state, viscosity, density, entrained solids, aggressive fluid characteristics, pressure, temperature, etc. Example: Two-phase flowing conditions can render most flow measuring devices meaningless.

- *Rangeability* (Turndown) is the ratio of the maximum measurable value to the minimum measurable value of a measured variable. Example: For differential head flow meters, the rangeability is limited. Measurements beyond the available rangeability produce unacceptable values requiring the use of alternate technology or multiple meters.

- *Accuracy/Precision/Reproducibility.* All sensors have specific limitations. The selected field measuring device must satisfy the application requirements. Example: Standard thermocouple tolerances can result in significantly different readings at the same temperature.
- *Sensor Response Time* is the inherent lag or deadtime exhibited by a sensor. Example: The thermal lags associated with protective sheathing can increase the lag time by a factor of twenty relative to a bare thermocouple.

The above problems of accuracy and response time can be additive. The cumulative effects of errors and delays may lead to the inability of a system to respond to a process upset in sufficient time.

- *Materials Compatibility:* Materials used for process piping may not be suitable for field sensor construction. Example: Thin-walled tubing used in some Coriolis flow sensors may have no corrosion allowance and may require the use of higher alloys for some applications.
- *Failure Modes Safety* considerations will require knowledge of the most probable failure direction. Example: A pressure transmitter losing its capillary pressure sensing fluid will give a false low signal.
- *Reliability:* Special ruggedness designed into the instrument or successful application experience in the field may be required. Reliability may also be achieved with multiple sensors or diversity of sensor technology. Example: Electronic components specially designed to accommodate short-duration temperature extremes.
- *Fire Rating* is the consideration of the behavior of a sensor or cabling in a local fire situation. Example: Aluminum will melt in a hot fire, while hot cast iron shatters if quenched by firewater.
- *Maintainability* is the ease with which repairs and maintenance can be carried out in a field environment. Example: Can a device be repaired, tested, and replaced without a process shutdown?
- *Robustness* is the mechanical ruggedness, overrange tolerance, and recovery. Example: The ability to sustain physical abuse in the field or overpressure during process upsets.
- *Installation Requirements* are mechanical and stress considerations when installing field sensors. Example: The piping stress exerted on directly coupled devices or the use of thick-walled piping to provide protection against breakage of the device connection.
- *Environmental Considerations:* Fluid properties must be compatible with the process conditions required by process-lead lines and wetted sensor surfaces. Devices must be tolerant of all ambient conditions (e.g., heat and vibration) to which they are subjected. Example: Freezing/solidification as a result of ambient conditions.

- *Electrically hazardous locations* Example: Instruments must meet the NFPA National Electrical Code requirements for the area classification in which it is installed.

Additional considerations for the installation criteria include:

- *Representative sampling technique:* Is the measurement truly representative of actual conditions? Is the immersion length sufficient? Examples: A thermocouple installed in a stagnant fluid or pressure measurement downstream of filters.
- *Process-lead-line considerations:* Ensure that sensing lines do not inhibit sensor response. Example: Self draining (continuously sloped) taps in wet-gas connections or process-lead lines with regulated heat tracing to prevent freezing/condensing of process fluids.
- *Minimize transport lag:* Ensure that a measurement is representative of process conditions at all times. Examples: Use the shortest possible sample lines or highest possible bypass flow rates in sample loops. Temperature measurements reflect the active process.
- *Serviceability:* The installation should reflect the needs to remove, calibrate, and test the device. Examples: Isolating/depressuring capability. The ability to purge/vent/drain connecting lines safely.
- *Functional Testing* is verification of a field sensor by simulating process conditions with a portable testing device. Caution: The bench calibration of a field sensor and electronic signal simulation do not verify the actual performance of connected devices.

This does not constitute an exhaustive list but should help the control engineer recognize some important issues.

4.3.3 Flow Measurement

Flow measurement is one of the most widely utilized measurements in chemical plants and refineries. It may account for up to 70% of the transmitters. Before selecting and installing a flowmeter, it is necessary to understand the chemical and physical properties of the fluid, the accuracy and rangeability requirements, the materials of construction, the piping system, and whether the meter will be used for batch or continuous operations. In most flowmetering applications, it is essential that the meter body be installed so that it is either full of liquid or gas. Liquid meters that contain gas or vapor will give false readings, or not work at all. Gas or steam meters that contain liquids or solids may be destroyed by the high velocity particles hitting the sensor. It is important to have a method of calibration and verification agreed upon by operations and maintenance, before the meter is installed. Meters installed in piping systems carrying hazardous materials may require flanged connections as opposed to wafer-style bodies.

4.3.3.1 Head Type
The most widely used method of volumetric flow measurement is the differential-pressure (head) meter (i.e., d/P transmitter and a restriction such as an orifice plate, flow nozzle, etc.).

Although the introduction of new volumetric and mass electronic flow measuring methods offer significant alternatives, the orifice plate and d/P transmitter are still in common use because of the ease of verifying calibration.

4.3.3.2 Vortex
This flowmeter is an in-line-mounted device having no moving parts. It provides linear signal characteristics to flow with improved rangeability and accuracy. Some versions allow sensor replacement without removal of the body from the line. Vortex meters are considered rugged, forgiving flowmeters in difficult liquid applications. Because this meter is limited to a minimum fluid velocity, flow is electronically forced to zero at flow rates below 10–15%. Hence there is the possibility of undetected low flow rates. Process vibrations/pulsations can affect certain designs, impairing their performance.

4.3.3.3 Magnetic
Magnetic flowmeters are suitable for metering electrically conductive slurries or liquids that are corrosive, dirty, viscous, or otherwise difficult to measure. Special care must be taken to ensure that the metal electrodes are totally inert to the process liquid. The meter should be installed in such a way that the meter body is full of liquid to ensure accurate measurement. Rangeabilities well in excess of 10 to 1 are typical. It is important to properly ground the meter and electronics.

4.3.3.4 Turbine
Turbine flowmeters may be used on clean streams where high accuracy and/or high rangeability is required. Viscosity must be within the manufacturer's recommended range. Inlet strainers ahead of a straight run of pipe are required to prevent turbine damage. Process conditions that can cause flashing are known to cause turbine blade overspeed which can damage rotor bearings. Liquid impacting the rapidly rotating blades of a turbine meter in gas service can cause rotor failure.

Insertion-type turbine meters are available for use on clean noncorrosive fluids (such as steam, water, nitrogen, or air). These units are especially useful where shut down of the line is not possible and a block valve can be "hot-tapped" on the pipe. In addition to having the same requirements and limitations of in-line meters, the positioning of the inserted turbine may also impact the metering accuracy.

4.3.3.5 Coriolis
Coriolis-type meters are high accuracy mass flowmetering devices. Each manufacturer has its own flow path configuration to achieve the desired accuracy, sensitivity, rangeability, pressure drop, and special features. This class of meters is subject to piping stresses and vibration. Care must be taken in installing the sensors in properly fabricated and supported piping systems. In high-risk applications, secondary enclosures may be required to contain failures of the thin walled vibrating tubes, welds, or brazing.

4.3.3.6 Ultrasonic
Nonintrusive ultrasonic flowmeters have their sensors mounted on the outside of the pipe which eliminates the concerns about materials of construction, mechanical wear and leaks from and around the meter body. Meters can be added without shutting down and draining the system. Some types of ultrasonic meters have sensors that are in contact with the process fluids. Accuracy and repeatability can be affected by gas bubbles and slurries which may make one or more of the different varieties of meters useless. Because this technology is very complex, the manufacturer should be advised of all essential physical characteristics of the flowing liquid and the piping system (such as pipe size, schedule and materials of construction). Though there are many successful applications of this type of flowmeter, it is generally considered complex and technically challenging. Success in one application does not guarantee success in a seemingly identical application.

4.3.3.7 Thermal Mass
Thermal-mass meters are available in in-line body configurations and insertion styles. Their greatest potential is in applications with high rangeability and low pressure-drop requirements. Probes are affected by imperfect velocity profiles or incorrect insertion depths. Changing thermal properties of the measured fluid also impacts the accuracy of the measurement. Many installations utilize this meter in flow threshold applications, i.e., flow/no flow. Meters designed for liquid insertion can cause potentially dangerous, high temperatures if the piping goes dry.

4.3.4 Pressure Measurement

Many technologies are available for field measurements of gage or absolute pressures and of vacuums. The choice of wetted materials, the use of isolation diaphragm seals, and elevation or suppression are some of the considerations in selecting pressure transmitters. The effect of pressure overrange, especially in very low-range models, needs to be considered, since very high thermally induced hydraulic pressures can be generated in isolated equipment and piping.

Pressure gages that are used on process fluids should be of the industrial type with blowout protection (or NPT connections are preferred), pigtails or siphons (on steam), and blocks and vents similar to transmitters. Special designs are available to prevent gage failure from vibration, pulsation, and overrange.

4.3.5 Level Measurement

Level transmitters are available in a variety of different operating principles.

4.3.5.1 Head Type

One of the most common types of level transmitters uses a differential-pressure transmitter (d/P), which measures the weight of the liquid between the two connections.

Calibration of level transmitters requires a knowledge of the fluids and their specific gravities. The connecting lines to the transmitter are at ambient conditions, and since bench calibration is usually done with water at ambient conditions, a conversion from the calibration fluid is required. This is often a source of error.

The level transmitter often involves suppression of elevation techniques to compensate or null out the static head of reference legs or the process fluid.

Sealed-diaphragm capillaries are often employed to simplify process connections or to isolate the process fluids. Care must be taken in these installations to minimize differential temperature effects on the sealed system and to account for diaphragm-sealed elevation differences.

Condensible fluids or liquid carryover, which can contaminate the reference leg, can invalidate level readings. Reference-leg fluids must be chosen with process compatibility in mind. Ethylene glycol, a commonly used fluid, may be a contaminant in some processes. There may also be environmental concerns with the use of ethylene glycol. The use of food-grade propylene glycol may be an acceptable alternative.

Another variation of the d/P method is to install a "dip pipe" down into an atmospheric vessel and bubble small quantities (one SCFH) of an air or nitrogen down into the liquid. The dimensions of the vessel and the length of the dip pipe, along with the specific gravity of the liquid, are critical to the calibration. Once the unit is in place and operating, the major concerns are dip-pipe pluggage and leaks which will give false readings.

4.3.5.2 Capacitance Probes
Capacitance probes and other related sensors (e.g., radiofrequency) are normally Teflon-coated, stainless-steel rods that are installed on to the vessel. They are dependent on the conductivity and dielectric constant of the fluid and are independent of liquid specific gravity and pressure. Their major problem may be the method used to calibrate them. The ideal method would be to install them into the vessel, zero the transmitter, and begin filling the tank with the process liquid under actual conditions while the calibration is proceeding. Unfortunately, these transmitters are normally used on hazardous chemicals which precludes looking into the vessel. It is therefore important for the operating and maintenance people to agree on a calibration and verification method before the unit is placed into service. Problems may also be encountered with packing-gland leaks, vibration, coating, and mechanical damage to the probe.

4.3.5.3 Displacer (Buoyancy) Type
Displacer (buoyancy)-type level instruments measure changes in level, specific gravity, or interface. A change in the measured fluid causes a change in the forces on the displacer, which is transferred to the transmitter mechanism.

Displacer-type level instruments are practical on clean, noncorrosive, noncoating chemicals. The multiple connections of external chambers can be sources of leaks. Mechanical components and physical size/weight make repairs and overhaul a complex procedure.

4.3.5.4 Nuclear
Nuclear-level measurement is used primarily where there are no other reliable techniques available. Because a nuclear-level device is external to the process vessel, it has found acceptance in measuring viscous materials and in agitated vessels where other types of transmitters would create installation problems. With proper design, installation, and maintenance, the units are no more dangerous than other types of level measurement. However, governmental licensing is required for some applications. Use of these units requires a designated safety engineer responsible for them. Disposal of used sources may be a problem. Personnel protection may be required. Effects of background radiation and half-life deterioration can cause shifts in calibration. The detector is sensitive to moisture and ambient-temperature changes. Administrative procedures are required to ensure that personnel are not exposed to radiation.

4.3.5.5 Level Switches
Level switches are used where continuous level measurements are not required or to back up transmitters. Switches are available utilizing the same types of measurement principles as transmitters and require care in selection

and installation. Usually level switches are simpler mechanical devices giving reliable on/off signals for alarming or interlock applications.

4.3.5.6 Gauge Glasses

Liquid gauge glasses should be of the armored reflex or transparent type with top and bottom connections. However, these types should be used with caution on highly hazardous materials. Internal mica shields should be provided where the contained material is damaging to glass. Armored magnetic types should be considered where glass, gaskets, or threaded connections cannot be used because of corrosive environment or toxic process streams. Block valves and excess flow check valves should be used. Thermal stresses and bolting torques are important considerations. In some applications gauge glasses should be isolated when not in use.

4.3.6 Temperature Measurement

Temperature transmitters are liquid, vapor, or gas filled systems that operate on thermal expansion principles or use thermocouples (T/Cs) and resistance–temperature detectors (RTDs). The major advantages of T/C and RTD transmitters are as follows:

- They can be calibrated in place using electronic sensor simulation equipment.
- They do not require long lengths of protected capillary from the thermal element bulb to the transmitter.
- The transmitter can be located in the control room, equipment room, or in the field (mounted on or away from the thermal element).
- They respond faster to temperature changes.

RTDs or T/Cs can be wired directly to the temperature input card of some systems. This reduces the number of wiring terminations and the need to mount and calibrate a transmitter.

Three- and four-wire RTDs tend to be more accurate than T/Cs. However two wire RTDs can have the advantages of more signal than T/Cs without the added failure modes with the added wires. Twisted, shielded instrument wire is used instead of special thermocouple wire. There is no concern about cold-junction compensation errors when using RTDs. T/Cs are more common in the CPI and can be used for higher temperature applications like furnaces. RTD elements are vibration sensitive. In both cases, care must be taken to make sure that the element, wire, and transmitter types all match. RTDs have several standard α (alpha) coefficients (DIN, ISA, and SAMA). Care must be taken in calibration and substitution. In running thermocouple wire, care must be taken to maintain the correct polarities all the time. If the wire is switched twice, the error will be the difference between the temperature of the field junction box and the place where wire finally terminates. This error could be small in the summer and large in the winter, but difficult to detect. The failure modes of RTDs are not always predictable, and this must be considered in safety sensitive applications. Local temperature indicators should be of the industrial type and installed in wells similar to those used by RTDs and T/Cs.

4.3.7 Failure Modes

Not all instrument failures are predictable, or normal, and many unpredictable failure modes exist. Because thermocouples may open circuit, an electronic system can detect the opening and force the signal high to simulate a high-temperature condition or low to simulate a low-temperature condition. This is commonly referred to as "up-scale or down-scale burnout." While most electronic failures result in down-scale indication, some pressure transmitters have been demonstrated to fail upscale. Since the failure mode of a level transmitter may be down scale, indicating a low level, the control system may continue to add liquid to the vessel. A high level switch may be used to alarm or shut down the feed.

Failure modes for some equipment components are impossible to predict absolutely. In highly reliable designs, it is common to use multiple sensors and voting logic to provide fault tolerance. Another technique is to employ diversity in field measuring devices. Plant experience can also be gained if failures are investigated, analyzed, and kept in a database.

Rigorous, on-line simulation and loop-function testing, combined with routine and preventative maintenance, can enhance the reliability of critical systems, often uncovering covert faults. It is recommended that rigorous testing be done and documentation kept for all safety-critical equipment and that their performance be analyzed formally to ensure continued high levels of availability. See Chapter 6.

Techniques using artificial intelligence can be used to validate sensors. BPCSs and SISs can be designed to validate sensors in the process on line by calculating acceptable rates of change and other considerations.

4.3.8 Use of Sealed and Purged Instrument Process-Lead Lines

For installing pressure and d/P transmitters, the standard procedure is to install the transmitter below the tap in liquid and steam service and above the tap in gas service. In liquid and steam service, the process-lead line fills with liquid and should be zeroed this way to give an accurate reading. The problem is that with some chemicals (e.g., water), the process-lead line may freeze or plug as it cools to ambient temperature. Heat tracing and insulation can be used to prevent the freezing; however, care must be taken not to overheat the liquid and start a boiling process. Electric tracing is available with thermostats or self limiting systems; however, a fault-detection system should be provided to determine if every circuit is operational when required.

Another method of keeping process-lead lines clear is to purge them with an inert or compatible fluid. In this case a small quantity of fluid is flow-controlled into the sensing line using a purge regulator with a needle valve and check valve to keep the process from backing up.

An alternative to process-lead lines is the use of chemical seals. These are available in a large variety of styles, connections, materials of construction, and fill fluids. Fill fluids need to be compatible with the process fluid in case a leak develops. Silicone fluid, which is the standard fill, is relatively inert, but will react with chlorine. Chemical seals should be purchased from the transmitter manufacturer, who will then be responsible for the proper design, filling, and calibration. Great care should be taken in refilling seals. This is a specialized operation, usually done successfully only by the original supplier.

4.3.9 On-Stream Process Analysis

Process-analytical instrumentation can be defined as unattended instrumentation that continuously (or semi-continuously) monitors a process stream for one or more chemical components. A process with continuous on-line analysis is contrasted with one in which "grab samples" are run in a control laboratory.

The purpose of many process measurements (temperature, flow, etc.) is to indirectly control the composition of some product or intermediate. Periodically, the performance of the unit is checked by the control laboratory which runs samples taken by the operator. The problems with this procedure are: exposure to obnoxious or toxic materials when sampling, composition changes in the sample bottle, time lost before results can be used to make corrections in the process (on batch operations, there may not be sufficient time to resample after a correction is made) or, conversely, time lost while this is done.

Chemical plants produce X number of pounds of a material per hour or day with Y composition. We can now measure X to better than ±0.5% in a matter of seconds. It takes hours or days to measure Y with the required precision using the chemical-analysis laboratory.

The faster a composition deviation can be detected, the easier it is to take corrective action. When reactors are charged with new or fresh chemicals from a large storage tank, there is time to do a laboratory analysis ahead of time to ensure that the ingredients are within specifications. But, when recycle materials are charged from other process stages or buffer tanks, there is not enough time to ensure that the composition is correct. There is also the case of batch reactors in series. A batch may drop from reactor "one" to reactor "two" even though there was an undetected problem in step one.

On-stream analysis can detect potential problems so that corrections or transfer to an "off-grade" tank can be made without unnecessary delay. All of these benefits translate into substantial savings in yield, throughput, material balance control (recycle), energy, and potentially, safety. Much of the benefit of process control systems is missed if the "ultimate" process variable needed to optimize the process is not measured.

Analytical instrumentation increases the margin of safety in operations by giving timely warnings or leaks and spills. Assuring that materials remain in specs, reduces the likelihood of corrosion or plugging and keeps reactions within stable limits.

Well applied analyzers are intended to control or monitor operations to reduce off-grade material, control by-product formation, and hold effluent compositions and quantities to some limit. They can lead to better controlled incinerators and scrubbers, reducing breakthrough to the atmosphere. If necessary, analyzers can also monitor final gas and liquid effluents along battery limits or fence lines.

Composition analysis equipment for process streams, ambient air, and waste water requires specific application engineering. This begins with the perception of the need, the merits, and potential problems of the various analytical methods, testing of promising methods, final design, and commissioning of the working analytical system. The final system must include elements for handling the process sample and generating a calibration standard.

Design of process analytical systems should be furnished by a group of specialists. Because the analyzer signals are integrated into the basic control system and analyzer subsystems require their own instrumentation, the analyzer specialists require a strong background in conventional instrumentation.

One important characteristic of process analyzers is that their measurements are related to a standard, not the process. As such, calibration is an essential component of analytical field measurement. Process analyzers are slow and suffer considerable periods of unavailability; consequently, process analyzers are seldom used for critical safety applications.

General characteristics of on-line (on-stream) process analyzers include:

- The sample systems require at least as much time and effort to design as the analyzer itself.
- The majority of the maintenance requirements and problems may be in the sample system.
- The accuracy of the analyzer system may be affected by the condensers, filters and other hardware in the sample system which may remove chemical components and otherwise modify the sample before it gets to the analyzer.
- In many cases, analyzer systems are complex, miniature chemical plants with their own complex control and interlock systems which may require dedicated maintenance.

- Analyzer systems take long periods of time to design, build, test, and turn over to the plant analyzer specialists.
- Many systems require dedicated, air-conditioned houses to hold the sample systems, complex sensors and electronic systems.

For a detailed discussion of the various analyzers used in the process industry, see Ref. 4.21.

4.3.9.1 Transport Lags

Transport lags are related to the type of analyzer used, its location in respect to the sample tap, and the design of the sample system. Many of the problems associated with an analyzer's ability to track a process can be traced to the location of the letdown valve that reduces the process pressure for the sample system, the length and diameter of the sample tubing, and the flow rate of the sample fluid. One must not forget that the analyzer is not seeing the process stream. Temperature and pressure changes in the sample conditioning system can alter the sample substantially.

These factors are all part of the sample system design; however, a few general rules are listed:

- Locate the pressure letdown valve close to the sample tap.
- Keep the sample lines as short and as small as possible.
- Purge significantly more fluid through a bypass line than the analyzer actually needs.

4.3.9.2 Failure Modes

Sample systems may be the source of more failures than the analyzer itself. Plugged filters, lines, or exchangers can be detected by the installation of low-flow switches. The sample system is a small chemical plant that can have some of the same problems as a big one. Therefore, all the safeguards that would be applied to the main plant should be used in an analyzer system if reliable performance is expected.

The analyzer itself will require periodic maintenance depending on the type. Many of the mechanical parts of the older analyzers have been replaced by microprocessor based electronics; however many parts still need to be cleaned and many sensors have limited life cycles and need to be replaced.

4.4 FINAL CONTROL ELEMENTS

No matter what the control system, from manual to sophisticated PESs, some action has to be taken by the operator or the control system to change the flow of some material. This usually means manipulating the position of a valve or changing the speed of a pump or compressor driver. In batch or sequencing operations, the majority of valves are open/closed and the motors on agitators, pumps, blowers, etc. are running/not-running. There also may be some throttling control valves, control dampers and variable speed drives. In continuous plants, the majority of automated valves in the BPCS are the throttling control type. The on–off type are used mainly for the shutdown systems. The majority of the motors still tend to be the on–off type, although variable-speed controls are becoming more commonplace.

4.4.1 Automated Valves

Automated valves can be divided into two main categories: on–off (or block valves) and throttling control valves. On–off valves are typically ball, plug, wedge, butterfly, or gate valves with a pneumatic, electric or hydraulic operator. They are usually line size which means there is no special sizing procedure (other than the actuator). The major requirements of these valves are that the initial leakage rate through the seat must be acceptable for the application; essentially zero leakage through the packing (or atmospheric seal); and the valve must go fully open or closed when a signal is sent. Absolute shutoff may require double block valves with a purge (bleed) valve between them. The actuator (pneumatic cylinder or spring and diaphragm, electric motor, hydraulic system) needs to be mounted on the valve in such a way that removal of the retaining nuts, does not cause the valve body to come apart. In many cases, on–off valves are equipped with limit switches to provide a signal to the control system or operator that the valve is fully opened or closed. Since in many installations these limit switches have caused more problems than the valve, it is important to properly select, install, set and test these devices before operations begin. It is important to properly design the linkage, couplings and brackets and to use hermetically sealed or proximity switches.

Control-valve selection should never be a hit or miss operation. Improperly selected valves and actuators will not be capable of regulating the flow over the required range with the proper sensitivity. They may leak, corrode, plug, cavitate (which can destroy the valve and piping), make noise beyond the acceptable limits, and become a major maintenance problem.

Control valves are required to regulate the flow of fluids over a wide range of conditions and are therefore very sophisticated with many varieties and options available. The first requirement for the design specialist is to have available all the operating conditions and abnormal conditions that the valve will be subjected to, the characteristics of the fluid, and all process data. This includes the phase or phases (gas, liquid, slurry, or steam); temperature limits; viscosity; vapor pressure; minimum, normal, and maximum flows with the corresponding pressure drops; specific gravity or density; etc.

The importance of accurate data *cannot* be overemphasized. Assigning ten pounds per square inch of pressure drop to every valve is not acceptable and leads to incorrectly sized valves. It has been estimated that more than 50% of the valves currently in service are oversized. Operation near the closed position (less than 10% open) may result in irregular flow. Undersized valves will restrict the flow.

Automated valve body style selection requires an understanding of the fluid and the piping requirements. Globe bodies are not appropriate when liquids must not be trapped in the valve. Rotary valves with hardened trim are appropriate for hard slurries. Seals on rotary valves tend to leak less than the standard packing on sliding stem valves. Packless valves, bellows seals and double packing are all available for highly hazardous materials, where zero leakage to the atmosphere is required. "Firesafe" valves should be used on flammable, lethal and toxic services to minimize

internal leakage and leakage to the atmosphere in case of fire. These valves utilize backup metal and graphite seals that supplement the primary flexible seal if it deteriorates in a fire. "Firesafe" valves should have flanged connections with spiral wound gaskets, welded connections, or lugged design for butterfly valves.

A fundamental concept often missed in process design is the need for tight shutoff and isolation. A control valve is designed as the final control element with shutoff characteristics which satisfy the process. In some applications, automated isolation valves and shutdown valves are separate devices which are designed and applied for a specific purpose. Control valves may be used for shutdown if the valve leakage does not create a safety or quality problem.

4.4.1.1 Valve Piping

Automated valves must have connections that are compatible with the piping specification, which means that flanges or welds should be used for steam, hot fluid services, flammable and highly hazardous materials. Screwed or wafer style valves with long bolts may not be acceptable in these services. It is desirable to have the valve located so that it is accessible for maintenance and properly supported (both valve and actuator). This does not mean that all automated valves must be brought down to grade. Platforms are acceptable in some applications. Some valves have a design feature known as "in-line maintenance." Theoretically, this allows the valve to be repaired in the field by one craft (e.g., an instrument mechanic); however plant policy may not allow repairing the valves in place.

Automated on–off valves are normally line size. Many properly sized throttling control valves will have body sizes smaller than the piping system. In these cases the pressure drop in the block valve and reducers must be taken into consideration when sizing the other control valve. Some companies have policies of installing only line-sized block and line-sized control valves with reduced ports. There are companies with standards that require double block and by-pass valves around every control valve to facilitate maintenance. This can become dangerous if the valve is operated manually and incorrectly in the field by an operator that cannot position a block valve with the precision and speed of an automatic control system. If individual valves are critical to an operation and cannot be shut down for any reason, an installed spare should be provided with the proper safeguards. When block valves are used to isolate a control valve, depressurizing, drain and/or flush valves are required to clean the control valves before the piping is opened up. It is important that block and by-pass valves do not leak and that their packing does not leak fluid to the atmosphere. Depressurizing, drain and/or flush valves should be installed with caps or plugs.

4.4.1.2 Materials of Construction

The body material of automated on–off valves are normally the same as the piping system in which they are installed. However, in some cases, it is appropriate to use a different material. A common example is the use of cast iron valve bodies with steel piping. The problem may be the mating of cast iron flat-faced flanges with the raised-faced flanges of the carbon-steel pipe. Excessive force in tightening the flange bolts during installation can crack the valve body. Proper coordination of the piping flanges, gaskets and torque required will eliminate the potential problem; however this may be difficult to control in the field. A simple solution might be to not use cast iron valves. Cast iron, malleable iron, or semi-steel valves should not be used in steam, flammable, toxic, hazardous chemical service, corrosive chemical service, temperatures above 200°F, or pressures above 250 psig due to the fact that the material is brittle and could easily break if hit by a hard object or exposed to temperature shocks.

The trim or inside materials of automated on–off valves should be of a better alloy (i.e., more resistant to corrosion or erosion), than corresponding manual valves of the same type. An example would be carbon-steel ball or plug valves that come standard with plated balls or plugs. Even in relatively mild service, the ball or plug is not very smooth and will tend to pit. After a number of cycles, the valve seals will begin to deteriorate due to the rough metal surface rubbing against them. The use of a higher alloy, such as 316 stainless steel (316SS), will help to reduce this wear. If 316SS is used as the minimum acceptable ball or plug material, care must still be taken to ensure that it will be satisfactory (chlorides can destroy stainless steels faster than less exotic materials). Other materials used inside the valve cannot be neglected. Certain monomers will permeate materials like Teflon® and polymerize, causing a swelling. Other valves require "firesafe" construction, which will mean the seals and packing will require backup metal or graphite materials that can withstand the heat of a fire. Care should be taken in achieving tight shutoff with soft-seat materials, because loss of the seat through abrasion may give a poorer seal than hard metal.

Because of the higher velocities around the seat and plug areas of globe-style control valves and at the edges of the discs in rotary valves, many control valve manufacturers provide higher alloy materials as a standard. An example is the standard use of 316SS or other hardened stainless steels in almost all globe-style control valves. Another example would be the use of Monel®, Hastelloy®, tantalum, or Teflon®-coated stainless-steel stems, plugs and seat rings in carbon-steel valves used in chlorine service. In other cases, hardened stainless steel or Hastelloy® may be required to reduce the erosive effects of slurries or flashing.

4.4.1.3 Valve Positioners, Transducers, Position Transmitters, and Accessories
An operator sitting in a control room and monitoring a display may require control valve position status. Watching the output meter on a controller go from 0 to 100% does not tell the operator whether the valve is opening or closing unless the controller has a tag that indicates an air-to-open or air-to-close valve. The majority of control valves have either a spring and diaphragm or air-piston type actuator. These actuators can be obtained in configurations that will open the valve with increasing pressure (may be termed "fail closed") or close the valve with increasing pressure (termed "fail open"). Additional accessories can be provided to hold position on air failure by trapping the diaphragm air or the use of air-failure cylinders.

Electronic controllers have a switch or software that can be used to help the operator to know the direction that the valve is going. The output meter (sometimes called the "valve-position indicator") can be set so that at 0% the valve is supposed to be closed and at 100% the valve is supposed to be fully open. As an example, the controller will put out exactly 4 ma at 0% and 20 mA at 100% (or vice versa). The electronic signal will go to a field transducer (current-to-pressure—I/P) which will convert the 4 to 20 mA signal into a 3 to 15 psig or 6 to 30 psig air signal for the valve. Because the valve and actuator are a combination of mechanical parts with springs and hysteresis, and the pressure of the process stream fluctuates, the valve may not move from 0 to 100% as the signal varies from 0 to 100%. In some cases (especially with rotary valves), the valve may not start to open until the output is 25% or more and may be fully open at 75%. When calibrating control valves on the bench, the manufacturer's nameplate data needs to be followed.

Control output signals are relative indications of valve position, provided that the valve is mechanically operable and the accessories are in good repair. An increasing output signal does not mean the valve is moving. Position transducers are accessories available to be applied to valves and may be required where valve position and valve response need to be verified.

1. Valve Positioners. Valve positioners are mechanical instruments that mount on control valves to sense the position of the stem. The positioner receives the signal from the 3 to 15 psig air signal from the controller or I/P transducer, compares it against the actual valve stem position, and sends a signal to the actuator to position it in the proper place. In effect, it is the secondary feedback control loop of a cascade system with the added benefits of making the valve stem move from 0 to 100% linearly and with increased sensitivity and speed. Valve positioners and I/P transducers are available as combination units in one box. Separate positioners may have a bypass switch that combination units do not have. The positioners are normally mounted on the valve actuator by the manufacturer. In this way the valve performance, which is affected by the filter regulator, tubing diameter and length, solenoid valve and other accessories, can be verified.

Some reasons that are used to justify the use of positioners are:

- Must be used with piston-actuated operators.
- Use on slow loops like temperature.
- Use on pH or level loops.
- Use on large valves for faster response by increasing air pressure.
- Use on valves with graphite packing because of increased friction.
- Use to reverse air signal.

One of the most common uses is on split-ranged control valve combinations. With pneumatic and analog electronic controllers there is one output. If more than one control valve is required, positioners can be used to sequence their operation. One example is a temperature control loop that has a heating and a cooling requirement. From 0 to about 45% of signal the cooling valve goes from fully open to fully closed. From about 45% to about 55%, both valves are closed. From about 55% to 100%, the heating valve goes from closed to fully open. In this way the cooling valve would go open and the heating valve would close on a loss of signal or instrument air. It is also possible to close both valves by setting the output to 50%. Another application is a valve range requirement so high that one valve cannot do the job. In this case, a small valve would be fully opened before the large valve begins to open.

There are problems with the use of positioners that the BPCS can handle very well. Most of the problems with positioners are due to the inability to calibrate them and keep them calibrated properly. In split-range applications, an overlap in the calibration will have both the cooling and heating valves open at the same time. In split-range pressure applications, the venting and pressurization valves may be open at the same times. In pH control, there may be requirements for two types of split-ranging at the same time (e.g., acid and base valves and multiple valves of each type for increased rangeability.) In this case, some positioners may not be capable of being calibrated 3 to 6 psig input to get 3 to 15 psig output. In all of these cases, PESs have the ability to output multiple 4 to 20 mA signals, that can be scaled and reversed as required, to separate transducers on each valve. With this arrangement, any combination of final control elements can be operated in any order, with the programmer determining the splits.

2. Valve Transducers. Most valves that require current-to-pressure (I/P) transducers have them mounted on the side by the valve manufacturer. This saves field labor and keeps the air tubing short. The valve manufacturer can determine how the response time of valve is affected by the filter regulator capacity, tubing diameter, tubing length, solenoid valve, boosters and any other accessories that are mounted on the actuator. Most of these I/Ps can handle the normal vibrations associated with the piping systems. If the vibration is excessive, the piping system should be checked and/or the I/P relocated to some external location.

On up-grade projects, where pneumatic systems are converted to electronic, remote I/Ps may be of the rack design type located remotely from the valve (i.e. junction boxes). In these cases, the air capacity of the filter regulators, I/Ps and tubing length should be checked to determine how they will effect the response of the control valves. As an example, valve mounted I/Ps have capacities in excess of 5 SCFM and less than two feet of signal tubing; while some rack mounted devices are rated at one SCFM and have more than 50 feet of signal tubing. When the I/P capacity is low and/or the tubing length is long, positioners or boosters may be required for the valve to provide an adequate stroking speed.

3. Valve Position Transmitters. Position transmitters are essentially reverse valve positioners. They are connected to the stem of the valve and can output a 3 to 15 psig or 4 to 20 mA signal. Their use is limited primarily to critical valves. The control system or control room operator can determine that the valve is in the position that is required and also check the stroking speed of the valve. If just one or two positions, such as fully closed and/or fully open are required, a limit switch can be used. However, as described in the on–off valve section above, care should be taken in the selection, installation and calibration of all limit switches.

4. Filter-Regulators and Other Accessories. Transducers and positioners require filter-regulators to reduce the plant air supply down to the required pressure (typically 20 psig for a 3 to 15 psig I/P). These regulators are available with a pressure gage and are normally supplied as a package that is mounted on the valve by the manufacturer.

Solenoid valves are another accessory that is mounted on the valve assembly. The solenoid should be mounted in the control air signal to the actuator. In this way if the positioner or transducer is out of calibration, the solenoid will still either dump all the air or put full supply air to the actuator. Solenoid valves should be rated NEMA 4 and 7 and have continuous duty Class F or H molded coils for industrial use. Solenoid valves used in throttling control valve applications should not be larger than inch and cannot be pilot operated, because the control signal can go below 5 psig, which would keep the solenoid valve from reopening. The capacity of the solenoid needs to be large enough so as not to restrict the control air flow.

In some cases, boosters (which are pneumatic multipliers) can be used to increase the control-air volume and/or pressure to an actuator. Care must be taken in the selection, mounting and testing to ensure that the booster and positioner do not interact to the detriment of the system. It is advised, that all these accessories be mounted and tested by the valve manufacturer.

All the valve accessories need to be specified as to the operating requirements, especially on critical applications, in order for the manufacturer to verify that the valve will operate as required.

4.4.1.4 Valve Failure Modes
The loss of instrument air or control signal (pneumatic or electronic) should cause the automatic valves to go to a safe condition. Most valves that are feeding chemicals or steam for heating to a piece of equipment will be the fail-closed type. Cooling water, pressure venting, and steam for cooling are usually the fail-open type. However, each application must be evaluated. "Fail-in-the-last-position" is also an option that can be used; however, the instrument signal still needs to be designated as increase-to-open or increase-to-close.

Diaphragm, cylinder, piston, or electric motor valve actuators are available with a variety of "fail-safe" devices. The spring return feature is the most common. The spring will either close or open the valve on loss of instrument air or electrical power.

When an air failure occurs with piston operators that have no springs, lock-up relays and air bottles can position the valve in any position desired. This practice should be discouraged in safety applications. Electric motor operators, without springs, will keep the valve in the last position on a power or motor failure.

Final control element failures have received less attention than sensor failures but can occur almost as often and are frequently not recognized until too late.

While solenoid valves are especially vulnerable, they can fail due to moisture, low or high ambient temperatures and pipe scale in the orifices. Supplier MTBF data may be obtained and considered, and careful application engineering implemented. Thus heated enclosures for cold climates, air filters for plant air and high temperature molded coils are recommended.

Automated valves can fail from corroded stems and packing, rusted piston linings or springs, diaphragm fatigue, bent stems or linkages, and fouled seating surfaces. Miniature control valves are particularly prone to corroded stems from leaky packing and to bent stems from minor bumps.

Any valve will stick if the operator is marginally sized.

Plug valves may not fully open or close, due to the seal material cold flowing into plug crevices, if they have not been stroked for a long time or they are operating at a borderline temperature.

Fire proofing some valves, actuators, accessories and wires may be required, depending on the application.

4.4.2 Electric Drives

4.4.2.1 Constant Speed Drives

These drives utilize motor starters. The motor starter design should be reviewed to ensure proper reliability.

4.4.2.2 Variable Speed Drives

Adjustable speed electric motor drives provide the capability to continuously vary and control the speed of process motors for pumps, agitators, compressors, etc. They can save energy, eliminate control valves, dampers and provide a wider rangeability.

Improvements in power semiconductors, microcircuits and digital control technology have resulted in motor speed control electronics that are smaller, have fewer components and significantly improved diagnostics, reliability, and improved adjustability. In order to match the drive system to the application; it is essential to define the type of load (pump, agitator, fan, etc.), speed range and accuracy, type of control, and motor and controller environment. Variable speed drives have the ability to overspeed the motor. They can also run the motor at lower than design speeds which will cause overheating.

AC and DC drives are used in most applications. AC drive systems use a standard induction motor that is controlled by an invertor that converts AC line power to DC and then "chops" the DC power to synthesize variable voltage and frequency power to the motor. DC drives use a special DC motor with brushes that are controlled by electronics that simply convert and control AC power to DC.

AC drives use a simple motor but complicated electronics, while DC drives do the opposite. The main disadvantage of DC motors are the brushes that require replacement and an approved enclosure in hazardous locations. Current AC drive technology is a voltage source invertor using power transistors. AC drives have many advantages in sizes up to 500 hp, but DC drives have advantages in larger sizes. The final choice depends on the application and the environment.

4.4.3 *Steam Turbine Variable Speed Drives*

Electronic governors are replacing mechanical governors on steam turbines because of their superior performance and reliability. The input signal to the governor electronics comes from a magnetic speed pickup and the output is a 3 to 15 psig air signal that goes to the steam control valve. The speed is adjustable from a calibrated dial or standard process controller output, either electronic or pneumatic. Features of these systems include:

- Push-button or automatic start/stop.
- Fail-safe operation on loss of input signal.
- Underspeed and overspeed alarms and trips.
- Tachometer outputs.
- Battery backup.
- Dual pickups.
- Explosion-proof packaging.

4.5 PROCESS CONTROLLERS

This section discusses the safety issues related to the application of process controllers. Applications may require continuous, sequential, and discrete control. (Batch controllers are treated separately in Section 4.10.) Equipment-approval classification, management decisions, accepted standardization, maintenance, and operational factors may affect the process controller choice. The approach to choosing a process control system, of which the controller is a key component, is to first define the process control system requirements as discussed in Sections 4.0 and 4.1. Consideration must be given to performance, reliability, maintainability, and safety.

4.5.1 *Types of Process Controllers*

This section discusses some important considerations in the selection of process controllers.

4.5.1.1 *Analog Process Controllers*
"Bumpless" transfer between manual and automatic operating modes and set-point tracking are available with most analog controllers. These are important features which can help prevent bumping the final-control element and thereby disrupting an otherwise stable process during changes in controller operating modes.

4.5.1.2 *Digital Process Controllers*

How failures occur in digital process controllers and how the overall system reacts are very important. For example, the fail-safe values for outputs may be stored in memory for the input/output (I/O) processor to transfer to the I/O modules in the face of certain failures. A failure in the I/O processor must be detected, and an alternate means must be provided for fail-safe action.

Also be aware of the fact that third-party configuration software may not compile your program in the manner intended.

A digital controller typically has several modes of operation: manual, automatic, local/remote set point, supervisory, or direct-digital control. If the controller set point or output is to be controlled externally from the loop, then the controller must be in the correct operating mode.

1. Single Station Controllers. Single station controllers are available as single-loop controllers, multiloop controllers, and sequencing controllers. A single-loop controller (SLC) has the advantage that a single failure results in the loss of only one control loop. (An SLC may have multiple inputs, but it has only one output to a final-control element.)

In general, single station controllers have both multiple I/O and loops. Theoretically, this increases the risk associated with a single station failure because of the number of loops affected by a common-mode failure. In practice, however, single station controllers are usually used on closely related loops where common mode failure is not an issue. Depending on the application, some issues that should be considered include the failure modes of the controller and their effects, the sharing of a common database, the built-in HMI, the need for external watchdog timers, the need for redundancy, and the need for emergency-shutdown switches.

2. Distributed Process Control System (DCS) Controllers. The DCS controller is a multiple-input–multiple-output controller. Misapplication can cause common mode failures, affecting multiple loops. Controller capacity must be considered, and there must be a distribution of risk. The HMI interface and the I/O processor can be remotely located. The failure modes to evaluate must include not only internal failures, but failures in communication with either devices in the distributed control system or the I/O. The existence or absence of watch dog timers must be known along with their internal and external communication of failure and what controller failures may not be detected.

The DCS controller may have continuous, sequential, and batch control programming capability. With this capability also comes a multitude of limitations, such as total memory, total I/O, I/O by individual type, number of times individual algorithms may be applied, processor time, etc. Study the controller specifications carefully to be aware of the controller limits. Make sure sufficient spare capacity exists in the original DCS configuration. This will allow room for unanticipated control requirements. Another example might be minimum unused memory requirements for downloading new configurations into controllers. Without sufficient memory to accept a new configuration download, a controller may lock up, which would cause a serious problem during on-line programming changes. Controller capacity should also exist for future improvements and additions.

A DCS requires that special attention be given to its environment and its special electrical and grounding needs. Follow the individual vendor installation requirements and maintain those conditions to protect the DCS controller from premature failure and erratic operation.

3. Programmable Logic Controllers (PLCs). A PLC may also require that special attention be given to its environment. The vendor recommendations on environment and electrical and grounding requirements must be followed and maintained for PLC controller installations.

Limitations include those on total maximum I/O, maximum analog I/O, maximum discrete I/O, programming language(s), PID capabilities, motion control, types of communication interfaces, program memory size, data memory size, program scan rate per kilobyte, and type of LAN. All of these are important in specifying the most appropriate PLC controller for an application. In general, PLCs are best for sequential control, not regulatory (continuous) control. A PLC controller may not have an HMI. The PLC controller size, functions, and other specifications must be chosen to safely meet the control requirements.

The predominant programming language for PLC controllers is relay ladder logic. Care must be taken between vendors and even models to understand the functions used. Lack of careful study of each controller's individual programming language can result in unsafe situations, where incorrect functions are performed and different functions use the same memory locations. The logic solving path in a PLC controller must also be clearly understood to obtain the intended logic. For example, some PLC controllers solve network logic by row, while others solve by column.

4.5.2 Controller Algorithms and Features

The performance of the process controller depends on the configuration chosen, how it is applied, the process characteristics, and the control criteria for the most stable operation. Maintaining the process variables at set points or within certain limits minimizes demands on other protective systems and minimizes the risk associated with the process.

4.5.2.1 Analog

Most PID controllers allow the derivative mode to act on the process variable only, not the total controller error. This means an abrupt change in set point will not cause a sudden change in the controller output.

Optional features such as output limiting and anti-reset windup are available for some pneumatic and electronic controllers. Output limiting can be used to limit the travel of a valve or the range of set points to the inner loop of a cascade loop. Output limiting for safety purposes can be accomplished with physical limits on valve movements or flow restrictions (e.g., restricting orifice in the pipe line). Output limiting makes reset windup more of a problem. Analog controllers should be purchased with an anti-reset windup option when significant overshoot can affect the safety of the process.

Where additional signal manipulation is used in a control loop, care must be taken to maintain stable control and protect the safety of the process. For example, the square root of differential pressure measurements is used for flow by using a square-root extractor module. At low flows, the signal from the square-root extractor becomes very small and unstable. Sending both the linearized output and direct measurement through a high signal selector module maintains stability at very low signals. Square-root extractors should not be used with linear flow transmitters.

4.5.2.2 Digital

Digital process controllers vary in their programming or configuration choices. The safety implications of some of the more common elements and algorithms are discussed in this section.

1. Output Limiting. Digital controllers usually offer adjustable output limits. The use of output limits are to prevent the control valve from continuing to be driven open or closed once the valve has reached its stroke limits. When output limits are used, they should normally be set to match the valve stroke limits. Output limits less than the valve stroke limits prevent the valve from going completely shut or open. Output limits can also be used to limit the set point range transmitted to the inner loop of a cascade system.

If safety is a concern, do not depend on output limits, because placing a controller in manual may permit the output to exceed the limit and also limits can be too easily altered.

2. Restart and Fail-Safe Output Values. Some systems allow specifying an initial or restart output value. This allows the control engineer to specify a safe controller output value at the instant the controller is initialized or restarted. The ability to specify a fail-safe value for the controller output is an important safety feature. In addition to being able to specify an output value, the implementation of this feature should be reviewed. If the failure occurs between the stored failsafe value and the controller output, then the output could continue at the last value and create an unsafe condition.

3. Reset Windup. Reset windup occurs when a limit in the output transducer, positioner output, valve stroke, process output, transmitter output, or controller input prevents the controller from either reducing or seeing the error response. The long term error causes the reset action contribution to drive the controller output to an extreme. Reset action will not reverse the direction of its contribution to controller output until the process measurement crosses the set point. This will result in a significant overshoot in the process variable, which may be unsafe. Reset windup commonly occurs in override, limiting, surge, batch, and pH control loops. The duration of the windup is longer for processes with large ultimate periods and large process gains, because small proportional and integral actions prolong the reversal of controller output. Anti-reset windup should be used to prevent reset windup.

4. Set-Point Limits and Bumpless Transfer. Set point limits can be used to prevent either a primary cascade controller from driving a secondary-loop set point or an operator from driving any loop set point outside of desired limits. As mentioned above, consideration should be given to whether or not limits are maintained when there is a mode change. For critical applications, physical limits on valves should be considered.

The considerations of bumpless transfer in the discussion of analog controllers (Section 4.5.1.1) applies to digital controllers as well.

5. Rate Limiting. Rate limiting allows the rate of change in the controller output to be limited in order to avoid an unsafe situation that might be caused by too rapid a change in the controlled variable. For example, where multiple feeds are off a common header and a rapid increase in feed to one unit might starve other units, rate limiting can be used.

6. Filtering and Signal Characterization. Use of a filter on the controller input signal will in general degrade the performance of the loop. The process will not be as stable, and the input filter will also hide this fact from any display, alarm, interlock, and external communication that is using the filtered input value. The loss of stability may not be detected until a safety hazard has developed if excessive filtering is used on a critical loop. In general, for a safer, more stable control system, minimize all filter time constants. On some loops, the noise is so severe that the degradation in control because of the noise is greater than the degradation in control because of a filter. If the period of the noise is less than 5% of the process dead time, then a filter time constant can be carefully selected to give a negligible degradation in control (see Ref. 4.22).

Signal characterization improves control loop performance by linearizing either the measurement, as in pH control, or the valve gain, as in an installed, equal-percentage characteristic (see Ref. 4.22). It generally requires the use of a polynomial function to the nonlinearity. The more powerful PES-based controllers offer a sufficient variety of functions and calculation steps to easily implement either input or output signal characterization. If the measurement or valve nonlinearities are predictable, then the improvement in control loop performance and process safety can be significant.

7. Signal Selection. It is sometimes necessary to override a controller output in order to maintain safe operation or to protect process equipment. In addition, to protect against placing a plant in a hazardous situation, a selective control system may select the highest, lowest, or median signal from two or more transmitters. For example, if the consequence of an analyzer failure is unacceptable, then three analyzers can be provided and a median selector used. As another example, it may be necessary to always control the highest temperature in a fixed-bed reactor. The highest temperature location may be shifting throughout the bed. By having temperatures measured throughout the reactor bed and feeding all the temperature signals into a high selector, the highest temperature will always be used for control. Other examples would include discharge pressure control of a compressor with suction pressure override, reactor temperature control with pressure override, and compressor discharge flow control with discharge pressure override.

When controller signals are selected, two problems must be addressed. One involves reset windup in the unselected loops, and the other involves the controller algorithms. Reset windup must be protected against. See section (3) above. A velocity algorithm in a controller calculates a change in controller output as contrasted with a positional controller, which calculates an absolute, summed output. This protects against reset windup and bumping of the output during manual-to-automatic transfer and computer failure. Positional algorithms require initialization, but velocity algorithms preclude the safe use of override control utilizing controller outputs.

8. Cascade Control. The output of one controller can be used to manipulate the set point of another controller. The two controllers are then said to be in cascade control. Typical cascade control consists of two controllers, two measured variables and only one manipulated variable. The inner control loop is also known as the *secondary* or *slave* controller. The outer control loop is also known as the *primary* or *master* controller.

Cascade control can improve the safety of a process by correcting disturbances within the secondary loop before they can affect the primary variable, by overcoming gain variations in the secondary part of the process within its own loop, by reducing the phase lag of the secondary process to improve the speed of response of the primary loop, and by the secondary loop permitting an exact manipulation of the mass or energy flow by the primary loop.

Typical examples of cascade control are a reactor temperature control loop setting the exit coolant temperature control loop set point, a controller setting the set point of a valve positioner (controller), and product temperature exiting a heat exchanger control loop setting the steam pressure to the heat exchanger control loop set point.

Although cascade control improves performance, several problems can prevent its implementation. The outer controller output and the inner controller set point must be balanced to have a bumpless transfer of the inner loop from local set point control to remote set point control. Without bumpless transfer to cascade control the inner loop controller can become erratic and create a hazardous situation. Derivative action on inner loop set point changes can also bump the process. Derivative action on the total error can cause problems in some loops. The well known rule of thumb for cascade control is that the inner loop must be three to ten times faster than the outer loop. Where the inner loop is a slower loop, the outer loop must be detuned by reducing the gain (i.e., increasing the proportional band). If both the primary and secondary controllers have reset action, then both controllers may require anti-reset windup protection.

9. Other Control Algorithms. There are additional control techniques which can provide more stable and safer control in some applications. Some generic controllers include adaptive-gain, direct-synthesis, Smith predictors, model-predictive, and self-tuning. Self-tuning controllers require the ability to individually turn them on and off (since they may interact with each other when operated at the same time). Some specialized controllers are model based and require that the process be in a normal state. Specialized controllers, such as for compressor anti-surge control, also exist.

10. Other Considerations. Implementation of unsafe output calculations should be limited by both the algorithm and physical limiters on the field hardware. Correct the problem, not the symptoms. For example, use a smaller valve; do not limit the valve signal in the controller algorithm. Avoid division by zero and other illegal mathematical operations. Calculations must be realistic and constrained.

The ability to do extensive calculations in a digital controller is a two-edged sword. As additional calculations are performed in a control algorithm, additional dead time is added to the control loop. This dead time reduces the performance of the system and affects how tightly process variables can be controlled around their set points. This can place greater demands on the safety interlock systems.

4.5.3 Controller Tuning

Controller tuning criteria must represent the process objective. Maintaining the process conditions which allow the process to function safely is the most important objective to achieve. A clear understanding of the process technology is required to develop and translate process objectives into controller tuning criteria.

The tuning method must match the controller algorithm for which it will be used. For example, analog and digital controllers might have to be tuned differently in the same service. The parameters and their units must be understood and identified (e.g., gain versus proportional band, minutes per repeat versus repeats per minute, engineering units versus dimensionless values, and others). The tuning method must be carried out properly (using either open-loop or closed-loop tests). The tuning method must also match the process. Processes may be self-regulating, non-self-regulating, highly non-linear, and so forth. Controller tuning settings can be determined from process knowledge, plant tests, simulation, or automatically by self-tuning controllers (see Section 4.5.2). See any process control textbook for details.

Trend plot displays on the VDUs of consoles for process controllers in distributed control systems may be too slow for tuning some loops. Separate analog or digital recorders can be an aid in tuning control loops. Also, the compressed and averaged data collected by supervisory-control computers may also be unsuitable for determining controller tuning parameters.

4.5.4 Controller I/O

Controllers offer a wide range of input/output (I/O) options that impact communications (Section 4.7), redundancy (Section 4.1.2), and hardware and software technology. Each type of controller has the I/O necessary to perform their function (Section 4.5.1). Additionally, each controller type has its own unique characteristics, features, and range of options depending upon the vendor. I/O design considerations include understanding issues of (1) architecture, (2) I/O communication, and (3) I/O interfacing (i.e., discrete, analog, or special).

4.5.4.1 Architecture
Some actions to consider in implementing controller I/O properly in a BPCS include:

- Treat the I/O as a stand alone subsystem.
- Select I/O channel grouping, channels per card, etc. to match the process system architecture so that I/O failure, maintenance, and allocation do not impact system performance.

- Understand all the options available in the I/O subsystem prior to selection.
- Understand the power distribution (Section 4.8) and grounding (Section 4.9) required for the I/O subsystem.
- Select an I/O redundancy scheme to meet the BPCS reliability needs.
- Implement acceptable wiring methods.
- Determine acceptable software by-pass methods (Chapter 6).

4.5.4.2 I/O Communications
I/O communication issues involve the:

- Reliability and integrity of the I/O communication scheme. For example, a vendor may not offer redundancy in a packaged scheme, and the user may have to implement redundancy.
- Ability to isolate electronics between ground planes, such as that which is provided by fiber optic links.
- Selection of a local I/O communication scheme. Often local I/O communication integrity is limited, because it is assumed since the I/O is directly connected to the CPU that an electrically clean environment is provided.
- Selection of a remote I/O distribution scheme. A remote scheme has many possible topologies (bus, star, ring, etc.). Each topology has advantages and disadvantages, including susceptibility to common-mode failure, which should be analyzed before selection.
- Selection of the distributed I/O scheme (ring and star configurations are typical for these systems). Common-mode failures may cause special reliability problems.

4.5.4.3 I/O Interfacing
I/O interfacing issues for the discrete case include:

- Matching the input loading current to the sensor contact design (low current may allow oxidation of the contacts, allowing spurious trips).
- The use of DC sinking inputs (current sourcing input may turn on if a short to ground occurs).
- The use of DC sourcing outputs (current sinking output may turn on if short to ground occurs).
- Solid state output leakage current considerations (if output leakage current is too high, the load may not drop out in the "off" state, for example, a pilot-operated solenoid valve).

I/O interfacing issues for the analog case involve:

- The digital resolution (8 bits, 12 bits, etc.) may not provide sufficient digital resolution of the sensor input signal to control the process properly.
- Output resistance matching (load resistance driven with rated accuracy).

- Appropriate isolation (improper isolation voltage range, channel to channel, or channel to system will result in reduced reliability).
- Acceptable nonlinearity as a percentage of full scale (poor linearity may impact BPCS performance).

I/O interfacing issues for specialty I/O cases (RS-232, BCD input, smart transmitters, intrinsically safe barriers, etc.) include:

- Selection of the PID I/O card to match the PID functionality in the control processor for display, tuning, etc. (provides all the features available in the controller).
- The point-to-point specialty card selection and application do not degrade system integrity.
- The selection of high-speed boards (system response time analyzed to ensure that performance achieved is appropriate).

4.5.5 Overall Cycle Times of Digital Controllers

The overall cycle time of a digital controller consists of executing operating system tasks, reading inputs, performing control calculations, and outputting the results. The cycle time is determined by the design of the controller. Overall cycle times for PES-based controllers may give slower response than local, analog controllers. If the process is very fast, this may be important. But a PES based controller with cycle times in excess of 30 times a second will emulate the faster, more stable control of an electronic analog controller with a 0.001-second input signal filter. Cycle time may not be proportional to total loop response time because of dead time. This should be analyzed.

The control interval and the I/O scan rate of a digital controller, where adjustable, should be chosen to give stable control with minimal loading on the controller processor. The I/O scan time should not be slower than the control interval. One rule of thumb suggests that the overall cycle time should be approximately one tenth of the effective process time constant of the process (Ref. 4.23). Other simulations show that for optimal control loop performance and efficient computer usage, the overall cycle time should be set about four to ten times faster than the dead time of the process (Ref. 4.24).

4.5.6 Processor Loading of Digital Controllers

The processor loading is expressed in terms of how much of the cycle time is needed to complete all of the scheduled tasks (60% loaded, for example) or how much unused cycle time remains (40% free processor time, for example).

The safety implications of how different process controllers handle loading can be important. Processors have operating systems that approach loading limits differently. Some systems may make loading assignments to each

function used during the configuration process and limit the processor loading at the configuration programming. Other systems may allow assignments
during configuration that enable multiscan execution of code. Load shedding
priorities may also exist in some systems. For example, lengthy controller
programs that require several cycles to execute will give response times that
become difficult to estimate. Depending on how
    the scan is restarted, some software instructions may never be executed or
may take so long to execute that a simple PID algorithm would give better
control. Another example could be a load shedding priority that drops external communication as a preset loading limit is reached. Do not drop external
communication which could contain crucial data or coordination information.
The user must study the process controller specifications to be aware of the
processor loading limits and how the process controller performs as the
loading limits are approached and passed.

4.5.7 Digital Controller Memory

There can be several areas of safety concerns where configurable memory
exist. Spare memory capacity needs to be available as additional programming needs are identified to protect the environment, equipment, and process.
In controllers where program changes or new sets of instructions are required
to be transferred into the active process controller, a minimum amount of
spare memory capacity must be available. There should be diagnostics on
memory, such as checksum or parity for reliability.

4.5.8 Redundancy in Digital Controllers

There are several types of redundancy. $1 : N$ redundancy is where one controller is backing up N controllers, where N may be one or more. The safety
implications are in how fast the program, data, and control are transferred,
how closely the backup database matches the current database, and the fact
that redundancy no longer exists for the remaining controllers. A backup
controller may have a common-mode fault between the processor and the I/O
and does not offer full redundancy. A software common mode failure can
cause the redundant controller to fail as well. Assuming there was a hardware
failure, the failed controller must be replaced immediately and control restored to reinstate the $1 : N$ redundancy. There should be confirmation that
identical programs exist in backup controllers. Real-time databases may also
be corrupted if redundancy is not implemented properly.
    Manual backup, another form of redundancy, is where an additional con-
troller, analog or digital, or a manual loading station can be manually switched
into the control loop. The purpose of a manual backup for a process controller
is to compensate for a real or perceived lack of availability. A manual backup

must be carefully designed and connected. A manual backup can fail and must be constantly tested and maintained.

4.5.9 Failure Considerations

The user must understand the failure modes of the controllers in the BPCS. A failure mode and effects analysis (FMEA) or other appropriate tool discussed in Chapter 3 should be used to identify the failure modes. The Process Hazard Review must consider how the controller fits into the overall process.

4.5.9.1 Analog
Loss of an air-supply compressor in a pneumatic system or loss of electrical power in an electronic system causes a critical failure which can be prevented by having back-up power sources. In all types of analog controllers, most troubles come from the power source or environmental conditions such as temperature, vibration, and corrosion. If conditions do not meet the manufacturer's recommendations for any instrument, a high rate of failure will occur.

4.5.9.2 Digital
PES-based controllers have failure modes which may be difficult to recognize or are unpredictable. Failures can result from environmental electrical noise (e.g., welding machines, two-way radios, and other computers), failures in the air conditioning system, system modifications (e.g., replacement of boards or software with incompatible revision levels), and so forth. The vendor should provide lists of potential failure modes and lists of current software revisions and their effects on their controllers. The list is useful for selecting a controller for a given application and for developing recommendations to make the process remain safe in the face of these failures. See Appendix G for a list of failures.

In digital systems, the restart strategy must be analyzed for potential to cause unsafe operation—for example, is the last position available?

1. **Internal Diagnostics.** Internal diagnostics (including memory checking, indicating lights on power supplies or on I/O, internal watchdog timers, etc.) may be implemented in software and alert the operator or take corrective action through a system of vendor-supplied tools. This subsystem must be understood before the controller is used on a hazardous process. More than likely internal watchdog timer diagnostics will check all the functions and items that the manufacturer considers important to monitor. An internal watchdog timer system may provide user selectable options ranging from failsafe output values to the shutdown of the entire controller. The user must realize the internal watchdog timer can fail for the same reasons that the controller failed. The internal watchdog timer may not monitor functions and

items that the user considers important. The internal watchdog timer may not have external contacts or other means available to warn of the controller's operability status. The failsafe actions of the internal watchdog timer may be ineffective because of its failure or other failures. The controller vendor should provide a list of the functions and items monitored by the internal watchdog timer and the system actions taken on failure detection. Exclusions from the internal watchdog timer monitoring or difficulty in detecting internal watchdog timer system diagnostic status may warrant the use of an external watchdog timer. This is a factor of the application, the failure modes of the system, and the consequences of the failures.

See Section 4.11 for software for diagnostics.

2. External Diagnostics. External diagnostics (such as watchdog timers) should be considered for controllers where undetected failures could cause hazardous situations. The external watchdog timer can provide accessible contacts for interlocking, monitoring the application program and the complete controller. An external watchdog timer can give warning alarms, deenergize the controller or its outputs, or provide inputs into the SIS. See Chapter 5 for a more complete discussion of external diagnostics.

4.6 OPERATOR/CONTROL SYSTEM INTERFACES

The safe operation of most chemical-processing facilities requires that operating personnel play an interactive role in day-to-day plant operations. The operators' ability to quickly and confidently diagnose unusual situations and to solve problems is enhanced when they have a good understanding of the process, know the exact condition of the plant, and feel ownership of their work. The operator/control system interface, commonly referred to as the human/machine interface (HMI), is a major factor in an operator's ability to satisfy this role. The HMI must provide the operator with accurate, clear, and concise information in a prompt manner.

The HMI for the basic process control system is discussed from a safety viewpoint. There is a great need for proper organization in the HMI. This is true whether the operator's interfaces are conventional control panels or video-display units (VDUs). Although instrument panels are discussed, the emphasis is on modern, VDU-based operator consoles and work stations. Process alarms, which are an important subset of process information at all HMIs, are given special consideration in this section.

The HMI is especially important during abnormal plant operation, e.g., emergency situations. The HMI must provide a consistent and uniform set of controls for the process so that operating and maintenance people can take prompt and decisive corrective action in response to such situations. Primarily for this reason, multiple vendors and different generations of equipment in the same control room and even at the same plant site should be avoided whenever possible. This is particularly true of the HMI. See also Section 4.6.2 for specific remarks regarding consistency in graphic displays in DCSs.

4.6.1 Instrument Panel

Although analog control systems are being replaced by digital control systems, instrument panels are still common. Many of the safety concerns involving the instrument panel relate to specific features of the individual analog devices which are covered elsewhere in this chapter. In this section, instrument panels are considered from the standpoint of layout, indicator test features, and protection of switches.

4.6.1.1 Layout

Panel-mounted instrumentation must be arranged functionally and logically for operators to understand the control system and what it presents about the condition of the plant. A schematic of the process may be desirable above the control panel showing the precise arrangement of and relationships among the various devices. All controls and indicators for a processing area should be grouped together, and controllers and indicators in cascade loops should be next to one another. Controllers and indicators should be arranged to satisfy the operating staff's mental picture of the plant, for example, starting, with the feed-preparation section of the plant and terminating with the product-storage section. Recorders should be situated near the controllers for which process variables are being indicated. Process alarm annunciators should be located above or adjacent to the instruments to which they are closely associated.

The control panel should be well lighted avoiding glare on the faces of the instruments and located away from distracting equipment and machinery and from areas of high traffic. All controllers, indicators, recorders, switches, and annunciators should be easily accessible and prominently and consistently labeled for clear identification. Devices associated with parallel trains should be color coded or otherwise consistently identified and tagged for proper association.

4.6.1.2 Indicator Test Features

Test features should be available to verify that all process indicating lights and alarm features are functioning properly.

4.6.1.3 *Mechanical Protection of Switches*

Critical switches should be protected from accidental activation. This can be accomplished with switch covers, by recessing, or by other mechanical means. Pushbuttons and controls should be situated away from the edges of the panel. Panel-mounted, turn-to-activate (as opposed to push-to-activate) switches should be considered.

4.6.2 *VDU-Based Operator Workstations*

Most modern basic process control systems utilize PESs to provide the automatic control, data handling, information exchange, and status display required to operate a chemical-processing facility. The operator interface for this system is a VDU (video-display-unit)-based workstation.

Correct and timely action by control room operators can prevent many accidents and good engineering of the VDU-based work station can improve operator performance and help prevent operator errors. While a high degree of human-engineering features may be built into a VDU-based operator workstation, additional features are usually required.

While BPCS display consoles may allow hundreds of loops to be accessed by a single operator workstation, care should be taken to prevent an operator from being provided more displays and process data than can be assimilated during process upsets. Start-ups, shutdowns, and situations requiring emergency operating procedures are times when poorly designed HMIs are likely to lead to operator overload. For a case study, see Ref. 4.25.

Operator-workstation process displays typically include:

- Operating Groups.
- Alarm Groups.
- Point Detail.
- Overviews.
- Historical trends.

while control system displays include:

- Communication network status.
- System diagnostics.

Each operator workstation also has a keyboard with the following main functions:

- Process manipulation keys.
- Configuration keys.
- Control system keys.
- Special function keys.

Parameters that should not be changed during normal operation should be protected keylock; password, etc. Operator workstations are normally in the operational personality; however, a maintenance personality is often available. This mode is used to build the system configuration offline. A diagnostic personality is also often available and may be used for troubleshooting communication highway or operator workstation problems.

As the primary operator interface, the operator workstation allows the process operator not only to view the process but also to intervene by taking loops off computer control, controlling loops at the basic regulatory levels, or controlling them manually.

Pictorial displays showing the interconnections of the plant processing equipment, the state of the process, and the status and outputs of the automatic control system are developed by the user for specific applications. This display building effort generates the display database for the BPCS. Human-engineering principles should be applied to design user-friendly displays and keyboard layouts. The presentation of information on BPCS displays should reinforce the operator's conceptual model of the process being controlled. For example, the indication of all valve positions should use the same convention. In one convention, an increasing indication implies an opening valve. In another convention, the output changes the process variable in the same direction.

The following subsections are concerned with several important topics pertaining to VDU-based operator workstations.

1. Configurable Display Hierarchy. A display hierarchy is a collection of displays arranged in successive order or classes, each of which is dependent on the one above it. This hierarchy usually starts with the most detailed display, the point display, e.g., a controller faceplate, and proceeds upward to the overview display, where an entire plant, or major subsystem, may be shown on one display. In between, there are usually group displays, which show a smaller section of a plant or unit, and graphics (custom).

The hierarchy of displays must be structured to minimize the number of operator requests for replacement displays as an operating problem is addressed. A logical organization of displays makes it easier for the operator to make rapid decisions under process upsets and multiple alarm conditions.

Some operator interfaces used in BPCSs utilize a fixed hierarchy of operator displays. See, for example, Ref. 4.26. Pushbuttons on the keyboard typically allow the operator to move rapidly between related displays with a minimum of effort. In these cases, however, the number of levels in the hierarchy is usually limited. The user must divide and subdivide the plant to match the vendor's hierarchy.

A configurable display hierarchy allows the user to assign the relationship between displays. The lowest level in the hierarchy is frequently the point display. All the other displays are custom designed and assigned a unique position in the hierarchy. Although the displays can be made to emulate the traditional group, graphic, and overview displays, the freedom is available for the user to define his or her own standard formats. This may be combination displays, e.g., group/graphics, or displays that mimic existing hardware such as alarm annunciator panels.

In one example (Ref. 4.27), each operator display has an associated set of display numbers that can be assigned for each of seven keys on the operator's keyboard: PLANT, AREA, GROUP, GRAPHIC, LEFT, RIGHT, and UP. This gives the user almost complete flexibility as to which display the operator can access from a given display. This flexibility is needed with processes such as batch, where flexibility in graphic layout and content is very necessary. These types of processes tend to defy the standard OVERVIEW–GROUP–POINT hierarchy.

Some advantages offered by a configurable display hierarchy are (Ref. 4.27):

- The user can define the display hierarchy to more closely match the actual operation of his or her plant or unit.
- If the number of levels in the hierarchy is not limited, the process can be subdivided into as many levels of detail that are necessary to allow the operator to move smoothly and logically through the process.
- If two or more operator's consoles, within the same BPCS, can be configured with different display hierarchies, then one console could mimic the entire plant, another console might display a specific process area, etc.
- If uneven distribution of display types is allowed, the user can then define the number of displays at each level. This allows him or her to achieve the balance of display types required for the particular application.
- A more intuitive operator interface is provided, because the hierarchy is constructed based on the relationship between displays. This eliminates the need for the operator to recall operator display numbers or descriptors.

Combined graphical and text displays can be created that offer a representation of the plant which exceeds that directly observable by an operator. Such displays can reinforce the operator's mental image of the process and can lead to safer plant operation. Only a limited amount of information about the process, however, can be displayed on a single VDU at any one time. Therefore, it is necessary to build families of displays that are consistent with the operator's understanding of the plant. These displays must be organized for ease of navigation through operating tasks. Keyboard arrangement and display should be consistent and compatible with the operator's expectations. Multiple VDUs and keyboards may be required for simultaneous displays and for redundancy. These are incorporated into a single operator workstation.

Functionally, displays required by the control-room operator at a BPCS workstation can be categorized as:

- Control displays showing process variables, outputs, and status information, including interlock alarms.
- Decision-support displays providing plant overview, plant status, trending of variables, and help to the operator.
- BPCS hardware status displays.

Control displays will differ between continuous and sequential processes. In continuous processes, critical displays should provide a detailed view of a plant with all significant process variables and controller outputs provided when operator attention is needed. Set point and manual valve position changes are made from this display. When multiple control display screens are needed for a large or complex plant, a selection path should be designed to allow minimum keystroke movement among the related control displays.

For sequential processes, such as batch plants, unit-operations displays should clearly show the physical relationships of vessels, lines, and valves and indicate the status (active or passive) of both the equipment and flow paths in the operating sequence. Alarms and interlocks should be indicated. The unit operation graphic may be supplemented with controller and switch faceplate displays to permit operator manipulations of individual valves and motors.

Decision-support displays are essential for the safe and effective utilization of a BPCS workstation. There are two important display formats that provide an immediate and overall description of the state of the process in sufficient detail for the operators to anticipate significant changes in plant operation:

- Information-dense, current plant status displays, perhaps utilizing simple process flow graphics, show the operationally significant variables in a plant.
- Trend displays, showing changes in plant conditions, are necessary for monitoring several operating tasks, such as tank filling, and for assessment of dynamic plant conditions.

The operator should be able to select the trending of virtually any continuous or discrete process variable, set point, or controller output in the system. Hardcopy should be available. Upon request, help displays should provide instructions to the operator. Both text and graphical help displays may be provided.

2. Consistency within and across Graphics. The presentation of displays which communicate efficiently with the operators requires consistency within and across graphics. This consistency should extend over the entire plant site and include:

- Display architecture and operator access strategy.
- Color coding (to draw attention to critical information, to show relationships among similar or related items, and to aid grouping or separation).
- Use of highlighting and signalling devices (blinking and reverse video).
- Data formats for data presentation, prompts, messages, trend-graph labeling, etc.

For example, a pipeline for a particular stream should be uniform no matter which graphic screen is displayed. If green indicates a running motor and red a stopped motor on one screen, then the same color coding should be used on all other displays to avoid confusion and possible operator errors. All numerical values must include engineering units.

3. Establishment of Symbols, Codes, and Dictionaries. A standard set of symbols, codes (including line widths and colors), and dictionaries minimizes the operator's need to memorize information such as tag designations and process variable units. The use of common dictionaries of both names and abbreviations is a step toward this goal. This covers process stream names, status labels, and action words for commands, prompts, and help messages. A standard set

of symbols and conventions for process-flow and signal connection lines and shapes for vessels and equipment is also important. See, for example, Ref. 4.28 as well as the general discussion of standards for documentation in Section 4.13. This should be applied plant wide to facilitate operator movement from one plant unit to another. It promotes safety by increasing the correctness of operators' interpretations of data and of their resultant actions.

Finally, it is desirable to have almost all control-center, basic-control functions performed through a common set of BPCS workstations. This eliminates the need for the operators to be familiar with different displays which could lead to confusion during times of plant upset.

4. Operator Input Means. The means by which operators enter commands to a BPCS workstation also have safety implications. Keyboards, panel switches, touch screens, and track balls are all examples of such means. In most cases, operator command entry should require two distinct steps: one action to select a general function and a second to enter the specific command. Keys and switches performing related functions should be grouped together and arranged logically and according to their use. Consistent switch position and functional relationships should be maintained for all operator workstations in a control center.

Proper design of the operator workstation can reduce the instances of human error. Software acknowledgment of entered information and selections, with rejection of inappropriate inputs, is important. The number of keystrokes necessary for the operator to change displays should be minimized. This is an important feature when the operator is responding to an emergency.

Changes to the control, display, and communication databases should be controlled by appropriate security and administrative procedures, including documentation and approval. See Chapter 6.

Touch screens provide one means for quick entry of commands. Touch-screen entries should require the two-step procedure cited above—one to select the action required and a second to enter the command. A track ball or mouse for positioning the VDU cursor for function selection offers the advantage that high display density can be realized, since selection target area can be smaller. While this may reduce the number of screens necessary to cover a unit of the plant, displays may become too dense.

4.6.3 Process Alarms

This section discusses techniques that can be used to make more effective use of process alarms. Some of the techniques discussed apply to both analog and PES-based control systems.

Alarms are used to call attention to abnormal process conditions by the use of individual illuminated visual displays and audible devices. The visual displays usually flash and are color coded to distinguish between different alarm priorities (e.g., white for WARNING, red for CRITICAL). Various audible tones are also used to distinguish between different alarm priorities. An alarm indicates a change in the condition of the basic elements within the BPCS (Ref. 4.29). Basic BPCS elements (e.g., PID loops, devices) can generate various alarms; these alarms affect the second layer of protecting the process, humans, equipment, and the environment (see Figure 2.2). SIS alarms are discussed in Chapter 5.

BPCS alarm management is concerned with the selection of alarms, their signaling, display, message content, and prioritization. BPCS alarms are used to:

- Alert the operator to abnormal conditions that may require manual intervention.
- Indicate the urgency for action.
- Help determine when corrective action is to be taken.
- Require some form of acknowledgment.
- Indicate a return to normal conditions.

4.6.3.1 Alarm System Requirements
When an alarm condition is presented to an operator, it should be (Ref. 4.29):

- Highly visible to the operator.
- Recognized easily from previously acknowledged alarms.

- Quickly identifiable.
- Unmistakable as to what caused the alarm

In small installations, these requirements can be met by the standard backlighted lightbox annunciator. Lightbox annunciators are used in a traditional, panel-based BPCS. Alarms in PES-based BPCSs are typically displayed on the VDUs at the Operator Workstation. Lightbox annunciators may also be used in these systems for redundancy.

Meeting these requirements becomes more difficult as plants become increasingly complex and many variables have to be monitored. If an annunciator window is assigned to every point, the alarm display becomes unmanageable and confusing to the operator. Operator confusion can result if there are many simultaneous alarms.

Therefore, alarms should direct the operator's attention to the most serious conditions. Operators should be able to make use of the alarm information, in conjunction with other process information, to quickly and confidently diagnose process problems and prioritize their actions.

Integrating alarms into advanced PES-based BPCSs can expand the capabilities available to operators for detecting and reacting to events. These systems offer the ability to implement large numbers of alarms at low cost. They also can enhance the effectiveness of alarms by alerting the operators to situations requiring special action and by providing alarm management capabilities. For example, the use of computerized voice messages and logic systems can give the operator detailed alarm information and full diagnosis.

Processes must be properly monitored, and all process upsets and equipment malfunctions must be adequately reported. But the danger is that there will be so many alarms that the system can be overloaded or the operator can be totally confused. Because alarms are easy to configure and cost virtually nothing with a PES, there is a tendency to add too many alarms. However, too many alarms can be as dangerous as too few. For example, an analysis of several applications indicates that many of the alarm conditions generated during a process disturbance are (Ref. 4.30):

- Interrelated alarms that trigger subsequent nuisance alarms.
- Generated by inputs that cycle in and out of alarm.
- Nuisance alarms that are generated by process equipment that is known to not be operating.

It is difficult during a process upset for operators to detect critical alarms and differentiate them from less serious alarms. Because this over abundance of information is provided to the operator in a short period of time, a considerable amount of interpretation and attention is required by the operator, and misinterpretation is a possibility.

1. Alarm System Design. Alarm system designs should be based on (Ref. 4.31):

- Using flashing displays accompanied by an audible warning signal as the best way to convey the urgency of alarms to operating personnel.
- Providing plant operators with sufficient data to take appropriate action when alarms occur.
- Logging all alarm and event data (see Section 4.11).

Importance, grouping (priority), and context are important criteria in managing alarms. These criteria closely reflect the early questions an operator would ask when confronted with an alarm (Ref. 4.32):

1. How important is the event? How much time should I devote to it? Not every alarm has the same importance, so the operator needs to know how important an alarm is relative to the presence of other alarms.

2. What is the alarm? The logical grouping of alarms has always been one of the tools available when providing alarm management. The first implementations used physical groupings of flashing lights (annunciators). For example, the alarms associated with a particular reactor could be grouped together in one section of a window-type annunciator. Later generations of BPCSs offer a much more configurable grouping capability (e.g., by displaying a graphic of the process for the same reactor with the alarms displayed in their actual location around the reactor).

3. What was the state of the process when the alarm occurred? The process state, for example, the phase in a batch process, provides important context information. Depending on the process state, an event may or may not be important to an operator. The role of the process state in alarm management is very significant, because it allows the operator to quickly identify the importance of an alarm.

2. Annunciator Sequences. An annunciator sequence is "the chronological series of actions and states of an annunciator after an abnormal process condition or manual test initiation occurs" (Ref. 4.33). Three basic annunciator sequences are defined in ANSI/ISA Standard S18.1 (Ref. 4.33):

- Automatic reset (A).
- Manual reset (M).
- Ringback (R).

The AUTOMATIC RESET (A) sequence returns to the normal state automatically once the alarm condition has been acknowledged and the process condition has returned to normal. The MANUAL RESET (M) sequence is very similar to the A sequence except that the sequence will not return to the normal state until the alarm condition has been acknowledged, the process condition has returned to the normal state, and the manual reset pushbutton has been pushed. The RINGBACK (R) sequence provides a distinct visual or audible indication (or both) when a process condition returns to normal. The sequence that is appropriate for the application must be selected.

First out (first alert) is an auxiliary feature that is used with the A, M, and R sequences when one alarm can trigger another alarm and some method is needed to determine which alarm tripped first. For example, a compressor shutdown and low oil pressure shutdown alarm may appear to occur simultaneously. This may occur so fast that it is difficult to determine whether low oil pressure caused the compressor shutdown or whether the compressor shutdown caused the low oil pressure. The first-out feature indicates which of a group of alarm points operated first.

3. Separation of BPCS and SIS Alarms. The SIS is the backup device that intervenes and takes a system to a safe state when the BPCS fails to do so. To effectively function as a backup device, the SIS must be physically separated from the BPCS (see Section 5.1.2.5). Anytime the SIS actuates, either to shut down a part of the process or to prevent the process from proceeding in an unsafe sequence, the operator must be notified of this action. This is usually accomplished by providing alarms that are actuated at the same time that the SIS actuates. The BPCS also has alarms. However, the operator must always be able to see what is happening within the SIS, even if the BPCS fails. Therefore, the alarms for the BPCS and SIS may require physical separation to minimize the possibility that a common-mode failure in the alarm subsystems disrupts the operation of both the BPCS and SIS alarms.

4.6.3.2 *Implementing Alarms in Non-PES-based BPCSs*

When non-PES-based BPCSs are used, alarms are usually indicated on an annunciator, consisting of a number of alarm points powered by a common power supply. Each alarm point consists of:

- A process-alarm switch (trouble contact).
- A logic module.
- A visual display.

Annunciator points share some components (e.g., an audible signaling device (horn or bell), a flasher, and acknowledge and test pushbuttons). With some manufacturers, the neutral side of the power supply is switched to activate the audible signaling device. This can present a safety problem for maintenance personnel unless they are aware of this condition. It is safer to switch the hot side of the power supply line than the neutral side.

4.6.3.3 *Implementing Alarms in PES-based BPCSs*

There are many alarm features available in PES-based BPCSs that can be used advantageously.

1. Types of Alarm Displays. Video display units (VDUs) can be used as alarm visual displays. Alarms can be presented by worded alarm messages or by words or symbols that appear near the related equipment on CRT process graphic displays. The alarm indications can flash or be colored or both until acknowledged; they can also activate audible signaling devices. Displays should be designed so that an operator with color blindness can distinguish among them.

Alarms in BPCSs are often displayed as a list or table on the VDU. The individual alarms in these tables usually are time-stamped to show when the alarm was actuated, acknowledged, and returned to normal. Displays of this type allow a large amount of information about each alarm point on one BPCS display page. However, the operator may have trouble associating each alarm point with its actual location in the process. For this reason, alarm tables (lists) are usually supplemented with graphic displays where the alarm is displayed at its actual process location.

When VDUs are used, conventional annunciator illuminated visual displays are often installed to indicate the more important alarms. Since they are dedicated displays and can be located near the related process controls, they may direct attention to the trouble area more rapidly than VDU alarm messages.

2. Inferred Alarms. PES-based control systems can be used to generate inferred alarms. These are alarms based on calculated variables or combinational logic rather than direct measurement or contact sensing.

An example would be of inferring the rate of reaction by measuring the flow through and the temperature differential across a continuous reactor. An alarm can be generated based on this calculated value. Another example involves the inference that a vessel is not properly sealed, based on indications that a locking mechanism has been activated and a block valve is closed, but the internal pressure is below a minimum value after a predetermined interval.

3. Alarm Diagnostics. Advanced control systems can be used to help the operator diagnose problems. The objective is to inform the operator of the cause of the problems, rather than just the symptoms. A chemical reactor may have alarms on several temperatures, pressures, and flows. A combination of alarms on these points would indicate that something is wrong, but the situation could arise from a number of problems, both internal and external to the reactor. PESs have the logic and calculation capabilities needed to examine all of these possible combinations. This may be difficult or impossible for an operator to do, especially during a major upset.

Process models have been suggested as one basis of alarm diagnostics (Ref. 4.34). A computer model, operating in a simulation mode, could determine how the process differs from its baseline behavior. Another proposed method for alarm diagnosis involves expert system technology (Ref. 4.34). In this case, a knowledge base would be built using a set of rules provided by experienced process engineers, based on the way they would normally interpret alarm and other conditions to determine faults.

4. Advisory Information. The alarm management system can also be used to provide the operator with suggestions about actions that might be taken in response to alarms. This advisory information would probably be in the form of text. The advisory information may be important enough in some cases that its presence could be indicated to the operator with the same urgency as an alarm. For example, if the system detects that a valve is stuck open, the advisory information might be a message to tell the operator to have it closed immediately.

4.6.3.4 *Alarm Management Techniques*

Alarm management systems can enhance the safety of an operation by ensuring that problems are presented to operators in an understandable fashion. Response choices can be automatically displayed and given priorities. Alarm summaries or equipment histories can be reviewed quickly as an aid in selecting the best response to a new alarm. A system could monitor the status of plant equipment and emergency systems and modify its recommendations accordingly (e.g., by incorporating expert system technology).

Alarm management should provide methods for critical alarm interpretation and nuisance alarm suppression. Alarm management techniques should address the following areas (Ref. 4.30):

- Alarm suppression for nonoperating equipment.
- Alarm suppression during shut down of a plant or portions of a plant.
- Alarm suppression except for the most important alarms when one alarm is the direct cause of several others.
- Alarm segregation by priorities to ensure the most important alarms receive first attention.

These objectives can usually be achieved through alarm prioritization, suppression, and the use of process alarm states. However, caution must be exercised to ensure that a critical alarm condition is not suppressed. Implementing these techniques is generally only possible in PESs.

1. Alarm Prioritization. An alarm severity or priority level should be assigned to each data point. Each plant should have guidelines for selecting and prioritizing alarms for its specific processing operations. Below are general guidelines, with the goal of maximizing the significance of critical alarms.

For example, the alarm management system may provide four different priorities (i.e., LEVEL 1, LEVEL 2, LEVEL 3, and LEVEL 4). The alarm priority determines how alarm information is presented and how the operator is required to acknowledge the alarm. A single process variable may be assigned one or more priority, depending on the state of the process.

Based on the assigned priority, the operator may be required to acknowledge the presence of only the most critical alarms. Two of the alarm priorities require operator acknowledgement while the remaining two are informational. Alarms requiring acknowledgment should remain alarmed until acknowledged.

LEVEL 1. Critical alarms should be assigned LEVEL 1 priority. These alarms require prompt operator action to maintain the unit on-stream and to protect personnel, the environment, and major equipment. These alarms should be displayed in a distinctive color (e.g., red) and require acknowledgement on a one-by-one basis. This helps the operator see the alarm and take appropriate action. Audible signaling devices should be used to alert the operator to the actuation of a LEVEL 1 alarm. Different audible tones should be used to distinguish LEVEL 1 alarms from other, less important, alarms.

LEVEL 1 alarms should be designed to facilitate on-line testing and should be routinely tested. For fail-safe operation of critical alarms, use contacts that are closed during normal operation. The documentation and identification of LEVEL 1 alarms on process and instrumentation diagrams (P&IDs) and in operating manuals are essential. The documentation should include a separate listing of alarms, their associated SIS (if any), and methods for testing.

LEVEL 2. Alarm conditions that are important, but not critical (e.g., warnings), should be assigned this priority. Prompt operator attention is required to maintain the unit production rate or protect equipment. These alarms should be displayed in another distinctive color (e.g., white) and require acknowledgment by an operator. More than one LEVEL 2 alarm may be acknowledged at the same time. A different audible tone helps the operator distinguish LEVEL 2 alarms from others.

LEVEL 1 and LEVEL 2 color coding is usually desirable. The addition, deletion, or modification of LEVEL 1 and LEVEL 2 alarms must only be done according to formal plant procedures.

LEVEL 3. Alarm conditions not detrimental to the process may be assigned LEVEL 3 priority. The first stage of the normal annunciator sequence (i.e., flashing displays and audible signals) may be suppressed. If an audible signal is actuated, it may be automatically silenced after a specified time. Corrective action may be required, but not immediately. Time is available for the operators to investigate the problem. This priority requires no operator acknowledgment. These alarms are presented to the operator on the VDU. They should be steadily illuminated until the condition is cleared.

LEVEL 4. Alarm conditions requiring correction in the near future, but not affecting the process at present, may be assigned LEVEL 4 priority. This priority requires no acknowledgment and should only be recorded in an alarm log and not automatically displayed on the VDU.

In a PES system, each alarm (at least in the alarm summary) should contain a time-stamped message providing the:

- Description of the alarm.
- Current value of the process variable.
- Limit that was violated.
- Priority (or seriousness) of the alarm.
- Status (e.g., the time triggered, acknowledge, and cleared).

A single (perhaps dedicated) keystroke or action is desirable for accessing the display associated with the most recent alarm event.

A means for summarizing and logging each alarm event (i.e., triggering, acknowledging, and clearing) dynamically, in order of occurrence, is desirable. Not all PES vendors incorporate the capability of recording alarm acknowledgment.

Malfunctions in all redundant or backup system components should be audibly alarmed and their level designated (LEVEL 1 or LEVEL 2), depending on the consequences of the failure. These alarms should be designated as either warning or critical , depending on the consequences of the failure in the redundant or backup components. The operator should have a procedure to notify the maintenance department of the problem even though it is not causing an immediate process shutdown. Similar malfunction alarms should also be used wherever possible to monitor operating and backup processors. In the case of redundant processors, it is expected that the backup processor will bumplessly take control in the event of a primary processor failure. An audible alarm should report this malfunction as well as any problems with redundant processors. Should an undetected backup processor malfunction occur, the backup processor would not be available when needed.

In addition to the above alarms, the operator needs status information (e.g., pump running or not, valve open or closed, etc.). These do not require corrective action by the operator, and the VDU is probably the ideal place for this information. Operators also need trip analysis information that is related to, or leading up to, a shutdown of the unit. This information should be recorded so that a permanent record is available.

2. Alarm Suppression. This technique suppresses excessive alarms that tend to confuse the operator. As an example, assume that the occurrence of one main alarm triggers multiple alarms, which in turn activate additional lower level alarms. Alarms are propagated to lower levels and at the same time increase in quantity and create an operator nuisance. This problem may be corrected by alarm suppression techniques.

PESs allow alarm reduction through the use of logic blocks. A typical situation might be where a number of alarms are linked together so that the failure of one unit causes an alarm and triggers other alarms in the group. An example is a series of conveyer belts feeding each other that alarm when any belt stops. Each alarm is interlocked to shut down all belts if one stops (to prevent material pileup). In this case, when one unit goes into alarm, it can be used to inhibit the alarm of the others, minimizing nuisance alarms.

Two types of suppression techniques that can reduce operator burden without eliminating valuable information or increasing the plant risk level are (Ref. 4.34):

(a) Conditional Suppression. This is where an alarm is prevented from occurring. Conditions upon which the suppression is based include sensed or inferred states or variables, other alarms, or their acknowledgment. Conditional suppression is appropriate when an alarm does not represent a dangerous situation and is a problem symptom that can be readily deduced from the remaining active alarms.

(b) Flash Suppression. This is where the first step of the alarm sequence is bypassed. The message is shown steadily illuminated, with no audible warning (as if a standard alarm had been presented and acknowledged). This minimizes operator distractions while still reminding the operator of problems. It may also be of value in analyzing process conditions. This type of suppression is appropriate when an operator expects the condition based on other alarms, but the alarm still represents a potential hazard or malfunction.

Alarms may be suppressed separately or in groups. Summaries of PES-suppressed alarms are desirable for reference by the operator.

The value of suppression is illustrated by the alarms that may occur when an operator stops a pump. This may cause low discharge pressure and low flow alarms, even though the operator expects these conditions. It might be good to suppress those alarms if they occur when the pump is stopped. If the only purpose of the alarm is as a symptom of pump failure, conditional suppression might be appropriate. If the low pressure and flow indicate process hazards (e.g., if the fluid were emergency cooling water), flash suppression might be the better choice.

3. Process Alarm States. A technique lending itself to the alarm management solution in the batch environment (and which may also be applicable to the start-up and shutdown of continuous process units) is the ability to enable and disable alarms according to a process state. A process state is a normal state of a process (e.g., a START-UP, RUN, or FLUSH state). In the alarm management system, each alarm group in the system could be assigned several different process states. Each individual alarm (e.g., High–High, High, Low, etc.) for each tag in the alarm group could be enabled or disabled selectively for each process state. These alarms are enabled and disabled by the application program and not by the operator.

4.6.4 Dynamic Parameters for Display, Communication, and Control Functions

In a PES basic process control system, the speeds with which the system performs control calculations, updates display screens, responds to commands, and communicates with other devices are important to the safe operation of the plant. Although many of these response and communication times are fixed by the equipment manufacturer, some dynamic parameters are user selectable during the software configuration of the system. The dynamic requirements of a PES are determined by the specific needs of the process.

Possible parameters (generally response times) that the user may have to specify either in selecting a PES or in configuring the system include the:

- Data acquisition parameters.
- Graphic display build and update times.
- Operator-command response times.
- Control scan times.

PESs have finite resources (e.g., memory and CPU capacity). The extent to which these resources are utilized may have impact on these parameters.

Data acquisition parameters include: (i) the report period between transmission of field measurement signals from the input module to the control processor (not necessarily displayed on the operator's console, but present in the database for display upon operator request), (ii) the exception deadband reporting window (i.e., the change in measurement signal as a percentage of transmitter span that is necessary before a new transmission from the input module to the control processor occurs), and (iii) the alarm event reporting time between the time of violation of the alarm limit at the input module and the reporting of an alarm event on the VDU.

The graphic display build time refers to the time to display a graphic image of the process on the console. The display update time is the elapsed time required to change the value of a display point in engineering units. The operator-command response time is the delay interval between command entry on the console and the response of the field device. The scan time (or process control interval) is the time between execution of the control algorithm calculations in the control module.

Table 4.1 offers a brief checklist for evaluating operator interfaces and serves as a summary of this section.

Table 4.1 A Checklist for the Operator Interface in the BPCS

1 Are the VDUs user friendly (easy to read and simple to use)?

2 Are adequate display types provided (overview, loop, group, trending, database, graphics, batch sequencing)?

3 Are there sufficient VDUs per workstation and in the system for good operator access?

4 Do the displays conform to standards for graphical user interfaces?

5 Are universal (any display on any VDU) and dedicated displays applied properly?

6 Is the alarm annunciation technique appropriate for the application?

7 Are displays configured for good understanding?

8 Are the displays properly backed up?

9 Is remote display downloading required?

10 Is on-line display information downloading from a remote location required?

11 Do the operator and engineer interfaces (special displays and keyboards) meet the application requirements?

12 Is there good human engineering?

13 Is the console display speed response adequate for the application?

14 Is an appropriate interface to the control functions (e.g., tuning) provided?

15 Is the resolution of the VDUs adequate for the application?

16 Is operator login required?

17 Is alarm response convenience acceptable (number of keystrokes)?

18 Is set point changes convenience acceptable (number of keystrokes)?

19 Are tag names, labels, and units recognizable?

20 Can the user generate customized displays?

21 Are there sufficient displays (standard and custom) allowed per console?

22 Is the precision of values displayed appropriate for the application?

23 Are security features (keylocks) provided appropriate for the application?

24 Are accidental commands adequately prevented?

25 Is the keyboard spill proof?

26 Does the keyboard buffer have a configurable timeout period appropriate for the application?

27 Is the (color) hard copy appropriate for the application?

28 Ensure a display change during a screen copy does not alter the screen copy.

29 Does the form of variable displays (bar charts or curves) convey sufficient information?

30 Are enough variable plots allowed per display screen?

31 Are operating instructions and other text displays allowed?

32 Is sequence-of-events recording available on alarms and abnormal operating conditions? Is the time interval appropriate?

33 Are the alarm indication levels and attributes (blinking, color change, reverse video) appropriate?

34 Are keystrokes required to place control system in manual and open and close the loop appropriate for the application?

4.6.5 *Control-Center Environment*

Most PES operator consoles are situated in centrally located and environmentally conditioned control centers. Consideration should be given to making the control center a "safe haven" (appropriate protection from fires, explosions, toxic releases, etc.) with emergency stop buttons and power disconnecting means located at exits and away from potential hazards. See the NEC of the NFPA. Control centers must meet all relevant fire codes and building requirements (adequate fire detection, alarming, and extinguishing systems must be provided) and must be located in a building with adequate grounding and lightning protection.

Serious common-mode failures can occur from deteriorated workstation electronics when foreign substances enter the control-center environment (e.g., trace H_2S, NH_3, SO_2, and oxides of nitrogen, NO_x). Other potential problems that can affect the control center include area flooding, acoustic noise, electrical surges, poor grounding and static discharge, poor physical security, vibration, electrical interference from large feeders and motors, lack of secure storage for backup software, and lack of accessibility for equipment and supply deliveries. Emergency power for lighting and communication is essential. For more information on control centers, see Ref. 4.35.

4.7 COMMUNICATION CONSIDERATIONS IN TYPICAL PES ARCHITECTURES

This section presents an overview of data communication with emphasis on the BPCS. It includes both the hardware- (architecture) and software-related aspects involved in the transfer of information from one PES to another.

Three general aspects involved in connecting devices for the exchange of information are the interfaces, protocols, and topologies of the connections. Interfaces are concerned with electronic hardware items such as the types of connectors, signal characteristics, and their compatibilities. Protocols define the functional aspects involved in exchanging information. This includes both a set of messages, and rules for exchanging messages. Protocols control the establishment of dynamic links between devices and provide a structure for the exchange of information. Topologies describe how a system is organized and configured. As the number of nodes becomes large, topologies that offer fault tolerance should be considered.

4.7.1 *Classifications*

This section classifies communication systems according to physical distance involved. Note that it may be difficult to strictly categorize a given implementation into just one of the above classifications. The five groups of communications systems are as follows (see Ref. 4.36):

1. Microconnections (Component-to-Component). The analysis of proper micro connections for PESs implemented in the BPCS is transparent to the user when the rigors of classifying equipment as approved per Chapter 3 (Section 3.4.1) is implemented. Users typically become aware of changes to BPCS microconnections with the issue of an engineering change order (ECO) by the BPCS supplier. Microconnection problems common to chemical processing facilities can be caused by particulates, corrosive vapors, moisture, temperature variations, and vibration.

2. Intrasubsystem Connections (Board-to-Board via Backplane). The issues previously raised with ECOs are relevant with intrasystem connections as well. Noise caused by the electrical environment or by inadequate shielding and grounding is the main problem. Insulation integrity and ruggedness may also be problems.

3. Intersubsystem Connections (Device-to-Device). Intersubsystem connections are typically direct-wired and point-to-point. Cable integrity and shielding are concerns.

4. Local Area Networks (Multiple Devices Physically Nearby). Considerations in the application of local area networks (LANs) are similar to intersubsystem connections, except that greater distances are involved. The possibility of lightning and the issue of access (restricted or unrestricted) should be considered in the design of these systems.

5. Wide Area Networks (Physically Distant Devices). Wide area networks are also known as long haul networks. These communication systems are designed to connect diverse systems in widely separated locations. Wide area networks are not generally used in PES-based BPCSs.

4.7.2 Common Data Communication Issues

The following presents a number of general issues which are inherent in communicating information automatically from one location to another.

4.7.2.1 Physical Connections

Physical connection problems involve the manner in which connectors fit together both mechanically and electrically. The reliability of the physical link is dependent upon the mechanical and signal characteristics of the communication system, as well as the protocol used. The use of robust (highly reliable) protocols on a communication link can often mask the true condition of the interconnection media. Protocols supporting automatic retransmission or correction of erroneous data should provide error reporting, so that effective maintenance of the transmission channel can be provided. The conditions under which the parameters used to characterize reliability of the physical connections are measured are often not consistent; while one reported measurement may only characterize the reliability of the hardware involved, another reporting of that same measurement may also include the robustness of the associated protocol and supporting software.

4.7.2.2. *Establishing Logical Connections*

A logical connection is the means by which the sender of a message identifies the recipient of the message and the mechanism and media by which the message is to get from the sender to the receiver. Establishing logical connections may be simple: once the physical connection is in place, the connection is established. However, this is typically true only for simple connections, and most commonly occurs between components on a single board (micro connections). In many cases, there are switches to be set in hardware, and addresses, vectors, and other parameters which must be specified for the software. Establishing a logical connection often includes a series of "handshakes" (exchange of predetermined messages in a particular sequence) and may also require strict timing of messages and responses. Establishing a logical connection between two entities may require the use of several different physical paths, as well as establishing subsidiary logical connections. This is often true of local and wide area networks, so that they may provide the logical functionality of a mesh topology without actually requiring the large number of physical interconnections which would otherwise be needed. A single physical channel may support more than one logical connection simultaneously. There are two primary mechanisms for sharing a single physical channel, time-division multiplexing and frequency multiplexing; these two mechanisms may be used separately or together on a single physical channel. The BPCS utilizes the concepts noted above in communicating between its controller, HMIs, peripherals, and foreign devices. The need to determine real-time message response of the BPCS communication network is critical when defining the suitability of the BPCS for any application.

4.7.2.3 *Security of Data*

Access to the BPCS data may be required by foreign devices such as supervisory computers, management information systems, PLCs, single-loop controllers, and miscellaneous peripheral devices. Care should be taken to ensure proper security to prevent corruption of BPCS data. This is especially important during periods of maintenance and system modification.

The BPCS may be required to communicate with the SIS. This places a unique demand on this communication link. See Chapter 5 (Section 5.6.5) for acceptable methods for communicating between the BPCS and SIS.

Data can be corrupted by external influences such as electric and magnetic fields, capacitive coupling to electrical energy sources, different ground planes, and lightning. Wiring, raceway, and installation practices should follow the BPCS supplier's installation instructions. Where variances are required, BPCS supplier concurrence should be obtained during the design stage. The installation scheme should be documented in such a way that maintenance and future modifications can be accomplished without compromising data security. Internal diagnostics, such as watchdog timers, are important. See Appendix C.

4.8 ELECTRICAL POWER DISTRIBUTION SYSTEMS

This and Section 4.9 discuss electrical power and grounding aspects associated with chemical plant control systems. Process operations and maintenance define the functional criteria for the electrical power distribution system. Guidelines for design of reliable control systems power supply and distribution systems are presented in this section; electrical grounding considerations for personnel safety and to assure control signal integrity are discussed in Section 4.9. Additional electrical power distribution and grounding consideration for SISs are presented in Section 5.6.2.1. Detailed technical discussions of power distribution in process control systems are available in the literature (Refs. 4.37–4.40). Good electrical design practices will be emphasized in Sections 4.8 and 4.9 through the presentation of typical block diagrams. Single-phase power distribution systems are shown to avoid the additional complexity found in multiphase power distribution and grounding systems.

Faults within the electrical power distribution and signal grounding systems are perhaps the largest source of major failures in chemical plant instrument systems. Attention to power supply, electrical grounding, and control system signal protection are fundamental design and maintenance activities to achieve safe and reliable operation in chemical plants that process hazardous materials.

Power distribution system design requires detailed knowledge of electrical technology and strict adherence to codes (Refs. 4.41 and 4.42). Consequently, this technical work is generally done by electrical professionals. The importance of the electrical utility to safe chemical plant automation cannot be overemphasized. Definition of the electrical distribution system should be a stand-alone consideration with multidisciplinary review when establishing the design basis for a new processing facility.

The power-system availability needed is established by process operational requirements. Some processes may shut down but not suffer negative consequences during a power disturbance while others may become unsafe or be severely impacted by these events. Process control systems can be divided into two groups from a power supply perspective: (1) instrumentation systems that can be subjected to millisecond-duration power supply disturbances without loss of critical control functions, and, (2) those which cannot continue to function when these transients occur. Control systems which employ electronic and mechanical technologies or those with PESs which are not used for primary regulatory functions fall in the first category; and, control systems which use PES technology for primary BPCS functions generally are in the second. Power supply transients in the millisecond range will usually cause abrupt stoppage of many PES modules used for automatic monitoring and control of process facilities.

Uninterruptible power supply systems (UPSs) are required for the second type of controls. There are many types of uninterruptible power supply technologies (e.g., motor–generator, DC/AC invertor systems which employ solid-state switching between AC sources, batteries, etc.). See Ref. 4.43. The predominant sources of sustained power for BPCSs are distribution systems which incorporate UPS technology. Typically these power supply systems consist of redundant, alternating current feeders plus a battery charger, batteries, and a static invertor which provides a source of AC power isolated from most transient disturbances in the plant electrical system. Since power system design must facilitate the safe, controlled shutdown of the plant when all power is lost, UPSs may be provided for both types of control systems when a total outage of utility power makes safe shutdown of the plant difficult.

Typical power distribution block diagrams for control systems with and without a UPS are presented in this section; Section 4.9 provides a corresponding overview of important design considerations for control system grounding in these two types of power distribution systems. The power system illustrated by the block diagrams are for control systems which utilize a deenergize-to-trip philosophy for fail safe operation. When implementing an energize-to-trip system, additional power system design factors must be considered (Refs. 4.39 and 4.44).

4.8.1 Distribution System without UPSs

A wide variety of instrumentation systems can be safely and economically powered using a power distribution system which automatically selects between two plant feeders as is shown in Figure 4.4. This power system passes plant electrical disturbances to the load instruments even though isolation transformers are installed. These disturbances typically occur during line switching when a fault develops in the selected feeder.

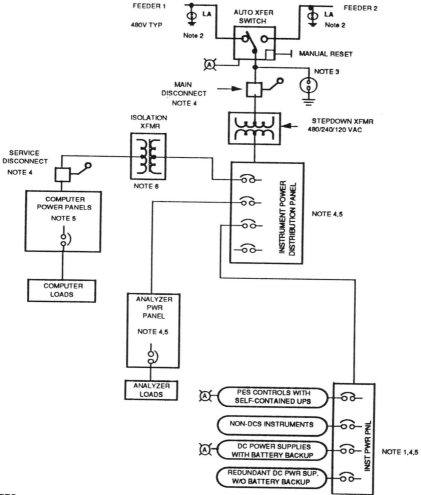

NOTES

NOTE 1 — Redundant DCS Modules should be fed from separate, individual breakers. Use separate Power Panels if possible.
NOTE 2 — Lightning Arrestors should be installed.
NOTE 3 — High-speed surge arresters should be installed.
NOTE 4 — Short circuit and overload protection not shown. See applicable Codes.
NOTE 5 — BPCS, Analyzer, and Computer Vendor power line conditioning and grounding requirements vary. Design for compliance with specific requirements.
NOTE 6 — Computers may require a single-point ground system development at this point.

Figure 4.4 Block diagram of typical power distribution system for computers, analyzers, and miscellaneous instruments that do not require an uninterruptable power supply.

The following instrumentation systems may be sufficiently robust to ride through typical power disturbances found in the power distribution system of Figure 4.4:

- Electromechanical panel instruments and relay-based SISs which can filter the switching transients.
- PESs which have an integral, battery-backup supply and can operate for several minutes during the loss of supply AC.
- Supervisory process computers which can take an abrupt halt without loss of plant control.
- Auxiliary instrumentation systems which are used for monitor only functions such as many process analyzers.

Recovery to normal operating conditions may require special restart procedures for some instrumentation systems.

Important features of the power distribution system include the following items:

- Two reliable, separate and diverse power feeders from separate plant substations for maximum power-supply reliability.
- Each feeder should be selected to provide high quality power (including minimum noise, good voltage regulation, good frequency control, and high reliability) for the instrumentation system.
- The electrical distribution system shown contains minimum power line conditioning. Installations often will require line voltage regulators, filters, frequency regulation, etc. Technical design specialists are required to determine the need for this equipment.
- A suitable automatic transfer switch with manual reset should be supplied to provide transfer from the selected feeder-to-backup source on loss of supply power. An alarm message should be provided for the operator to indicate both (1) a transfer to the secondary source and (2) the loss of the secondary source.
- High-speed voltage surge arresters should be installed at the entrance to the instrumentation power distribution panel. Lightning arresters are required at other points in the power distribution system (Refs. 4.45–4.47).
- The electrical distribution system must contain means to isolate equipment necessary for maintenance and future modifications.
- Distribution of power at the branch circuit level should be consistent with the redundancy philosophy of the BPCS. Power-supply connections for redundant modules should be made to separate branch circuits to avoid a common-mode failure point at one circuit breaker.
- Loss of power at any level should not result in a process hazardous event.
- Power is wired to a series of distribution panels, each provided with appropriately sized over-current circuit breakers, for further distribution

to power control system modules (e.g., operator workstations, AC-to-DC power supplies, and individual instrumentation subsystems.
- Good electrical circuit overload design practice requires the use of properly coordinated circuit breakers and fuses.

4.8.2 Distribution System with UPSs

Most PESs have a limited reserve of electrical energy and are designed to initiate an automatic power down sequence if the supply power voltage falls below a minimum value for more than a few milliseconds. Typically an uninterruptible power supply (UPS) as shown in Figure 4.5 is provided to assure that a reliable, stable power source is available for control system modules. When assigning PES modules to power sources (UPS or other sources), the time required for a module to restart after a power outage is an important consideration. Restart times for PES modules differ widely; consequently, valid information may not be available immediately on power restoration to the PES. Therefore, sufficient time for re–booting of the PES modules during power recovery should be provided by the design of the PES power distribution system.

Additional important power-distribution features include those listed in Section 4.8.1 as well as the following:

- A manual transfer switch for selection of the AC power feeders is recommended.
- Two sources of power for the UPS are needed to allow maintenance of the electrical supply system without sacrificing control system power reliability.
- A manual bypass switch with make-before-break switching action is necessary for maintenance of the UPS system.
- The UPS batteries should be housed in a protected and environmentally controlled utility area with controlled access and should be installed in a dedicated room. Batteries may be the source of several operating hazards arising from hydrogen-gas evolution, flash heat and sparks from high-energy short circuits, and acid spills. See Refs. 4.48 and 4.49. These potential hazards need to be considered during battery system selection and design.
- Battery capacity should be sufficient to allow for safe operation during a power outage and for orderly shutdown of the process if required. Typically, battery capacity is supplied to support operations for 30 to 240 minutes.
- An alarm message should be available for the operators to indicate (1) failures within the UPS control circuitry and (2) an automatic transfer from the UPS to the backup supply.

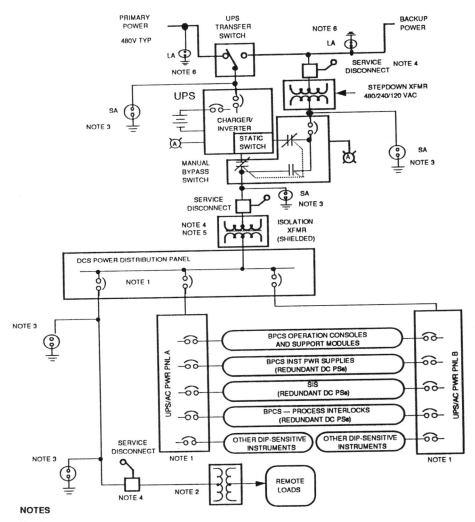

NOTES

NOTE 1 — Redundant DCS Modules to be fed from separate Power Panels.

NOTE 2 — Isolation Transformer recommended at the load anytime UPS Power supplied to a location with a different ground plane.

NOTE 3 — High-speed surge arresters should be installed.

NOTE 4 — Short circuit and overload protection not shown. See applicable Codes.

NOTE 5 — BPCS, SIS, and remote load Vendor power line conditioning and grounding requirements vary. Design for compliance with specific requirements.

NOTE 6 — Lightning arrestors should be installed.

Figure 4.5 Block diagram of typical power distribution system for BPSC and SIS when an uninterruptable power supply is required.

- The selected UPS must be able to handle high inrush current loads resulting from inductive loads. In addition, sizing of the UPS must consider nonsinusoidal, harmonic loads. (See Ref. 4.50.)
- Supply of uninterruptible power to loads remote from the control building (i.e., to areas with a different ground plane) is not recommended. If this must be done, then installation of an isolation transformer and appropriate arresters is required. See Section 4.9 for grounding details for isolation transformers.

4.9 CONTROL-SYSTEM GROUNDING FOR PERSONNEL AND SIGNAL PROTECTION

Detailed technical discussions of electrical grounding and signal protection in process-control systems are available in the literature (Refs. 4.51–4.54). Typical electrical grounding systems for the control systems described in Section 4.8 are shown in Figures 4.6 and 4.7. The discussion in this section is only an overview of the detailed system grounding shown in these diagrams.

4.9.1 Grounding Systems

All power and grounding systems must be designed in accordance with all applicable codes, standards, and practices. In this section, those additional requirements and practices for PESs and instrumentation that address signal integrity are described.

In today's DCSs, two electrical reference "ground" systems are needed in each building that house control and communication modules. One ground system, referred to as the *building ground* (BG), is used to assure personnel protection from electrical shock hazards. The BG is often referred to as the *safety ground*. (See Ref. 4.55). The BG is established by electrically connecting all building structural steel, utility piping, and foundation steel members to a low impedance metal grid buried in the earth. The following items of control-system equipment are typically connected to the BG:

- Conduit systems.
- Instrument-panel frames.
- Housings of electrical equipment and apparatus.
- Metal enclosures of switching equipment.
- Neutral connections of instrument transformers.

Equipment connections to the BG are often made using uninsulated wire conductors between the equipment to be protected and a convenient, grounded, steel-building member.

The BG also has a number of subsystems which require single-point ground connectivity. (A single-point ground is a system with a star topology with only one path to the BG.) Single-point grounding subsystems should be provided for cable shields, intrinsic safety barriers, and in some cases raised computer floors. See Ref. 4.56. Each subsystem should have its own uniquely color-coded, insulated conductors to facilitate maintenance.

The second electrical grounding system is the *electronic reference ground* (ERG). It is needed in many PESs to (1) minimize internal faults and (2) to maintain control signal integrity (Ref. 4.51). The following PES connections are generally made to the ERG:

- Supervisory control computer.
- DCS and SIS modules (signal common, logic ground, network ground, etc.).

All electrical conductors and bus bars which are interconnected to form the web of the ERG must be insulated from building ground paths. Insulated connectors from the voltage reference terminals of the noise-sensitive control system components should be individually wired to isolated ground buses (ERGs), and these remote ERGs should be star-connected to a main ERG bus to minimize voltage shifts caused by ground currents within the plant electrical system.

The main ERG bus is typically connected both to the BG and to a high quality earth ground which is identified as the *master reference ground* in Figure 4.7 and which is established by installing ground rods to assure a stable voltage reference. The ERG-to-BG connection is made at only one point to a substantial BG tap. This connection is sometimes temporarily lifted to facilitate noise diagnostics. This should only be done with the greatest care and with a safety permit.

The preparation of detailed grounding drawings, complete testing prior to commissioning, and maintenance of "as-built" ground system drawings are important to the safe automation of chemical-processing facilities.

The following are additional, important considerations in the grounding of PES control systems:

- The NEC of the NFPA requires that all electrical equipment enclosures be grounded for personnel protection.
- The sizing of grounding conductors (wires and buses) must be adequate for the maximum possible fault currents as established by the NEC.
- The planned use of existing grounding systems for new process facilities requires a thorough analysis and testing to determine suitability. A grounding system upgrade may be required.
- Isolation transformers are available with and without shields between the primary and secondary transformer windings. Shielded transformers are

installed to reduce the passage of electrical noise present in the supply system. This type of isolation transformer should be provided for noise suppression when PES subsystems, e.g., the supervisory computer, are part of the control system. An isolation transformer operating at low loads may lose noise rejection capability.

- Isolation transformer secondaries are used to establish a dedicated, separately derived, single ground reference point in AC circuits. Typically the center tap of the instrument transformer secondary (see Figure 4.6) is grounded to the BG to establish an AC supply voltage reference, and the grounded AC neutral is isolated from the BG at all other points in the power distribution system. Codes require that this single ground point is established at the transformer or at the first disconnect in the secondary side of the power distribution system. Both methods of isolation transformer grounding are illustrated in Figures 4.6 and 4.7.
- Isolation transformers (with shields between primary and secondary windings) are installed to reduce the passage of electrical noise present in the supply system. This type of isolation transformer should be provided for noise suppression when PES subsystems, e.g., the supervisory control computer, are part of the control system. Isolation transformers may lose noise rejection capability when operated at low load levels.
- A shielded isolation transformer is recommended on the outlet of the UPS to provide a single ground reference point for PES since AC power may be supplied from either the UPS or from backup feeder power.
- Supervisory control computers frequently require special ground system treatment for correct operation even when this equipment is not connected to the uninterruptible power distribution system. Computer equipment often requires that all cabinet housings be electrically isolated from the BG except for a single point connection to the building ground. This results in the installation of special isolated ground outlets and necessitates special mechanical installation procedures to prevent the casual grounding of a computer housing through electrical conduit, floors, or walls of a building.
- Signals originating from instruments powered by sources external to the control building (e.g. a gas chromatograph in the analyzer building) often are electrically referenced to another ground potential. These signals must be electrically isolated from the remote ground plane when brought into the PES. See Ref. 4.57.
- Vendor site preparation guidelines for power distribution and grounding should be followed.

Design of a power distribution and grounding system for a PES-based control system requires technical expertise. A formal review of detailed, control-system grounding plans is recommended for each project in which a

NOTES

NOTE 1 — Isolation Transformer grounding per
        manufacturer's specifications.
NOTE 2 — Computers and electronic equipment
        may require electrical isolation of
        frame and chassis from Building Ground.
NOTE 3 — Short circuit and overload protection not
        shown. See applicable codes.
NOTE 4 — To insure "single-point" ground, IGB ground
        wire must be insulated.
NOTE 5 — Ground wiring sizes to be calculated for short
        circuit capacities, in accordance with
        applicable standards.
NOTE 6 — Power and Grounding Systems must be
        designed in accordance with applicable
        Codes and Standard Practices.

LEGEND

· LA — Lightning Arrestor
· SA — Surge arrestor
· IGB – Ground Bus insulated from enclosure.
· — Connection to Building Ground System

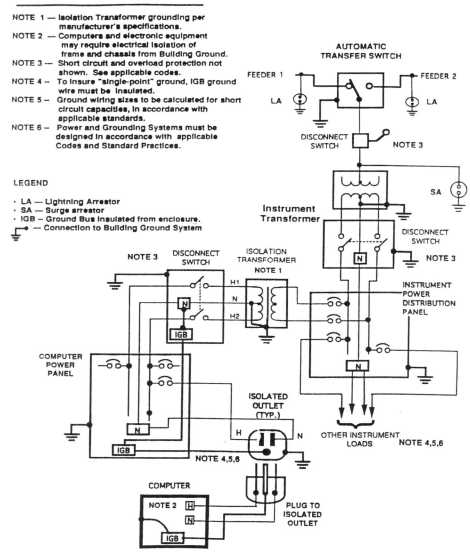

Figure 4.6 Typical electrical grounding system for computers and miscellaneous instruments that do not require an uninterruptable power supply.

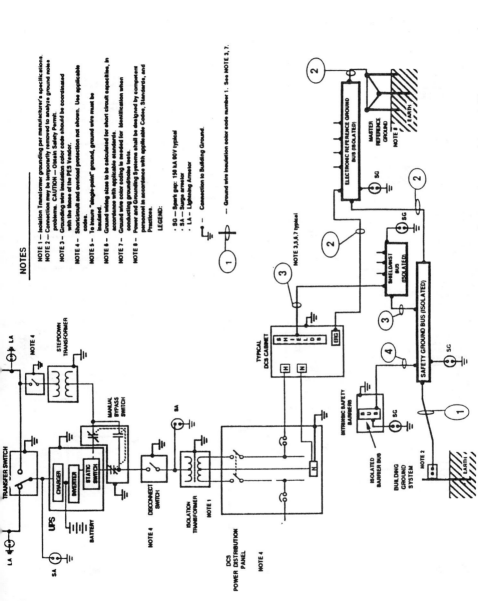

Figure 4.7 Typical electrical grounding system for PESs and instruments that require an uninterruptable power supply.

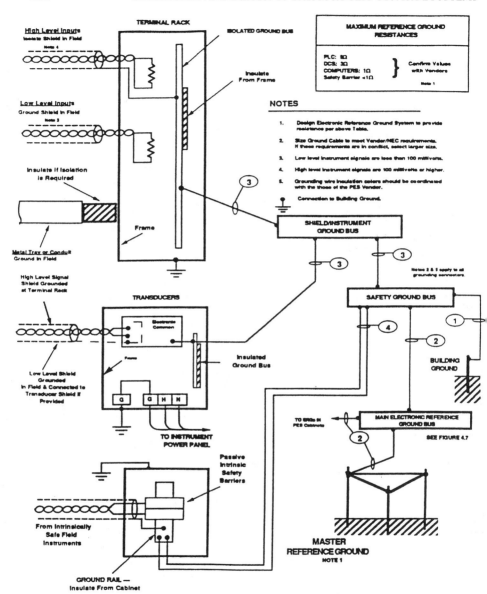

Figure 4.8 Typical instrument system signal shielding and grounding practices.

PES is to be installed. The PES manufacturer's technical consultant on system grounding should review the detailed grounding drawings before the PES is installed.

4.9.2 Instrument System Signal Shielding and Grounding Practices

Protection of control system signals from electromagnetic and electrostatic noise sources also requires a disciplined use of cable shielding and shield-grounding practices. In addition, the need for physical separation of control system signals from electromagnetic noise sources is discussed in section 4.2.3. Many of the good practices in this area are shown in Figure 4.8. Note that individual signal pair wires are shown as twisted pairs with a grounded signal pair shield. Twisting of the signal pair greatly reduces electromagnetic noise, and properly grounding the pair shield significantly reduces RFI noise. Grounding of a low-level signal pair shield is done as close to the source of the signal as possible (typically at the transducer). Thermocouples are an example of low level signals (typically less than 100 millivolts). Grounding of high-level signal-pair shields should be done to an isolated ground bus at the terminal rack.

Each cable shield must be grounded at only one point to provide maximum noise reduction. Shield continuity must be maintained from the signal source to the terminal rack. And in the case of some low-level signal transducers, the signal-pair shield may be continued to the transducer input circuit by connecting the pair shield to an instrument shield (guard) connection.

4.10 INTEGRATION OF BATCH CONTROL WITH REGULATORY AND DISCRETE CONTROL

4.10.1 Introduction

This section discusses the safety issues related to the coordination of batch control (i.e., sequencing) with regulatory and discrete control for processes that are sequential. Sequencing is also a part of the start-up and shutdown of some continuous processes (e.g., a burner management system used to start up and shut down a combustion process). This discussion will concentrate on batch process applications. Much of this discussion parallels the work going on in ISA's *Batch Control Systems* standards committee SP88.

Coordination between the regulatory and discrete control subsystems and the batch control subsystem is essential. The batch control subsystem manipulates set points, tuning parameters, etc. of the regulatory and discrete control subsystems as a function of various logic variables in the batch control system.

Batch control systems need the ability to check the status of large groups of devices and make logical decisions based on the result.

4.10.2 Comparison of Response Times

Regulatory, discrete, and batch control have different signal processing requirements. Control requires consistent, predictable update times. Typically these update times are 1 second or less. Discrete (on/off) control often requires a faster update time, possibly from 10 to 50 milliseconds. Because of their different requirements, it makes sense to have separate regulatory and discrete (on/off) processors that can be optimized independently. Batch control usually does not require the fast update times needed by the discrete control processor. Therefore, sequential logic for batch control could be implemented in the same processor as the control loops or in a general purpose computer that is handling supervisory activities.

BPCSs designed for continuous control may not be suitable for implementing batch control if high scan speeds are required (e.g., a fast response time is needed for monitoring external events). This may be true when implementing discrete control in a DCS or regulatory control in a PLC.

4.10.3 Control Activities

Figure 4.9 defines the control activities that are important in batch control systems. This drawing shows the control activities in a hierarchical fashion starting with the safety interlocks and proceeding up to higher level activities such as recipe management, scheduling, and information management. This section discusses those areas that are shaded in Figure 4.9.

- *Safety Interlocking.* These control functions prevent abnormal process actions that would jeopardize personnel safety, harm the environment, or damage equipment. A more detailed discussion of safety interlocking is given in Chapter 5.
- *Equipment-Related Control.* This control activity includes process control and unit management. Process control includes those loops and devices that perform sequential control, regulatory control, and discrete control. Unit management is responsible for coordinating the activities associated with the batch units (e.g., allocating resources within the unit, ensuring that batch sequences proceed in the proper order, etc.).
- *Process Management.* One of the main functions of this control activity is to select a master recipe from the *Recipe Management* control activity, edit that recipe and transform it into a control recipe suitable for downloading to the *Equipment-Related Control* activity, downloading the recipe (i.e., initiating the batch), and then to supervise the execution of the batch.

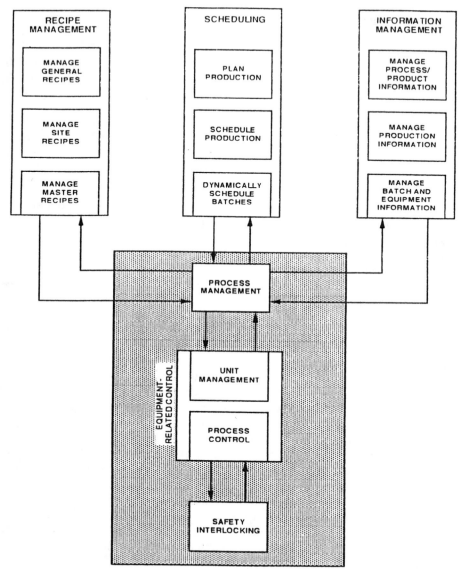

Figure 4.9 Batch control activities.

- *Recipe Management.* This control activity includes creating, editing, storing and retrieving three levels of recipes (e.g., general, site, and master recipes) and interfacing with the *Process Management* control activity. Some interfacing (not shown in Figure 4.9) is also needed between this control activity and with the *Scheduling* and *Information Management* control activities.

- *Production Scheduling.* This control activity is primarily concerned with determining what products will be made in the batch plant and when those products will be made. This requires the control activity to interface with the *Process Management, Recipe Management,* and *Information Management* control activities.
- *Information Management.* This is the control activity that maintains a history of data associated with previously executed batches, equipment operating rates, etc. and provides this information to higher levels in the organization. This requires the control activity to interface with the *Process Management, Recipe Management,* and *Scheduling* control activities.

A level in Figure 4.9 (e.g., the *Safety Interlocking* level) does not depend on any higher level for the performance of its functions. It might receive commands from an upper level (e.g., open valve A), but this level will continue to function even if the upper level fails. The higher level cannot be used unless lower levels are operational.

For example, the batch process should not run, even in the manual mode, unless the SIS is in operation. These rules apply to the interface between all levels.

Figure 4.10 shows the interaction between the various levels and the role of the human-machine interface (HMI). Although this drawing implies that there is an HMI at each level of the model, these are not independent HMIs. It simply means that the operator must have the ability to interact with each of the control activities. The HMI allows the operator to stop the command from the higher level and enter direct information. This capability is necessary in case the upper level fails and the lower level continues to run. For example, at the Equipment-Related Control level (which includes process control), the operator may need the ability to switch a regulatory control loop from "remote set point" to "local set point" and to enter a specific set point. This may be because the Process Management level fails, and this level is supplying set points to regulatory control loops at the equipment-related control level based on an optimization routine. As Figure 4.10 shows, a lower level has the ability to reject a command from a higher level if it considers that command illogical (e.g., reject unsafe commands at the safety interlock level).

Note also that at the safety interlock level, the HMI is a "read only" device. It can only monitor the operation of the safety interlocking level. It does not give the operator the ability to bypass the safety interlock system.

Figure 4.10 also shows that data can come into the system at any level. However, if that data is needed at more than one level, it must enter at the lowest level where it is needed and then be transmitted up to the higher level. Outputs can be generated at any level.

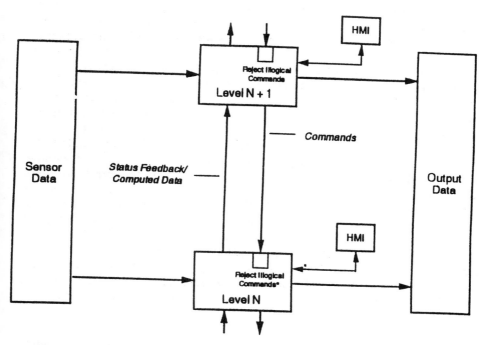

Figure 4.10 Relationship among levels. Note: if Level N is the *safety interlocking* level, then the HMI is a "read only" device, and the arrow only points toward the HMI. "Reject Illogical Commands" also changes to "Reject Unsafe Commands."

4.10.3.1 *Safety Interlocking Level*

The safety interlock level (Figure 4.9) is for safety systems that protect personnel, equipment, and the environment. The sensors, interlocks, and output devices communicate with the operator through the use of indicating lights, switches, pushbuttons, and annunciators. This level includes redundancy and fault tolerance as necessary and independence of safety functions from control functions to ensure that the process is maintained in a safe condition regardless of failures or errors at higher levels (Ref. 4.58).

This level should be separate from all others; the software should also be separate from the BPCS software (see Section 5.1.2.4 and Appendix B). This means that the hardware/software, operator interface, of the SIS should meet separation requirements in Appendix B. It also means that all the critical inputs and outputs, along with the safety interlock logic, should be totally contained in the safety interlock system. It does not mean that the safety interlock system should not be integrated in with the rest of the batch control system (e.g., the basic process control system), such as by a data highway. However, when the safety interlock system is integrated into the basic process control system, special precautions must be taken to ensure that the communications network cannot corrupt the integrity of the SIS. This means that other devices on the network can only *read* information from the safety system; *write* capability must meet separation integrity requirements outlined in Appendix B.

In some applications, the same SIS logic may be suitable for all recipes. But this is not always true; it may be necessary to change set points of critical interlocks, deactivate some interlock logic, and/or activate some additional interlock logic. When this situation exists, then a PES should be selected for the SIS. This greatly simplifies communications between the BPCS and the SIS. However, the BPCS cannot be allowed to corrupt the SIS logic. Therefore, special techniques are required for implementing these communications. For example, set points could be fed into the SIS using direct-wired interconnections between the BPCS and SIS. Or the BPCS could download information into a buffer in the SIS communications interface. Then the SIS could read the contents of that buffer. More discussion on this topic is given in Chapter 5.

When critical information is downloaded to the SIS, the BPCS should read back this information from the SIS and compare this feedback information with what was downloaded; this should be repeated for every batch. If there is a discrepancy, the batch can be put into a hold state, and the operator can be alerted.

Even when a PES is used for the above reasons, it may still be desirable to direct-wire (e.g., with relays) those interlocks that don't change. Then things that are changed frequently are implemented in the PES.

It is not unusual in a batch reactor to provide for the shutdown of complex equipment and chemistry (e.g., an active rapid quench of a chain stopper may be needed to stop the reaction). This makes the design of the SIS more complex. For example, if a quench is required, the agitator must normally be working to distribute the quench material; the chain stopper may have to be added before the agitator stops turning. This puts additional speed demands on the SIS.

The engineer or technician configuring the batch control sequence should not have configuration access to the SIS. Although this level does have an operator interface, it does not provide the operator with the ability to bypass safety interlocks.

The safety interlocking level prevents the operator from placing the plant in a dangerous condition. For example, in the manual operating mode, an operator may have a switch that can be put into the "HAND" position to open an automatic block valve, but that "HAND" position must not bypass the SIS. However, it would be able to bypass the signal from the batch control level (batch sequencing logic) that calls for the valve to be closed (see Figure 4.11). For additional details on the safety interlocking level, see Chapter 5.

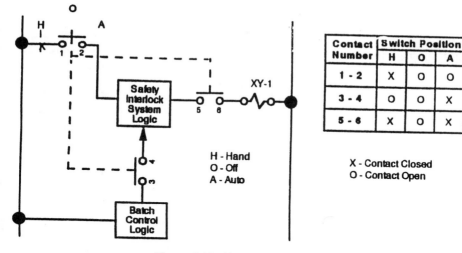

Contact	Switch Position		
Number	H	O	A
1 - 2	X	O	O
3 - 4	O	O	X
5 - 6	X	O	X

X - Contact Closed
O - Contact Open

Figure 4.11 Manual operation.

4.10.3.2 Equipment-Related Control Level

This level provides the ability for the operator to run the batch process in manual. But the plant will still be operated safely, because the safety interlock system must be fully operational. This manual mode may be implemented through a VDU-based workstation, individual stations, or a combination of the two.

This level also has interlock logic, but at this level the system is protecting product quality or preventing minor equipment damage; this level is not intended to protect personnel, the environment, or major equipment. At the SIS level, a reactor may be shut down if its temperature exceeds a high limit value, because this could cause a runaway or decomposition that could eventually injure personnel, damage major equipment, or harm the environment. At the equipment-related control level, the pumpout from a reactor may be prohibited if the valves on the pump discharge are not lined up to the correct destination. This could cause the product to be contaminated if the transfer is made, although it probably doesn't pose a safety hazard (Note: combining some materials can cause a safety problem, and those systems would be included in the SIS level).

Automatic control is also implemented at this level, but now controllers are placed in automatic mode so that they attempt to control the process to some specified set point.

4.10.3.3 Process Management Level

Batch control is added to the plant at this level. The operator often has the ability to drive the execution of the recipe by initiating the individual phases.

A phase is a process-oriented action (e.g., add ingredient A). The logic for implementing the phases is built into the control system; the operator is simply initiating phases. This also means that the operator accepts some responsibility for making sure that phases are executed in the proper order. The operator also usually has the ability to run the batch process under automatic control, where the system is simply asked to "make a batch of product B."

This *Process Management* level is also concerned with the management of batch production. In its simplest form, this means providing information to an operator that all the materials and equipment required to process a batch are available. A primary function at these levels is to maintain some form of equipment history logging for preventive or predictive maintenance, not so much to prevent downtime, but to prevent it from occurring while a batch is in process. However, it also provides the ability to switch production processes efficiently between products or grades of products.

This means that a recipe is tested for validity (i.e., a recipe cannot be downloaded to equipment that is not suitable for making the product or that does not have the correct interlock integrity level). Once a recipe is downloaded (e.g., to a unit controller), it may be necessary to read the recipe back from the unit controller and compare it to the original recipe to see if the download was correct. If not, the batch should not be initiated.

4.10.4 Batch Control Safety Aspects

Batch control is required to step the batch process through its prescribed sequence. Organizing these batch sequence events is a very important design issue. These events should be grouped into manageable and definable objects such as operations, phases, control steps, etc. (e.g., as defined by ISA's SP88 *Batch Control Systems* standards committee) and should follow the natural operations of the process.

Batch control strategies involve more than minimizing the set point deviations caused by steady state disturbances. There are often alternate sets of control actions needed to respond to events such as major ambient upsets, equipment failures, or major process load changes. Other strategies are needed because of the frequent start-ups and shutdowns that occur. Strategies that involve changing control actions based on events generally fall into the classification of sequence control strategies.

Batch processes use large quantities of two-state devices, such as automatic block valves. Other devices typically found are two-state pumps and motors. These on/off control requirements of batch control contribute to the difficulty and complexity of the control system application since many on/off functions must be performed during each phase of the process. This can lead to problems, such as:

- Processing steps not performed in correct sequence.
- Processing steps not performed at the correct time or for the correct time duration.
- Incorrect material addition.
- Incorrect quantity of material added.

Whether a sequence control strategy applies to the simple start-up and shutdown of a motor or to a sophisticated batch process, the control actions and status of the process equipment change with time. Time and event sequencing, whether fixed or variable, is the underlying consideration in designing a batch control strategy.

Events that initiate changes in the state of the batch process are called trigger events. Trigger events are usually based on either the control actions themselves (e.g., such as the reaching of a temperature end point) or by the completion of a certain time period (e.g., a 1 hour hold). Several events may have to occur simultaneously to trigger a change in the sequence.

It is essential to verify that the system is in the correct operating phase before allowing certain operations to start. For example, pump-out is not started unless the reactor is in the cooling step *and* the batch temperature is below 100°F. This prevents pump-out from inadvertently beginning anytime the temperature is below 100°F. In this case, the temperature will probably be below 100°F when the reactor is not operating.

Off-line simulators can be used to verify the operation of the batch control system before a batch is actually run (Ref. 4.59). This allows extensive system checkout to be done prior to plant start-up. In addition to improving the safety of the system, this can also drastically reduce start-up time. The simulator can also be used for "hands on" operator training. Input can be received from operating personnel based on operating the simulator. Necessary changes can then be incorporated into the system prior to start-up.

The BPCS is used as a protection layer in a batch control system and has the same duties as it does for a continuous process.

4.10.5 Safety Aspects of BPCS Regulatory Control

Regulatory loops in the BPCS are needed to control operating parameters like reactor temperature and pressure. Flow and level controls are also examples of regulatory control. However, regulatory control consists of both continuous control (e.g., temperature control) and discrete control (e.g., depressuring a reactor when the pressure exceeds a predetermined set point).

Regulatory control of a batch process is basically the same as regulatory control for a continuous process. It involves the monitoring and manipulation of process variables in the same manner. The batch process, however, is discontinuous. This adds a new dimension to regulatory control because of frequent start-ups and shutdowns. During these transient states, control parameters such as controller gain adjustments may have to be changed for optimum dynamic response.

By the very nature of batch processing, it is inevitable that process equipment will have idle time between batches. The idle time could also be because the controller is not being used sometime during the execution of the batch (e.g., a flow controller may only be used during the feed of one of the reactants). During the idle time, control considerations such as reset windup must be considered. There may also be frequent changes in recipes, product grades, and in the process itself. All of these things put increased demands on the regulatory control loops in a batch process.

Different control strategies are often necessary for the same piece of equipment. This could involve changing to a different controller or changing the control algorithm. The choice of which controller or control algorithm to use may be product dependent and must be specified in the recipe.

Batch processes have optimum conditions that will yield the maximum product with minimum time and cost. This often involves a process variable versus time profile (common with batch reactors), and the control strategy must conform to and be tuned

for this profile. This can cause problems when product grades are changed and different ingredients are used. This could cause changes in reaction rates which will affect the controller tuning. It could also require a different profile, which means that the control system must be capable of generating the new profile.

It may also be necessary to monitor more variables in a batch process in order to take supervisory action (e.g., put the system on hold if a particular manual valve is not closed).

Reset windup is common to many control loops, both single loop and multiple loop. (See Section 4.5.2.2.) But with the emphasis today on improved control, improved product quality, energy conservation, etc., it is no longer possible, in most cases, to live with the offset inherent with straight proportional control. When the process must be controlled tightly at set point, reset (integral) action must normally be added to the controller. However, when reset action is present in a controller, any sustained deviation of the process from set point will cause the controller to wind up (saturate). This means that the controller output will go to its maximum or minimum value, depending on the direction of the deviation.

Reset windup leads to the process variable overshooting the set point. In many batch processes, controlling overshoot is critical to product quality. This problem may occur during the extended heating interval often required for starting exothermic reactions. As the reactor temperature slowly rises to initiate the reaction, the integral term winds up. Once the reaction begins, the integral term may not decrease fast enough to prevent temperature overshoot.

One way to prevent integral windup is to ramp or gradually increase the set point. This prevents the error from becoming large enough to allow the integral term to wind up. This technique only works as long as the control system is capable of following the ramp. If not, the controller will still wind up. Another technique that can be used to overcome the problem of reset windup is to use a batch switch. See Ref. 4.60.

An important consideration with batching stations is the capability of two-stage shutoff. This means that at some point before the final set point is reached, a switch is actuated that reduces the feed flow rate. For the last part of the charging, the meter will be fed at a lower flow rate. Because the final control element cannot close immediately, some material will pass through the meter between the time the switch actuates at the final set point and the time the final control element actually stops the flow. Feeding at a lower flow rate during the final stage of charging will minimize this overshoot. The batching station should continue to record the material that is fed through the meter after the final set point is reached, so the actual amount of material that was charged is known.

4.10.6 Safety Aspects of BPCS Discrete Control

BPCS interlocks are designed to alter operating conditions so that abnormal operating conditions are avoided. For example, a feed valve to a reactor cannot be opened if the reactor vent valve is not open because reaction in a closed vessel causes an undesirable pressure increase.

There are two basic types of interlocks: failure interlocks and permissive interlocks. Failure interlocks are continuous and are usually associated with equipment shutdown. A high level switch in a tank may be interlocked with the fill valve. This causes the valve to close when the level switch sees liquid and prevents a tank overflow. A permissive interlock is used when a specific condition must exist before some other action may be taken. An example of a permissive interlock is when the reactor sequence must be in the "add ingredient A" phase before valve A can be opened. The interlock conditions must be easily discernible to the operator, i.e., the operator must see what causes the interlock action to occur.

Many applications also require process variable monitoring and exception handling logic designed to deal with abnormal conditions (Ref. 4.61). This monitoring logic is different from safety interlocks because it may change as the batch sequence changes. For example, some interlock logic may have to verify that sequence actions occur as directed (e.g., the valve really did open). The monitoring logic must also determine if an unexpected change of state has occurred and alarm it if necessary.

Exception logic allows different actions to be taken for certain abnormal situations, for example (Ref. 4.61):

- *Abnormal situation, but not critical.* The control system must alert the operator to the abnormal condition. No corrective action is required by the control system.
- *Abnormal situation.* This means that the control system must either take automatic corrective action or drive the process to a predefined safe state.
- *Return to normal.* Recovery logic may be necessary that aids in the return to normal operations so that an operator does not have to finish a batch by hand.

In addition to interlocks and exception-handling logic, other types of discrete signals and actions must be considered. For example, start signals, or start permissives, are discrete signals. The start signal may be initiated by a manual pushbutton station that is interfaced directly with the discrete control subsystem. It may also be derived from an analog measurement in the regulatory control subsystem, based on the analog signal being above or below a specified set point, being in a range between two set points, etc. In PES-based BPCSs all changes of state of discrete signals, such as start permissives, may be included in the event log. A sequence-of-events recorder may be needed in some applications. Similar logic holds true for stop signals, which may be used to shut down a particular piece of equipment or to put the sequence into a hold state.

Overrides are used to prevent the process from reaching the interlock state.

Many situations arise where the conditions for starting a process are considerably different from the conditions needed in normal operation. This may mean that some interlocks may be active in the start-up state, but not in the normal state, or vice versa. Control modes or controller algorithms may need to be changed when moving from one state to another. Good communication is needed between the various subsystems to handle these changes of state in the process.

4.10.7 Process Operating States

The sequential nature of batch process operations results in many potential equipment operational states. These operational states must be identified and functionally defined so that the system can handle abnormal conditions during the production of a batch of product. The following operational states are typical of many batch processes (Refs. 4.62 and 4.63):

- *Start-up.* A batch is ready to start. Any necessary auxiliary equipment is brought to operational condition.
- *Normal (Run).* The process is operating according to the prescribed procedure.

- *Hold.* This is a partial shutdown because of one or more abnormal conditions. Operating conditions are maintained at a safe level.
- *Normal Shutdown.* This is a planned shutdown, e.g., at the end of a batch or the end of a campaign. All equipment is emptied out to save as much of the raw materials and product as possible and to minimize safety problems while the equipment is idle.
- *Emergency Shutdown.* This is an unplanned shutdown that occurs because of a hazardous condition in the process. All equipment is stopped immediately. Usually, no effort is made to empty out the contents unless required for safety considerations. An emergency shutdown can also be manually initiated because of a fire, an accident to personnel, etc. The restart logic may have to be built into the recipe.
- *Restart.* This is exception logic that enables a safe transfer from an interrupt condition (e.g., a hold) back to the normal state (i.e., the state the process was in prior to going to the hold state).
- *Maintenance (Idle).* The equipment has been emptied, and cleaned if necessary, and is ready for the next normal processing.

Control is transferred within these states as a function of process conditions. For example, the actuation of a high pressure shutdown switch could result in a transfer from the normal to the emergency shutdown state. An operator might detect some unusual condition and decide to stop normal operation and transfer to the hold state.

Or a laboratory analysis may indicate a need for additional catalyst to keep the reaction going. The return to normal from other control states should always be initiated by the operator.

The interaction of operational modes become increasingly complex with increased numbers of process equipment units (Ref. 4.62). Taking a unit to the HOLD mode can be a very straightforward operator action in a process with only one unit. However, as the number of units under the control of a common control system increases, more mode issues must be considered. Because of the programmable nature of most BPCSs, it is possible to build safeguards into the system to protect against operator error (e.g., the operator tries to initiate a phase that is not allowed). Integration between the safety interlocks, regulatory loops, and batch control programs is essential to effectively control the batch reactor process.

4.10.8 Manual Operating Modes

Manual control requires access to the individual inputs and outputs that interface the control system to the process equipment. Under manual control, discrete and analog inputs to the control system may be read, and discrete and analog outputs may be manually changed. Manual control should ideally be accomplished through the BPCS so that the full monitoring capabilities of the BPCS are available to verify correct operation. In addition, the operator may be required to verify that the manual actions have resulted in the desired process change (e.g., flow was initiated, so the valve was not plugged). Feedback monitoring of external devices should not be reduced because the system is in a manual operating mode.

Manual control is open ended. The operator may make the change manually, but the feedback normally goes to the BPCS which is then responsible for verifying that the change was actually executed or that the change does not put the process or equipment in an unsafe condition. When the BPCS makes the change, it monitors the feedback to verify that everything occurred as planned.

4.10.9 Application Programming

The application program should be written in modular form. For example, use recipes instead of hardcoding the sequence into logic, break the system down into units, equipment modules, loops, etc. (i.e., partition the process). This makes the system more flexible, makes debugging and troubleshooting easier, and allows these modules to be used in other batch programs.

New recipes must follow the rules established for existing recipes (i.e., there should be documented ground rules for software development). For example, if timing is done with a timer in one recipe, don't do timing with a counter in another recipe. Each recipe should be completely documented.

It may be desirable to mirror the SIS logic in the BPCS (i.e., take the same sensors into the BPCS as into the SIS and duplicate the logic). This allows sensors to be checked (e.g., calibration of analog transmitters), allows discrete sensors to be compared, and allows the logic to be checked.

4.10.10 Operator Displays

The role of the operator in the batch environment is continuing to receive attention. This is because of the complexity of the plants that can be placed under the supervision of an operator, because of the large amount of information available, and because of the number of possible decisions make the job more demanding in a batch than a continuous facility.

The entire batch process must be reviewed from the standpoint of normal and abnormal operation to determine the type of information required and its organization for operator display, and to ensure that the operator is provided with sufficient controls. Standard console designs and display formats that are used in continuous systems are of limited value since batch plants are unique. The analysis should also determine that input and output signals are available to satisfy the control requirements, and the means for coping with a power failure, air failure or equipment failure have been provided. By the end of this analysis, the size of the system in terms of inputs and outputs, the operator console, and the required batch language features should be fairly well defined.

Special operator's functions are needed to enable an operator to communicate with the batch control system (e.g., how to select a recipe, change recipe parameters, assign batch size and number, and start a batch). With a VDU display and data entry console, the operators and engineers should be able to display process schematics with dynamic data and control loops, monitor modify and, if necessary, intervene in the automatic control of the plant equipment. Batch systems tend to complicate the HMI problems because of the many different operator actions that are necessary. For example, the operator may be involved in resource allocation (e.g., selection of the proper hose needed to make a raw material transfer). This is in addition to his needs in relation to recipes, monitoring the operation of the batch as it executes, and a myriad of other activities.

It is necessary to give an operator the ability to intervene in a batch process. However, a situation should be critical before an operator is allowed to halt it at the instant he or she requests a hold. Operations are divided into phases, where each

phase is made up of one or more steps that perform equipment-oriented functions. Phase boundaries provide valid unprogrammed hold points in the step sequencing. They also provide a convenient way to subdivide an operation. A phase boundary does not have to be a valid hold point; however, a valid hold point may occur only at a phase boundary.

4.11 SOFTWARE DESIGN AND DATA STRUCTURES

This section discusses safety issues pertaining to the design and selection of software for BPCSs. Data reliability topics are also covered.

4.11.1 Software Design

Software for the BPCS should be designed (or selected) for reliable operation throughout its life. To be reliable, software should:

- Perform in accordance with its specification.
- Be written in a logical, modular, and structured manner to allow it to be readily modified to accommodate changing user requirements.
- Be robust (e.g., tolerant of user data-entry errors) and secure (immune to sabotage by computer viruses and inadvertent changes).

- Include sufficient documentation to allow it to be used and maintained properly.

Reliable software design requires that the following be specified:

- The phases of the overall design program (see below), the responsibilities of all participants in the design (or selection) process, and the major milestones in the development of the software.
- The documentation to be produced during each phase of the program and the standards for documentation to be met.
- The standards or procedures to be applied in the implementation of each phase of the design.
- The quality-control procedures to be applied at each phase.

Reference should be made to established guidelines in each phase of the software design program. These may include:

- Formal specification techniques.
- Techniques for program design relating to control flow structures, modularization, data partitioning, and concurrent processing.
- Standards for programming languages.
- Techniques for program analysis, verification, and testing.

The quality control procedures which are to be applied should be established before the start of software development. These may include:

- Design reviews during each stage in development.
- Checking and approval procedures at the end of each phase in the development and the acceptance criteria to be satisfied before proceeding to the next phase.
- Test procedures and acceptance criteria.

It is essential that throughout the software development procedure all changes be implemented in a controlled fashion to ensure that:

- All possible implications of a change are considered and that the proposed change is authorized before implementation.
- The changes are communicated to all those involved in the design.
- The documentation relating to each phase of development is kept up to date.

The actual phases of a software development program will vary from system to system, but typically the following phases are necessary:

1. Software-Requirements Specification. This phase begins with the user's specification of requirements and culminates with a specification, agreed upon between the user and the supplier, of the functions to be carried out by the PES, the required performance, and the facilities and interfaces to be provided. This phase forms the basis of the detailed specification of hardware and software and normally forms the basis for final system acceptance. However, since errors or omissions may not be detected until the PES fails in operation and since it may be difficult to implement changes retroactively while maintaining the safety integrity of the system, it is particularly important that the requirements specification be complete, correct, and unambiguous.

Some of the items that should be considered when developing a specification for software are given in the following checklist:

- Are there standards and procedures for the writing of the software specification?
- Is there supervision to ensure the application of standards and procedures?
- Is the specification written in a form understandable to both system designers and programmers?
- Is use made of a formal specification language or some other means of ensuring a precise and unambiguous specification?
- Is there a procedure for generating and maintaining adequate documentation?
- Is there a procedure for the control of changes in requirement?
- Are design reviews carried out in the development of the software specification involving users, system designers and programmers?
- Is the final specification checked against the user requirements by persons other than those producing the specification before beginning the design phase?
- Are automated tools used as an aid to the development of the software specification in documentation and consistency checking?
- Within the software specification, is there a clear and concise statement of each critical function to be implemented, the information to be given to the operator at any time, the required action on each operator command including illegal or unexpected commands, the communications requirements between the PES and other equipment, the initial states for all internal variables and external interfaces, and the required action on power down and recovery (e.g., saving of important data in nonvolatile memory)? The different requirements for each phase of plant/machine operation (e.g., start-up, normal operation, shutdown)? The anticipated ranges of input variables and the required action on out-of-range variables? The required performance in terms of speed, accuracy and precision? The constraints put on the software by the hardware (e.g., speed, memory size, word length)? Internal self-checks to be carried out and the action on detection of a failure?

- Is there a software test specification? If so, is it written to the same standards and procedures as the requirements specification?
- Are all the requirements testable?
- Is the test specification produced independently of the requirements specification so that the probability of common errors is minimized?

2. Software Specification. This phase defines the techniques, procedures, and standards to be employed in the production of the software, and the functions to be implemented. The functions to be implemented should be specified in a form which can be checked against the requirements specification but which allows precise translation into program design and coding.

3. Software Design. This phase defines the required program modules, the functions performed by each, and the data and control flow interrelationships between modules.

4. Software Coding. In this phase, the program modules are coded using a suitable programming language.

5. Software Test. In this phase each program module is tested against the specification for the module in terms of its functions and its interfaces to other modules, and the complete integrated software system is tested against the software specification.

6. System Test. This is the testing of the PES as an integrated hardware–software system against the requirements specification. This phase may comprise an acceptance test program agreed to between the user and the supplier.

Each phase is the software development program should be documented sufficiently to allow:

- Adequate communication between all participants in the design program.
- The design to be reviewed as it proceeds.
- Each stage in the design to be checked and approved on completion.
- The operation of the software to be fully understood in the future.
- Changes to be incorporated in a controlled manner.
- The full implications of changes to be evaluated.

Well tried and tested standard software packages are available for many applications and should be used wherever possible. However, although individual programs may be standardized, the program package may be unique, so adequate checking and testing is still essential.

4.11.2 Types of Software

Software in a PES-based system can be classified as either system (embedded) or application.

Typically, system software is that software provided by the hardware vendor and is necessary for the hardware to be utilized properly. It is generally transparent to the user and resides in the PES as firmware or as disk-drive software. Occasionally system software resides in memory, either battery-backed-up or capable of being refreshed from some fixed source.

A vendor's techniques used to develop system software should be examined. The following is a checklist of some of the items that should be evaluated:

- The manufacturer should provide evidence of an independent safety assessment of all the system software supplied with the PES.
- Identical PES-based systems should have been in satisfactory use in other similar applications for a significant period.
- The manufacturer should have significant experience in the manufacture and maintenance of similar PES systems.
- The system software must be sufficiently well documented for the user to understand its operation and resolve problems in programming or installation in the plant.
- The manufacturer must be able to provide competent technical support.
- The software must be used with sensor inputs wholly within the ranges previously experienced.

Application software is normally implemented by the end user. This software can involve various levels of code from machine-language programming all the way up to the configuration of a PID loop by specifying values in a database. Since this software is application dependent, it does not have the ability to be tested over a long time period or at very many sites and therefore requires very thorough testing and analysis.

The following is a checklist regarding some areas that should be examined before developing an application software package:

- The method of programming must be readily understood by those carrying out the programming. For example, if a ladder diagram PLC programming technique is used, the people involved must be familiar with and experienced in the design of relay logic.
- Existing application modules which have been tested should be reused whenever possible.
- Procedures for generating and maintaining adequate documentation must be used.
- Procedures to control changes to the program must be established.

- Design reviews carried out in the development of the program must involve users, system designers, and programmers.
- The final program should be checked against the user requirement by people other than those producing the program.
- The system must provide adequate tools for program development, for example: facilities for printing the program and I/O reference listing, for cross-referencing the inputs with the derived outputs, and for checking program operation by the simulation of input logic states.
- A means must be defined for testing the program prior to installation and commissioning.
- The test method must be designed to simulate exceptional conditions as well as normal conditions, thereby finding faults as well as proving correct operation.

4.11.3 Database Structures

BPCS PES database structures range from nonexistent to highly structured. Proprietary architectures of many BPCS PESs allow no user access at all.

Systems that lack a "traditional" database structure, such as programmable logic controllers, are typically enhanced with the use of third party programming/documentation software. The fact that this type of software is so popular says something about the needs of the user versus the understanding of the control system manufacturers. The user must not assume that the control system manufacturers have provided the proper tools and methods for control system development.

Systems that allow access to the database structure provide users with a way to manage the configuration data. This approach works well during initial stages of development, but may require the maintenance of two sets of data once configuration of the control system begins. This offers a potential source of error due to confusion of which system is the "master."

Data historians employ still another type of database which is typically a subset of the control system information. Care must be taken when dealing with this data, since the sampling rates can be several orders of magnitude slower than the rest of the system. Historical data are often stored in a compressed format. These effects can mask fast process oscillations or excursions and prevent data from being useful for such purposes as reconstructing sequences of events or loop tuning.

4.11.4 Data Sampling and Resolution

Due to the different subsystems contained within each BPCS system it is important to understand how each subsystem scans and reports its data to the rest of the system. Typical I/O channel scans are in the range of 5 to 200 msec.

Some systems even allow the user to vary these scans. These data are normally sent to the controllers in either a serial transmission or on a parallel bus.

Serial transmission of thsse I/O data can be either periodic or on an exception basis. Both methods can introduce a source of deadtime into the measurement or control system. Periodic transmissions typically cause a fixed deadtime each scan, while exception reporting can introduce a variable deadtime as a function of how fast the process is changing. In most process applications these delays are a small percentage of the total system deadtime, especially if control valves are being used as the final element. However, in fast loops, the amount of deadtime may become significant and should not be ignored.

Parallel I/O buses do not introduce the transmission delays as in serial systems. They are, however, susceptible to common mode noise problems. As reliable serial transmission speeds increase it appears that parallel I/O systems will be abandoned, except for backplane and board level systems.

The required reliability is often achieved by internal diagnostics, such as internal watchdog timers. See Section 4.9.2.1.

4.11.5 Data Reliability

In order to provide a software based system with an acceptable level of fault tolerance, it is important to insure that the program algorithms, the input data, and the output data generated is correct. The software can be tested to guarantee the correctness of the equations, but handling bad data requires some up front design considerations and system analysis.

When designing software for a BPCS certain assumptions must be made about the control systems ability to pass data internally without error. In general it can be assumed that the integrity of data passed through internal memory or across a backplane will be maintained. Errors such as these and those generated in the I/O hardware generally set system integrity flags which can be used by the programmer as required.

The sources of bad data vary, but can be classified into some general categories based upon their origin and impact on the control system. Table 4.2 lists typical sources of error and common methods for an applications programmer to handle them in a BPCS.

It is good engineering practice for each software module to verify the validity of any data that it uses or passes on to other modules in the system. Performing limit checks on data, passing through valid values and generating exceptions if invalid, is a normal technique for data manipulation. In the case of an exception, however, the applications programmer must make a decision based upon process requirements, system limitations, or some other criteria which has been decided as the result of a systems analysis procedure. It may for example, be acceptable to default the last known "good value" for data which has been determined as invalid. Other systems may require a halt until the problem has been resolved.

Table 4.2 Possible Sources of Data Errors

Data Source	Possible Error	Corrective Actions
Field Measurement	Noise	Filters
	Process upset outside sensor span	Limit check with alarm
	Instrument failure	Limit check with alarm Limit check with default Limit check with last good value Multiple field instruments with signal selection Data reconciliation by gross error detection
	Loss of field power	Hard zero detect with alarm
	I/O System failure	Integrity check with alarm
Operator input manual data entry	Bad data Key entry errror	Interactive limit check with alarm
External data system serial communication link	Bad transmission	Limit check with alarm Message validity check Limit check with default
	Bad data	Limit check with alarm Limit check with default
Application software	Program error	Rigorous testing Limit check with alarm
	Input data limits allow invalid calculation	Limit check with default

4.11.6 Execution Rates

Control devices with multiloop capability typically allow variable scan rates in order to minimize CPU loading. This allows the user to select the frequency at which the control algorithm is executed. The range of execution times normally runs from 0.1 to 60 seconds. While execution times of 0.2 to 0.3 second are adequate for most loops, applications such as compressor surge control may require scans less than 0.1 second. It is important for the user to recognize these applications and design for these requirements.

4.11.7 Event-Logging/Data-Historian Requirements

The ability to go back in time and trace a series of events and operator actions is an important feature required in most process situations. The following list represents the type of events that should generally be logged.

- Alarms.
  —Activated.
  —Cleared.
- Operator requested changes to:
  —Set points.
  —Modes.
  —Outputs.
- Operator messages:
  —System asks for operator input—printed.
  —Operator response—printed.
- Operator starts operation or sequence.
  —Operator status message for each step.

The user needs to understand where this logging function takes place and how the messages are generated in order to interpret the data generated correctly. The message logging may not, depending on the system, provide messages in chronological order.

Sequence of events recorders, or recording annunciators, provide a printed record of the alarm identification and the date and time that alarms occur and return to normal. This printout is used to identify the alarms and might be coded as alarm point numbers or as short-worded alarm messages.

Sequence of events recorders are used for determining and recording the times that annunciator inputs change state. Inputs can be analog in addition to contact types. They usually scan field contacts rapidly, store information when alarms occur in rapid succession, and print the information in chronological sequence. The time can be recorded to the nearest millisecond to allow accurate analysis of the sequence of events.

And finally, a note about automatic logging of operator actions to either a printer or mass-storage device: if possible, a DCS should be configured to dynamically log all operator actions which affect the regulation of the plant. This includes changes in controller and alarm set points and controller modes (e.g., automatic to manual or remote to local). This information should be time stamped and could aid in the evaluation of a safety incident and help prevent its recurrence.

A sequence of events recorder can be of considerable help for troubleshooting nuisance alarms. In addition, a complete history of the operation of the annunciator is provided. Several operating modes are available. In the standard mode, all input changes are recorded when they occur. In the snap-shot or trigger mode, one designated input acts as a trigger to initiate the recording of all inputs. In the contact histogram mode, one input is monitored continuously. Troublesome contacts (e.g., nuisance alarms) can be resolved easily using a contact histogram. Most sequence of events recorders provide 1 millisecond event resolution. As long as the contact does not change state any faster than the one millisecond resolution, all contact changes will be recorded with the contact histogram.

High-speed scan time (and time stamping within 1 millisecond of the actual occurrence) is generally only possible using dedicated sequence of events recorders (i.e., dedicated hardware/software dedicated to recording sequences of events). One millisecond resolution is not possible if the sequence of events recorder is implemented within the BPCS (e.g., within a unit controller or PLC that is part of the BPCS). In this case, it is limited by the scan time of the BPCS. This also means that the data is time stamped with an inaccurate time. See also the comments in Section 4.11.3 concerning historical databases.

4.11.8 Partitioning of Modules and Software to Support Operations and Maintenance

Today's distributed control systems provide the software engineer with significant flexibility. This flexibility is an important attribute of the control system allowing custom applications to be efficiently developed. However, as is true with most software applications, the greater the flexibility the greater the complexity.

A multitude of ways exist to design application software for performing a specific task. Four independent software engineers, given the same control problem, will generally develop four unique algorithms. All algorithms may work efficiently but will vary in the format of the program (readability) and, more importantly, the amount and organization of information provided to the operator for understandability and efficient troubleshooting.

Software design concepts such as "modular" and "top-down" design have been used in the computer science fields for over twenty years. The application of these concepts, however, varies. Application programmers too often approach system design as more of an art than a science. Computer-science methods are required to efficiently design and implement a control system.

Support and maintenance of the control software is simplified when consistent formats and techniques are utilized. Troubleshooting interlock logic which is not implemented in a structured format is frustrating and tedious.

Developing a control system which provides for ease of operation and maintenance requires that the control engineers, software engineers, production supervisors, and maintenance supervisors agree on:

- Design of the operator interface including overview displays, and troubleshooting/maintenance displays.
- Format of the control logic documentation.
- Partitioning of control code into simple modules which can be reused throughout the control system.

Agreement on a structured approach is imperative before any configuration of the software begins. This approach should provide the basis for a design specification document which defines the system development scope.

Typically, 80% of the configuration process can be partitioned into modules that are reusable throughout the distributed control system. Examples include interlock logic, control valve tracking, timers, and flow totalizers. Once a module is developed and tested using simulation techniques it can be reused with little risk. It is important that these modules be well documented so that they are easily understood by the person wishing to make use of the module. This is important to minimize the misapplication of modules.

The other 20% may represent custom logic which is unique for that area of the process. Through consistent formats, variable utilization, and effective documentation within the logic, custom logic can be made easy to understand and maintain.

Through the use of a structured and modular architecture the software becomes easier to read, troubleshoot, modify and extend. Many "war stories" exist with systems that were configured without using these techniques. The system may have been proven to function per the control specifications but when it came time to troubleshoot a problem, or to implement a modification, the task became nightmarish. It quickly becomes obvious that a structured approach is vital in these flexible yet complex distributed control systems.

4.11.9 System Access Security

All BPCSs contain some of, if not all, the security elements for the levels of data access and manipulation shown in Table 4.3. These levels are normally password or keylock protected.

4.11.10 Programming Utilities

Programming techniques vary from system to system between on-line and off-line. Extreme caution must be exercised by the user when utilizing on-line techniques to protect against accidental equipment operation. Off-line changes, though safer during the actual programming, can cause process upsets during installation of new programs if done while the process is operating. Some control systems allow changes to be made with little disruption to current operation.

Table 4.3   Levels of Data Access and Manipulation

Access Mode	Description
Monitor	Data can be viewed; however no changes can be made. Typical mode for remote console locations or backup systems in standby.
Operate	Normal operating mode allowing operator to manipulate set points, modes, and outputs, as well as start and stop higher level sequences and systems.
Tune	Changes to loop tuning parameters, alarm set points, overrides, and other register variables is allowed. User access to system parameters is not normally available. This mode is typically reserved for special situations such as loop tuning, troubleshooting, or some type of failure recovery mode.
Program	This mode may vary depending on the systems ability to support "On-line" and "Off-line" programming functions. On-line programming allows changes to the control system to be made while the BPCS is operating. Off-line programming requires that all changes be made remote from the operating system memory. After the changes are completed the new control program is swapped with the old program resulting in varying degrees of disruption to the control system operation. The effects of both methods must be considered before use on an operating system.

4.12 ADVANCED COMPUTER-CONTROL STRATEGIES

Advanced computer control strategies usually employ general-purpose, supervisory-control computers (as shown in Figure 4.1) having the capacity, speed and architecture for the development of complex applications of advanced control strategies. To accomplish advanced control, access to the entire database may be needed. Supervisory computers, though reliable, are not always on line. They are usually not redundant, since the BPCS is capable of controlling the process during computer outages.

With the emergence of distributed computing and the increasing power of BPCSs, the distinction between the BPCS and the stand alone supervisory computer is becoming less clear. In this section, we will consider the role of the supervisory computer with a dedicated VDU and keyboard.

4.12.1 Operator Interface

The operator's "window into the process" is provided by the operator's workstation, which consists of the BPCS VDUs and keyboards and the supervisory computer's display. The operator interface typically consists of multiple operator consoles, providing redundancy and multiple display functions and interacting keyboards. Printers for alarms and operating summaries provide hard copy. In addition, computer driven recorders, alarm panels, graphic mimic panels and alarm summary video displays are the typical components of the operator interface. Supervisory computers typically either report data to the operator for process manipulation or indirectly regulate the process by manipulating set points in the BPCS. Process computer applications realize advanced levels of control, such as economic optimization, constraint control, model-based control, override control, multivariable control, statistical process control, etc. With supervisory computers, any process variable can be linked via a computer application to any set point in the regulatory control system to which it is connected. This overcomes the database limitations in many BPCSs.

In addition to the primary operator display and input device described in Section 4.6, the process computer also has a display terminal and input device. This operator interface device delivers some of the most important information and features of the supervisory computer control systems. These include:

- Historical data collection/reporting over specified time bases, for example, hours/days/months for any point.
- Alarms/messages.
- The link between the process computer and the regulatory instrumentation (BPCS).
- System status/information.
- Help screens.
- Input signal diagnostics/suspension of computer control.
- Displays (profiles, plots, schematics complete with process measurements, status, etc.).
- Integral (reset) windup status.
- Computer application restart facilities.

The operator's supervisory computer display is normally not redundant, since loss of the computer facility can be tolerated with the process control fullback position covered by the redundant operator stations of the BPCS. Also there is usually a display for the application and process-control engineers which allows separate access to the supervisory computer.

4.12.2. Initialization

Initialization provides bumpless transfer between operational modes to avoid process upsets. Initialization is a procedure applicable to the BPCS and the supervisory control computer strategy.

Initialization of the regulatory controls in the BPCS is required in the following cases, among others:

- Single-loop initialization (bumpless transfer to automatic).
- Cascade initialization (primary and secondary set points, as well as windup status).
- Multilevel control structures (automatic-ratio and automatic-bias).

In addition to the transfer of operational modes for regulatory controls, the initialization task involves filtering, resetting accumulators, and collecting past values for use at the time of initialization.

Initialization of the supervisory computer is sometimes called "restart," and typically has three options:

- *Cold*. The supervisory computer can be initialized on the restart command without affecting the BPCS. In this mode the computer control link has not been reestablished to the BPCS. Therefore, each control loop has to be individually initialized.
- *Warm*. In this mode the computer has not been down long, and the process has not changed appreciably. The computer automatically follows predetermined initialization paths to accomplish bumpless transfer to computer control.
- *Fast (or hot)*. In this case the computer outage will have been very short (1–2 minutes). All BPCS control loops in computer mode will be taken over and returned to computer control directly.

4.12.3 Failure Modes

Process control computers must not be relied on to protect against unsafe process conditions; worse than that, they have the potential of suddenly or gradually driving the process into an unsafe operating condition. Applications often permit running closer to safety constraints, thereby decreasing the safety margin.

To avoid safety problems the supervisory control computer should only be able to manipulate the set points of BPCS loops that are in the computer mode. Applications must not prevent or inhibit the backup regulatory controls of the BPCS from performing their design functions in any way. The supervisory computer should not normally be capable of changing the operational mode of BPCS loops, except for going to backup mode on computer failure or to computer mode on system initialization. Clamps and limits imbedded in the BPCS loop configuration should not be changed by the application program.

Diagnostics should be built into the application. For example, instrumentation inputs should be monitored by the computer for instrument failure (bad process variable), overrange/underrange, frozen values, or a fault mode as reported by smart digital sensors. Failure mode strategy may involve continuation of computer control using the last good value and alarming bad process variable signals to the operator. Another strategy may be to ignore a failed sensor or sensors when redundant sensors are used. Alternately the computer may shed control from the computer mode to the BPCS.

4.13 ADMINISTRATIVE ACTIONS

4.13.1 Documenting the BPCS

All aspects of the basic process control system, including computer programs, should be carefully and completely documented. Known limitations and faults in the BPCS control strategy, including the effects of "bugs" in the control software, should be thoroughly documented and made known to the operating staff.

Specific personnel within the plant should be assigned the responsibility of keeping piping and instrumentation diagrams (P&IDs), operating manuals, and computer programs up-to-date. Careful consideration should be given to the personnel who will perform this task. If only one person is responsible for continuing to keep the process-control system up-to-date operation may become a problem. Plant safety, as well as operating efficiency, could be jeopardized if that one key individual were to become unavailable.

Complete documentation of any process-control equipment should be obtained from the vendor. Engineering personnel should furnish a complete documentation package to the operations and maintenance departments to enable proper operation and support of the system, and all documentation should be kept current throughout the life of the system. It is imperative to have documentation of procedures, sequences, alarms, and advanced, supervisory control schemes for the basic process control system.

Although software backups are normally associated with general-purpose computers, many of the same backup principles also pertain to a DCS. If backup procedures are inadequately documented, with only one copy kept in someone's desk drawer, there is a risk that this information will be lost. This type of information should be officially documented with the originals maintained in a formal filing system along with other project drawings and documents. The frequency with which backups are to be performed should be part of this documentation package. PES application programs and databases for control, display, and communications should be covered by a procedure which assures that current backups are made on a regular basis with at least one copy stored remotely from the building housing the DCS. Backups must be properly labelled with their dates and contents for future reference.

Documentation should be standardized whenever possible. Numerous standards and recommended practices cover the documentation of instrumentation and control systems. See, for example, Refs. 4.64–4.73.

Other organizations which offer standards pertaining to documentation include the International Electrotechnical Committee (IEC), the Institute of Electrical and Electronic Engineers (IEEE), the International Federation of Automatic Control (IFAC), and the International Organization for Standardization (ISO).

Additional topics relating to the documentation of the basic process control system are alluded to in Sections 4.6.2 and are covered more generally in the discussion of documentation in Chapter 6. In what follows, several specialized aspects of good documentation for the basic process control system are emphasized.

4.13.2 Documentation Required by Standards and Guidelines

Deviations from accepted, safe practices should only be approved by specific, designated authorities within a plant. A distinction should be made between standards (or standard practices) and guidelines. Standards for safety contain information that is so critical that deviations from them should only be approved by a designated high-level safety authority within the plant, for example, the chairman of the local process safety committee. Guidelines, on the other hand, are only recommendations, and deviations from guidelines can be approved at a lower level of safety authority, perhaps by an operations manager. Corporate safety standards and guidelines can be internationally, nationally, or locally recognized. Typically, corporate standards will simply cite international and national standards and codes as appropriate for their company.

4.13.3 Documenting Engineering Changes

In the construction of new plant facilities, corporate safety standards and guidelines pertaining to process control and instrumentation should be adhered to. With regard to existing facilities, however, retrofits to computing and control systems required by standards and guidelines adopted after construction of the facility may not be warranted when the risk of any safety problem is deemed low and the cost of compliance is high. In situations such as these, a careful review should be conducted, perhaps under the direction of the process safety committee within the plant. This committee would determine whether or not the upgrading of the existing instrumentation and control system is required. Consideration should be given to the technical requirements of the standards and guidelines, as well as previous loss experience. Even if partial compliance or no compliance at all is warranted, the basis for all decisions should be documented. Whenever deviations from standards and guidelines are granted, copies of the variances should be kept in a file maintained perhaps by a centrally located corporate process safety committee and should be accessible for review by safety auditors.

Corporate safety standards and guidelines should be reviewed periodically for relevance and updated as required. Revisions to standards and guidelines should be properly documented. Revisions should be properly issued to appropriate personnel so that the modifications are effective.

4.14 REFERENCES

4.1 Thompson, L. M., *Industrial Data Communications, Fundamentals and Applications*, Instrument Society of America, 1991.

4.2 Moore, J. A. and S. M. Herb, *Understanding Distributed Process Control (Revised)*, Instrument Society of America, 1983.

4.3 Lovuola, V. J., "Consider Your Choices," *InTech*, Vol. 36, No. 3, March 1989, pp. 38-41.

4.4 Calder, William, and E. C. Magison, *Electrical Safety in Hazardous Locations*, Instrument Society of America, 1983.

4.5 Magison, E. C., *Electrical Instrumentation in Hazardous Locations*, 3rd. ed., Instrument Society of America, 1978.

4.6 Magison, E. C., *Intrinsic Safety*, Instrument Society of America, 3rd ed., 1984.

4.7 ISA-RP12.1-1991, *Definitions and Information Pertaining to Electrical Instruments in Hazardous Locations*, Instrument Society of America, 1960.

4.8 ANSI/ISA-RP12.6-1987, "Installation of Intrinsically Safe Instrument Systems for Hazardous (Classified) Locations," Instrument Society of America, 1987.

4.9 ANSI/ISA-S12.12-1984, "Electrical Equipment for Use in Class I, Division 2 Hazardous (Classified) Locations," Instrument Society of America, 1984.

4.10 NFPA 493-1978, "Standard for Intrinsically Safe Apparatus and Associated Apparatus for Use in Class I, II, and III, Division 1 Hazardous Locations," National Fire Protection Association, 1978 (or later).

4.11 NFPA 497A, "Recommended Practice for Classification of Class I Hazardous (Classified) Locations for Electrical Installations in Chemical Process Areas," National Fire Protection Association, 1986 (or later).

4.12 NFPA 497M, "Manual for Classification of Gases, Vapors, and Dusts for Electrical Equipment in Hazardous (Classified) Locations," National Fire Protection Association, 1986 (or later).

4.13 ANSI/ISA-S12.10-1988, "Area Classification in Hazardous (Classified) Dust Locations," Instrument Society of America, 1988.

4.14 NFPA 496, "Standard for Purged and Pressurized Enclosures for Electrical Equipment," National Fire Protection Association, 1989.

4.15 ISA-S12.4-1970, "Instrument Purging for Reduction of Hazardous Area Classification," Instrument Society of America, 1970.

4.16 Krohn, D. A., *Fiber Optic Sensors—Fundamentals and Applications*, Instrument Society of America, 2nd ed., 1992.

4.17 Strock, O. J., *Telemetry Computer Systems—The New Generation*, Instrument Society of America, 1988.

4.18 ANSI/ISA-S7.4-1981 (R1981), "Quality Standard for Instrument Air," Instrument Society of America, 1981.

4.19 ANSI/ISA-S7.4-1981, "Air Pressures for Pneumatic Controllers, Transmitters, and Transmission Systems," Instrument Society of America, 1983.

4.20 ISA-RP7.7-1984, "Recommended Practice for Producing Quality Instrument Air," Instrument Society of America, 1984.

4.21 Kenneth J. Clevett, *Process Analyzer Technology*, John Wiley & Sons, 1986.

4.22 McMillan, G. K., *Tuning and Control Loop Performance*, Instrument Society of America, 2nd ed., 1990.

4.23 Corripio, A. B., "Tuning computer-based feedback controllers," *Control*, February 1990, pp 54-58.

4.24 Gerry, J. P., "Reader Feedback," *Control*, May 1990, page 79.

4.25 Spurlock, M. G., "Control Operator Overload During Emergency Situations," American Petroleum Institute Operating Practices Committee Executive Session, Seattle, October 6, 1987.

4.26 Tucker, T. W., "Guidelines for Organizing Displays of Process Information on CRT Monitors," *Advances in Instrumentation*, Vol. 36, Part 1, Instrument Society of America, 1981.

4.27 Goble, W. M., Frantz, D. J., Johnson, D. A. and L. C. Lewis, "User-Configurable Operators Console Fits a Variety of Specific Control Needs," Paper #85-0762, ISA 1985.

4.28 ANSI/ISA-S5.1-1984 "Instrumentation Symbols and Identification," Instrument Society of America, 1984,

4.29 Hanes, R. L., "Understanding Annunciators...Selection and Application Factors," *Plant Engineering*, May 1, 1980.

4.30 Shaw, J. A., "Smart Alarm Systems: Where Do We Go from Here?," *InTech*, December 1985.

4.31 Hanes, R. L., "Design Alarm Systems to Favor Critical Information," *Control Engineering*, October 1978.

4.32 Arnold, M. W. and I. H. Darius, "Alarm Management in Batch Process Control," Paper #88-1473, ISA 1988.

4.33 ANSI/ISA-S18.1-1975 (R1991), "Annunciator Sequences and Specifications," Instrument Society of America, 1986.

4.34 Schellekens, P. L., "Alarm Management in Distributed Control Systems," *Control Engineering*, December 1984.

4.35 ISA-RP60.3-1985, "Human Engineering for Control Centers," Instrument Society of America, 1985.

4.36 McNamara, John E., *Local Area Networks*, Digital Press (1985).

4.37 American Petroleum Institute, *API RP540 Electrical Installations in Petroleum Processing Plants*, 1982.

4.38 Institute of Electrical and Electronic Engineers, *IEEE 399 Recommended Practice for Industrial and Commercial Process Systems Analysis—1990* (IEEE Brown Book), (ANSI).

4.39 Institute of Electrical and Electronic Engineers, *IEEE 493 Recommended Practice for the Design of Reliable Industrial and Commercial Power Systems—1990* (IEEE Gold Book).

4.40 Institute of Electrical and Electronic Engineers, *IEEE 141 Recommended Practice for Electrical Power Distribution for Industrial Plants—1986* (IEEE Red Book), (ANSI).

4.41 National Fire Protection Association Code: *NFPA-70 National Electrical Code*, 1990.

4.42 American National Standards Institute, *ANSI-C2 1987 National Electrical Safety Code.*

4.43 Institute of Electrical and Electronic Engineers, *IEEE 446 Recommended Practice for Emergency and Standby Power Systems for Industrial and Commercial Applications—1987* (IEEE Orange Book).

4.44 Institute of Electrical and Electronic Engineers, *IEEE 242 Recommended Practice for Protection and Coordination of Industrial and Commercial Power Systems—1986* (IEEE Buff Book).

4.45 National Fire Protection Association Code: *NFPA-78 Lightning Protection Code,* 1986.

4.46 Robinson, M. D., *Power Plant Electrical Reference Series,* Volume 5, *Grounding and Lightning Protection* (Report EL5036), Electric Power Research Institute, Sept., 1987.

4.47 Institute of Electrical and Electronic Engineers, *C62 Guides and Standards for Surge Protection—1990.*

4.48 Institute of Electrical and Electronic Engineers, *IEEE 1145 Recommended Practice for Installation and Maintenance of Nickel-Cadmium Batteries for Photo-Voltaic (PV) Systems—1990.*

4.49 Institute of Electrical and Electronic Engineers, *IEEE 937 Recommended Practice for Installation and Maintenance of Lead-Acid Batteries for Photo-Voltaic (PV) Systems—1987,* (ANSI).

4.50 Institute of Electrical and Electronic Engineers, *IEEE 519—Guide for Harmonic Control and Reactive Compensation of Static Power Converters—1981,* (ANSI).

4.51 U.S. Department of Commerce, National Bureau of Standards (NIST)/Institute for Applied Technology, *AD No. 612 427 Instrumentation Grounding and Noise Minimization Handbook,* 1965.

4.52 National Fire Protection Association Code: *NFPA-77 Recommended Practice on Static Electricity,* 1988.

4.53 Institute of Electrical and Electronic Engineers, *IEEE 142 Recommended Practice for Grounding of Industrial and Commercial Power Systems—1982* (IEEE Green Book).

4.54 Institute of Electrical and Electronic Engineers, *IEEE 518 Guide for the Installation of Electrical Equipment to Minimize Noise Inputs to Controllers from External Sources,* 1982 (Reaffirmed 1990).

4.55 Institute of Electrical and Electronic Engineers, *Guide for Measuring Electrical Resistivity, Ground Impedance, and Earth Surface Potentials of a Ground System,* 1990.

4.56 U.S. Department of Commerce, National Bureau of Standards (NIST)/Federal Information Processing Standard Publication, *No. 94 Guideline on Electrical Power for EDP Installations,* Sept. 21, 1983.

4.57 Digital Equipment Corporation, *Digital Site Preparation Guide No. EK-CORP-SP-003,* Digital Equipment Corporation.

4.58 Rosenof, H. P., "Successful Batch Control Planning: A Path to Plant-Wide Automation," *Control Engineering,* September 1982.

4.59 McIntyre, C. and S. Monier-Williams, "Automation of Brewing Operations Using Modern Batch Control Techniques," ISA/90, Paper #90-419.

4.60 Fisher, T. G., *Batch Control Systems—Design, Application, and Implementation,* ISA, (1990), Research Triangle Park, NC.

4.61 Bradbury, R. K., M. C. Rominger, and J. D. Verhulst, "Promoting Modularity in Batch Control Software," ISA/88, Paper #88-1404.

4.62 Roerk, P. E., "Control, Scheduling and Optimization of Batch Processing, Batch Process Automation Difficult but Cost Effective," AIChE National Spring Meeting, March 25-28, 1985, Houston, TX.

4.63 Gidwani, K. K., "Batch Control Methodology," ISA/82, Paper #82-819.

4.64 ISA-S5.1-1984, "Instrumentation Symbols and Identification," Instrument Society of America, 1984.

4.65 ANSI/ISA-S5.2-1976 (R1981), "Binary Logic Diagrams for Process Operations," Instrument Society of America, 1981.

4.66 ISA-S5.3-1982, "Graphic Symbols for Distributed Control/Shared Display Instrumentation, Logic and Control Systems," Instrument Society of America, 1982.

4.67 ANSI/ISA-S5.4-1976 (R1989), "Instrument Loop Diagrams," Instrument Society of America, 1989.

4.68 ISA-S20-1981, "Specification Forms for Process Measurement and Control Instruments, Primary Elements and Control Valves," Instrument Society of America, 1981.

4.69 ISA-RP42.1-1982, "Nomenclature for Instrument Tube Fittings," Instrument Society of America, 1982.

4.70 ISA-RP60.6-1984, "Nameplates, Labels and Tags for Control Centers," Instrument Society of America, 1984.

4.71 ANSI/ISA-S61.1-1976, "Industrial Computer System FORTRAN Procedures for Executive Functions, Process Input/Output, and Bit Manipulation," Instrument Society of America, 1976.

4.72 ANSI/ISA-S61.2-1978, "Industrial Computer System FORTRAN Procedures for File Access and the Control of File Contention," Instrument Society of America, 1978.

4.73 ANSI/ISA-S51.1-1979, "Process Instrumentation Terminology," Instrument Society of America, 1979.

5

SAFETY CONSIDERATIONS IN THE SELECTION AND DESIGN OF SAFETY INTERLOCK SYSTEMS (SISs)

The fundamental process design should seek to eliminate all hazards. SISs should not be used as a substitute for either the appropriate chemistry or mechanical design. Safety interlock systems should only be used in well engineered designs where all other appropriate risk reduction items have been considered (see Chapters 2 and 3).

5.0 INTRODUCTION

This chapter primarily addresses the use of programmable electronics as the logic solving device in SISs [also commonly referred to as emergency shutdown systems (ESDs) or safety instrumented systems]. Other technologies used in SISs are discussed. Minimum design criteria is provided for applying SISs and integrating them with other protection systems.

The concepts discussed in this chapter should not be applied without a knowledge of the concepts presented in other chapters in this book (see Figure 5.1).

The complete SIS (see Figure 5.2) is addressed in this chapter, from the field sensors through the logic solver to the final control elements, and interfaces to other systems.

Guidelines for specifying and designing SISs are provided in accordance with three integrity levels (1, 2, and 3) described in Chapter 2, Figure 2.5. As used in this book, the integrity levels vary in availability from about 0.99 for Level 1 to 0.9999 for Level 3 (Ref.5.6).

Prerequisites for applying the guidelines developed in this chapter are:

- Identify process risks and estimate potential consequence
- Determine the need for safety interlock(s)
- Establish the integrity level requirements for each safety interlock (see Chapters 2, 3, and 7).

Process hazard analyses are integral to the identification of integrity levels of specific interlocks relating to safety and environmental protection. The information generated through these reviews is a necessary prerequisite to the

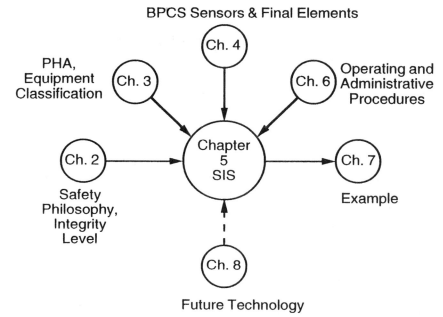

Figure 5.1   Relationship of other chapters to Chapter 5.

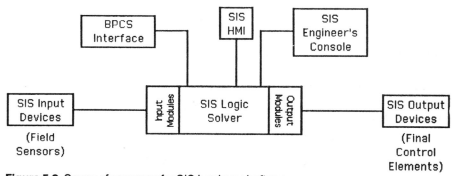

Figure 5.2. Scope of coverage for SIS hardware/software.

use of the techniques discussed here. Techniques for accomplishing process hazard reviews are discussed in Chapter 3.

Independent Protection Layer(s) (IPLs) may be involved in preventing hazardous events or in mitigating the consequence of a hazardous event. Establishing the need for an IPL(s) requires a systems design approach, based on results from the Process Hazards Analysis (PHA) (Chapter 3). The PHA includes consideration of all preventive and protection layers including the

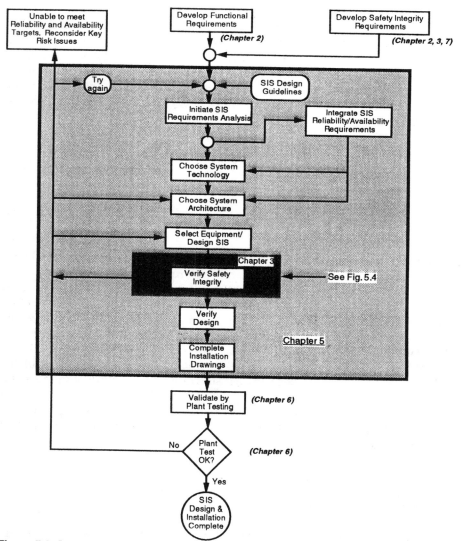

Figure 5.3. Scope of SIS design.

BPCS (Chapter 4) and administrative procedures (Chapter 6). Figure 5.3 shows those details involved in SIS Design that are discussed in this chapter.

Many of the BPCS design concepts discussed in Chapter 4 also apply to the design of SISs (e.g., software considerations, power supplies, and grounding, communications, and sensors and final control elements). This chapter builds on the base established in Chapter 4 (where applicable) and discusses additional requirements that apply to the design of SISs.

It is extremely important that SISs be properly tested and maintained and that unauthorized changes to the hardware/software are not made. This topic is discussed in more detail in Chapter 6. The example in Chapter 7 demonstrates methods that may be used to design an SIS. Chapter 8 discusses some future technologies that may affect how SISs will be designed and implemented.

SISs functions include:

- Detection of out-of-limit conditions.
- Solution of time and event-based equations to identify unsafe operating domains of the process.
- Taking of corrective/preventive action.
- Monitoring of the SIS to ensure it will function properly when a demand occurs.

The SIS performance objectives are to achieve the required integrity levels by reducing to the appropriate level, any failures that will cause the SIS to fail to a dangerous state [i.e., design for high availability (low PFD)]. SISs also need to perform reliably. Since these goals are sometimes conflicting, a balance needs to be achieved between high availability design and reliability while still meeting the required integrity level. This requires the identification and control of faults that cause SIS failures. Faults can be caused by environmental conditions or by problems with the design (including vendor design problems and component selection decisions), installation, or operation of the system.

The design of safety interlock systems should consider aspects of both equipment design (that provided by the manufacturer) and the end-user application. The principles discussed in this chapter apply to the total system design of SISs.

5.1 SIS DESIGN ISSUES

A basic set of information is needed before the design and selection of an SIS can begin. This information is provided in the Safety Requirements Specification (see Section 5.1.1) supplied by the Process Hazard Analysis (PHA) team. Included are the functional and safety integrity requirement of each safety interlock.

Design of an SIS must address:

- All fail-safe aspects, including deenergized-to-trip versus energized-to-trip operation
- Logic structures (Table 5.1)

Table 5.1 Logic Structures

Logic Structure Code	Logic Structure Diagram	Logic Structure Operation*	
		Elements Required To Operate	Elements Required To Shutdown
$1oo1$	—‖— (1oo1)	1	1
$1oo2$	—‖—‖— (1oo2)	2	1
$2oo2$	A / B (2oo2)	1	2
$2oo3$	A B C / C A B (2oo3)	2	2

- Fault prevention/mitigation
- SIS/BPCS separation
- Diversity
- Software considerations
- Diagnostics
- Human/machine interface (HMI)
- Communications (such as the interfaces to the BPCS)

There are many design choices to consider in each of the above items. These choices are discussed in Section 5.1.2 *SIS Design Basics*.

5.1.1 *Safety Requirements Specification*

The Safety Requirements Specification, is divided into Functional Requirements and Safety Integrity Requirements.

5.1.1.1. Functional Requirements
Typical SIS functional requirements include process parameters that should be monitored, actions to be taken when an interlock trips, status information, identified trip points, capability for full function test, minimum acceptable test intervals, and environmental conditions that the SIS will experience.

5.1.1.2. Safety Integrity Requirements
The safety integrity requirements, that is, the integrity level of each SIS, are used to establish an acceptable system architecture for achieving the level of performance, safety, and integrity required for executing the necessary SIS functions. The system consists of sensors, processors, input/output (I/O), final elements, communications, human/machine interfaces, power distribution, etc.

The number of interlocks and their integrity level has a direct effect on the SIS technology selected. For example, if a large number of interlocks are required, then PES-based SIS technology may be a good choice. For small SIS applications, electromechanical relays are often a good choice.

Target reliabilities and availabilities (PFD) of the SIS are needed if quantitative validation of the SIS is appropriate. The calculated reliability and availability of the SIS as designed are compared against the targets. If either of these checks fails, then the SIS should be redesigned, as shown in Figure 5.4.

In addition, this section of the Safety Requirements Specification should identify all IPL's associated with a risk that is to be controlled using an SIS. If other IPL(s) exist, understand their separation, diversity, independence, capability to be audited, risk reduction capabilities, and any if they have common mode faults with the BPCS and the instrumented IPL. All of these impact the design of the SIS.

5.1.2 SIS Design Basics

This section defines basics to consider in developing an SIS that is appropriate for a given application.

5.1.2.1. Deenergized-to-Trip versus Energized-to-Trip SISs
Safety interlock systems can be designed for deenergized-to-trip or energized-to-trip operations. SISs designed as deenergized-to-trip systems are most commonly used throughout the chemical industry. Deenergized-to-trip have the following advantages over energized-to-trip systems:

- Simpler in design
- Fail-safe compatible
- Design approach with the greatest operating experience

Many of the application rules are equally suitable for deenergized- or energized-to-trip systems. A brief overview of key features in each type system follows.

1. Deenergized-to-Trip SISs. In general, the equipment used in the SIS should fail to a safe condition. In most SIS designs, the outputs deenergize or a backup system takes over control when a component or circuit failure occurs in the logic associated with a particular output, if that failure would prevent the SIS from responding to a demand. Therefore, when there is a failure, the system will go to a safe condition. A system that is designed to operate in the "normally-energized" mode can be designed to fail-safe on loss of power. For example, a mechanically operated low flow switch with normally open (NO) contacts that close when the flow is above the *set point* (normal operation) and open when the flow drops below the set point (shutdown condition) is considered fail-safe in many applications.

The components used in the SIS should not only fail to a safe condition when that component fails, they should also be selected so that they fail in a safe condition when system utilities or signal continuity (e.g. electrical power, instrument air, etc.) are lost.

2. Energized-to-Trip SIS. The energized-to-trip SIS should have
- Diagnostics to detect failures in the wiring between the input sensors and logic solver and between the logic solver and final control elements.
- Battery backed DC power, and/or uninterruptible power supply system (UPS) for sufficient time to bring the process to a safe condition
- Alarms on loss of SIS power.
- Frequent testing and active diagnostics monitoring total SIS performance.
- Independent manual safety shutdown means and process shutdown means.

Properly designed energized-to-trip systems may be considered "safe" for loss of power when appropriate external means of taking corrective action are available and deemed suitable by the PHA team.

5.1.2.2. Logic Structures
SISs utilize specific terminology (Table 5.1) to reflect their design features, including varying levels of redundancy. The logic structure defines the number of elements required to operate, as well as the total number of elements required to take a control action. The code used in Table 5.1 defines typical logic structures. Each code is then provided with its logic diagram and a listing of the associated number of operate and shutdown elements.

Note that, the same level of redundancy is not necessary for all parts of the SIS. Figure 5.5 allows typical redundancy techniques for each part of the SIS (e.g., it may be necessary to have dual redundant sensors with a single logic

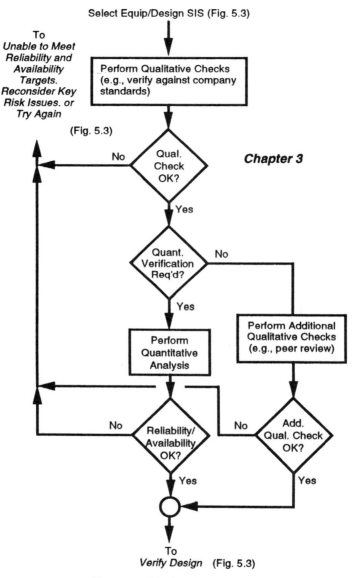

Figure 5.4 Verify safety integrity.

solver and single final control element). Each part of the SIS has its own level of redundancy based on application needs, reliability and availability (PFD) requirements.

A complete definition of the SIS logic requires specification of the structure and the type of logic required. Two types of logic may be specified: shutdown

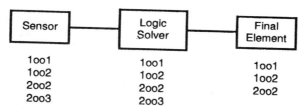

Figure 5.5 Redundancy possibilities in a SIS.

or permissive. The type of logic specified in this text is shutdown unless otherwise noted. Logic designed to restrict the start of a process operation unless defined conditions are met is permissive.

5.1.2.3. SIS Fault Prevention/Mitigation Techniques

The system consequences of component failures can be prevented or mitigated through application of fault avoidance, fail-safe, and fault tolerant design principles.

The selection of the following approaches is an application specific issue to achieve the desired reliability and availability performance.

1. Fault avoidance is the use of design techniques that avoid the occurrence of faults in the SIS. Fault avoidance is achieved by using devices with performance margins that minimize the chance of a detrimental failure (e.g., a safety relay is designed such that the fail-to-danger mode is minimized). This means using a User-Approved Safety relay (see Section 3.3.2.2) selecting the proper contact materials, using contact protection as needed, etc. (Ref. 5.2).

2. Fail-safe SIS design includes features that signal malfunctions of SIS components or of an energy source causing the corrective action to occur. When a component or energy source failure is detected, the SIS may:
   —Signal the failure and not be able to monitor the process until the fault is corrected, where this situation does not create a safety problem.
   —Initiate changes in process conditions to nonhazardous level while the SIS is in a failed state.
   —Initiate actions that automatically signal the failure, replace the failed component and continue monitoring the process.

3. Fault tolerance is that property of a system that permits it to carry out its assigned function even in the presence of one or more faults in hardware or software.

5.1.2.4. SIS/BPCS Separation

Two aspects of separation should be considered: physical and functional. Physical separation means that the BPCS logic solver functions and the SIS logic solver functions are performed in separate hardware. Functional separation is achieved through the elimination of common-mode failures in the execution of the BPCS and SIS functions. This may require the separation of sensors, final elements, and other I/O components; separation of the hardware used to perform the logic function; separation of the system software; and separation of the application program. Some communication between components and systems is allowed as long as no common-mode failures are introduced. Additional discussion of separation principle in SIS design is contained in Appendix B.

Separation is usually not a significant issue in traditional direct-wired interlocks and analog control systems because the typical assignment of one function to one module with interconnection using direct-wiring techniques naturally produces separate control systems. However, elimination of common mode failure can be a significant issue and separation is useful in controlling common mode problems.

In traditional discrete interlocking schemes, individual, stand-alone building blocks called relays (electropneumatic, electromechanical, and electronic) are utilized. Analog signals are typically converted to a discrete value through the use of electronic switches or electropneumatic switches. Communication between SIS and BPCSs is accomplished by interconnecting individual pairs of wires (i.e., direct-wiring) for each discrete function being communicated.

When industrial microcomputers and networking began to be used in regulatory control, the interlocking continued to be accomplished using relays and hard-wired techniques. As a result, separation of interlocking from regulatory control was effectively implemented, due in large part, to the inherent partitioning provided by the way relays are packaged, interconnected, and applied. The definition of "separate" at different plant sites depended on the method used to make interlocks distinct from regulatory control.

PESs are being used for interlocking as well as for basic process control. All of this control is done in software, which can be corrupted. Separation has become an important issue in SIS design, operation, and maintenance. This is because BPCSs can be easily integrated with fully automatic SISs to replace traditional SISs. BPCSs and SISs integration and/or interconnection can result in a system that may compromise SIS integrity.

As a result, personnel in design, process control, maintenance, safety, environmental, operations, and manufacturing should be aware of the importance of separation to maintain SIS integrity and security.

SIS separation from the BPCS is required to ensure that safety and environmental aspects of the SIS are consistent with user, manufacturer, local, national, and international standards and guidelines, because separation will:

- Minimize the effects of human error on the SIS from normal BPCS activities.

Protect safety system software from unintentional changes, by isolating the PES-based SIS from process-control-induced programming changes (i.e., in the BPCS).

- Provide access security.
- Ensure that SISs are maintained safely and correctly.
- Facilitate stand-alone testing and maintenance of the BPCS and SIS.
- Ensure security and integrity, allowing the PES-based SIS to achieve a level of security and integrity equal to or better than a direct-wired SIS.
- Minimize common mode faults (both hardware and software). Separation issues should be considered at the early stage of control system conceptual design. The SIS should have separate identification, documentation, programming, and maintenance.

Some aspects of separation, such as electrical power distribution and sensors, have been addressed for years by the electrical and instrument technologist. PES technology adds I/O communications, HMIs, and "smart" sensors to the list of separation considerations. Other technical aspects to consider when providing the needed degree of separation include:

- The possibility of multiple HMIs that may be used to access SISs. One may be used to display data needed for day-to-day operation [e.g., BPCS operator's console (Section 4.6)], another to provide critical SIS functionality and information to the operator (e.g., first-out information), and another to program and maintain the SIS [e.g., the programming panel or engineer's console (Section 5.6.7)].
- Modifications (authorized or unauthorized) in the SIS application programs, which are equivalent to rewiring or retubing in direct-wired or analog systems, may be easy to make at the engineer's console when the SIS is controlling (on-line) as well as when the SIS is shut down (off-line). Special steps should be taken in any modification of an SIS application programs to emulate the security, reliability, and predictability of traditional direct-wired and analog SISs (see Appendix A).

The following are suggested separation criteria:

- Make the application program as secure as possible (i.e., consider separate access paths for BPCS and SIS programming).
- Provide physical and functional separation and identification among the BPCS and SIS sensors, actuators, logic solvers, I/O modules, and chassis.
- Make SIS documentation distinguishable from that for the BPCS.
- Hardware write protect against sources other than approved SIS sources.
- Provide read-only hardware write protect communications for smart transmitters.

5.1.2.5. Diversity

When the BPCS and SIS are implemented in the same hardware/software (i.e., same controller) or when the same technology is used for both the BPCS and SIS (i.e., different controllers but same hardware/software technology), some issues of common-mode faults arises. Some separation schemes (see Appendix B) reduce the possibility of certain common mode faults. An important concern, however, is a common hardware and/or software fault in both the BPCS and SIS resulting from the use of identical hardware components (subject to common mode failure because of the same vulnerability to a specific environmental condition, design feature in both, etc.) and/or the use of identical software (having identical "bugs") which can be triggered by a common event). While possibility of these bugs occurring is difficult to quantify, a sound decision for the need to diversify can be made based on qualitative analysis.

Diversity refers to the factors that make two components (e.g., devices, subsystems, systems, software systems, communications systems, sensors, or final control elements) different in a way that minimizes common mode faults. However, the use of diverse components just to be different could defeat the intent of this effort by introducing unreliable components while trying to achieve diversity. Each diverse device should be User-Approved Safety.

Examples of how diversity can be applied include:

- Ensuring electrical service reliability by providing an uninterruptible power supply (UPS), with transfer capability to batteries and an alternate source. A major factor in this protection scheme is that the alternate source is separate and diverse.
- Monitoring the process with diverse sensor technologies to ensure reliable sensor information where the reliability of a single sensor technology is insufficient.
- Using two PESs with different hardware, system software, and application programs to minimize common mode faults in hardware and software.
- Providing two Independent Protection Layers (IPL), one being an SIS and the other being a non-SIS IPL.
- Measuring different physical parameters, either of which can detect the onset of the same safety-related event.

When a PES is used, and diversity is employed, it may be applied to only the hardware (e.g., two or more PESs from different manufacturers), to only the software (e.g., two or more implementations of the software by different teams using different design, coding, and testing techniques), or to both.

1. **Diversity in System Software for the BPCS and SIS.** System software is that software which is provided by the hardware vendor, and is necessary for

the hardware to function as specified. Utilizing the same controller, with the same system software, for both the BPCS and SIS may make it difficult to achieve diversity and may, therefore, result in a covert (e.g., concealed) common-mode failure.

The probability of a common-mode fault initiated by software in the BPCS and SIS can be identified, and then an analysis can be done to determine if diversity is appropriate. The use of diversity as a protection against the effects of software faults depends on the probability that the different sets of software do not each contain common mode software faults (i.e., faults that would cause similar failures of each set of software under the same set of operating conditions). This probability depends, in part, on the extent to which diversity is applied; in practice, it is difficult to achieve total diversity.

Caveats associated with the use of diversity, include support of two different systems and the increased burden placed on life cycle costs from design through training, spares and maintenance.

2. Diversity in Application Programming for the BPCS and SIS. Application programming refers to the user developed program (e.g., relay ladder programming for PLCs and configuration for DCSs) found in SISs and BPCSs.

When a single team develops the functional requirements for the SIS (see Section 5.1.1.1), the BPCS application program, and the SIS application program, the possibility of common mode faults in the resulting control system is high. If this situation is combined with common hardware and system software, the possibility of common-mode faults increases further. Therefore, applying diversity in functional requirement development and programming the SIS (separate individuals or teams) is an excellent way to reduce the potential of common-mode faults in application programs and should be considered.

5.1.2.6. Software Considerations
Section 4.11 addresses basic software considerations. This section addresses additional software issues relevant to SISs.

Problems (bugs) can occur in both the system software (supplied by vendor) and the application program (configured for user requirements). Bugs present in the system software may not show up during the configuration and checkout activities. In fact, they may not show up for a substantial period of time, until the exact combination of events occurs, causing them to finally surface.

Other problems can occur while the program is running (e.g., loss of parity checking capability, loss of the software watchdog timer, and undetected bit changes caused by external noise).

Issues to consider in software include the following:

- Storage devices (e.g., hard disks, floppy disks) should be checked for viruses with appropriate antivirus software before they are used in the SIS.

- When hardware components are changed (e.g., upgraded CPU), the vendor's system software or firmware may have also been revised. The user should work closely with the vendor to ensure compatibility and to ensure that adequate factory test procedures have been implemented (see Chapter 6).
- Users should be cautious about having the system software from a particular vendor modified to meet their specific needs. This makes their software one-of-a-kind and usually means that upgrades will be incompatible and that vendor support may be difficult to obtain.
- Problems can occur when changes are made to a program while the process is running (e.g., changing the preset of a timer). On-line programming is especially dangerous, although it is a very desirable feature during system checkout. Some type of key lock device or password protection should be installed to prevent dangerous on-line program changes. The PHA team should consider restricting on-line programming changes.
- The ease of programming in a PES also makes it possible for unauthorized personnel to change programs. Even if personnel know how to program the PES, they may not know the programming approach used for safety systems. The use of PROM or EPROM memory for program storage may add some protection against unauthorized tampering.
- Many PESs allow inputs and outputs to be forced on or off. This is equivalent to bypassing relay contacts in an electromechanical system. Forcing is a feature sometimes used during check out of a system, but it could result in defeating the action of the SIS and its use should be limited and managed (see Section 6.1.3.1).

1. Application Programming. In addition to the coding recommendations provided in Chapter 4 (4.11), program coding for safety applications should also

- Keep the program simple so it is easily understood.
- Use modular programming to break the program into smaller parts.
- Avoid use of NOR and NAND functions.
- Embed documentation within the program.
- Apply structured programming techniques.
- Verify design during the application program development process.
- Provide the proper amount of front-end work to thoroughly define the application program requirements and the application program development methodology.

2. Diagnostic Techniques. One of the methods to keep application software problems from creating unsafe conditions is to use diagnostic techniques that take application software decision- making for one system and compare it against:

- Measurements.
- Another system doing the same algorithms.
- Inferences based on artificial intelligence techniques. (This area offers tremendous promise and is discussed in Chapter 8.)

Another typical application of diagnostics is to provide for verification that unintentional changes are not made to the SIS. Diagnostic procedures are typically done periodically: they are also done after programming changes are made (e.g., provide a utility that allows a software compare of the current application program with a master program); and prior to start-up from a power-down condition.

3. **Verification and Validation.** Verification is the process of determining whether or not the product of each phase of the application program life cycle development process fulfills all the requirements specified in the previous phase.

Validation is the field test and evaluation of the integrated application program to ensure compliance with the safety, functional, performance, and interface requirements. The value of validation is:

- It provides for a disciplined and thorough approach to ensure performance.
- It can be reused throughout the life cycle of the process being validated.

A software validation plan should include the following:

- A list of individuals responsible for implementing each part of the plan.
- A list of documents that define the system.
- Assurance that appropriate software development standards are consistently followed.
- Identification of all systems that affect validation of the total system.
- The establishment and availability of change control procedures.
- A detailed list of backup and recovery procedures in case of system failure.
- Documented evidence that each module or section is performing its function.

5.1.2.7. Diagnostics
Diagnostics are required to detect failures in the SIS. This concept is fundamental to the application of PESs as SISs. However, diagnostics add hardware and software to the SIS that would not otherwise be required. The additional hardware and software may reduce the reliability, and increases

the complexity, of the SIS. Therefore, a balance between reliability and fail-safe features matching the process system needs is required to ensure desired SIS performance. Reference 5.7 provides guidance to determine the effective performance of diagnostics (e.g., coverage factor).

Diagnostics may require a test program to ensure that the diagnostics are functional. Some diagnostics are self-testing, which reduces the need for a separate test program.

This section discusses the reasons for using diagnostics, the types of diagnostics that are used, and the way diagnostics are applied in SISs. The need for diagnostics is greater for PESs than for many other SIS technologies.

Some unsafe failure modes of PESs (Appendix G) are well documented. Equipment suppliers are continuously upgrading their embedded diagnostics to allow self-diagnosis of internal problems as well as external wiring problems. As PES manufacturer's products continue to mature, inherently fail-safe PESs will reduce the need for users to add diagnostics.

However, the need for user-implemented diagnostics continues today to ensure detection of unsafe failure modes in:

- The logic solver
- Sensors
- Final control elements
- Energy sources

Diagnostics may be internal (embedded) or external. Internal diagnostics are provided by the manufacturer; external diagnostics are provided by the user.

Diagnostics may be active or passive. Active diagnostics continuously check the area of hardware/software being diagnosed to ensure satisfactory performance. Passive diagnostics only check the area of hardware/software being diagnosed on demand (e.g., when a command is given). It is recommended that some form of active diagnostics be utilized where high integrity SISs utilizing PES technology is applied.

Initially, designers of PESs provided internal diagnostics to provide a level of certainty that the PES was operational. These programs became known as "watchdog" or "heartbeat" diagnostics. The end-user had no access to the "watchdog" diagnostics and limited capability of executing corrective action on loss of the "watchdog."

To improve availability, users now provide external "watchdog" features to supplement vendor provided "watchdogs." Appendix C provides guidelines for the use of watchdog timers (WDTs) in PESs.

The BPCS may be capable of being programmed to duplicate the functionality of the SIS, a concept known as "mirroring." Mirroring is used to alarm abnormal conditions and/or shut the process down (Section 5.4.3). Mirroring can be a way to identify faults in the SIS. Mirroring by the BPCS is not considered an additional independent protection layer. BPCS mirroring diagnostics increases the complexity of both configuration and maintenance. The result can be increased availability, but BPCS reliability may be reduced slightly.

Appendix B discusses various secure communication schemes (i.e., a safety gateway). Diagnostic should be provided to ensure that this communicator link is capable of communicating the status of the SIS to the BPCS in a timely and secure manner without jeopardizing the separation requirements. Response time to perform these diagnostics and respond to the operator is a critical communication function that should be analyzed.

1. Passive Diagnostics. Passive diagnostics respond to a problem on demand, indicating a problem only when an event or system test occurs. As a result, the limitation of passive diagnostics is that they advise the user of abnormal performance conditions only when the user initiates a test or when the SIS is required to perform correctly.

An example is a solenoid that should deenergize on high pressure. When a high pressure occurs and the solenoid is not deenergized, passive diagnostics may be provided to indicate that the solenoid is not performing correctly (see Figure 5.6). This form of diagnostics is typically found in alarm systems where an indication is provided when a potentially hazardous condition or an out-of-range condition occurs.

Note that the user usually knows there is a problem only when a state changes. As a result, the use of passive diagnostics in safety applications requires a means to perform on-line testing, and/or an alternate means to implement the desired action (an example of this would be to turn off the power to the solenoid valve).

Passive diagnostics are excellent maintenance trouble shooting aids and are especially suitable for noncritical applications. Note that actuating a pushbutton (PB) to do a complete diagnostics check of the system may bypass protection during the test period (An example would be a test PB on a WDT that alarms only).

2. Active Diagnostics. Active diagnostics continually test the subsystem being monitored, even though an out-of-range condition has not occurred, to ensure that it is operable when requested to perform. Active diagnostics simulate out-of-range conditions, check the result, and then, if appropriate,

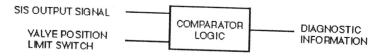

Figure 5.6 Passive diagnostics.

allow operation to continue. If the result is inappropriate, a decision should be made to shut down or take some other corrective action.

For example, active diagnostics may be used in the high pressure example of Figure 5.6. Assume a solenoid valve failed and stayed closed on a high pressure condition, when it should have opened. With active diagnostics, the signal to the solenoid would have been "bumped off" on a regularly scheduled basis (e.g., once a minute, once every five minutes, once an hour, or once a day), whatever is appropriate to achieve the necessary availability. The time interval of the "bump off" would be sufficient to ensure the solenoid valve is operating but short enough in duration to not affect the final control element. The solenoid valve would have tested "nonfunctional" prior to the high-pressure condition occurring. The system could have been shut down or some back-up form of safety support could have been provided until the system could be repaired. There may also have been a way to repair the system on-line.

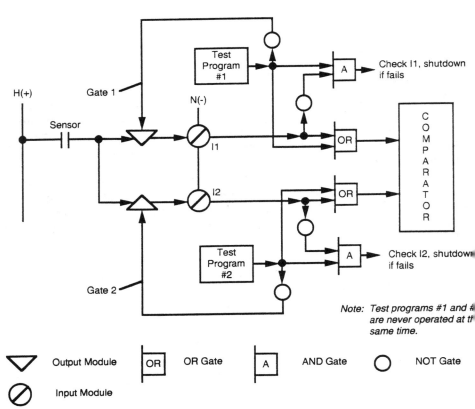

Figure 5.7 Active diagnosis.

Figure 5.7 shows an example of a technique that can be used for active diagnostics of an input circuit (deenergize-to-trip, therefore, the input should always be ON). This drawing shows a single sensor with redundant inputs to the PES. Note that this circuit does not test the sensor, only the input module. The sensor must be tested by other means (e.g., by a manual test or by comparison with other sensors).

The sensor input is wired through two PES output modules to the PES input module. If both test programs are turned off, their outputs will be at a logical 0, and the comparator will be comparing inputs I1 and I2 to verify they are the same. A miscompare will generate an alarm for the operator. This comparator circuit is an example of active diagnostics.

Only one test program is active at one time, so the comparator output is always based on at least one good input signal. When the test program is active, its output will turn on for a short period of time. Because of the NOT gate, gate 1 will turn off and deenergize the output, causing the input to go to a logical 0. The AND gate compares the gate signal to the input signal to verify that the input signal did turn off. If not, the output of the AND gate will go to 0 and trigger a shutdown. The AND gate must be bypassed when the test program is not active, since the output of the AND will be 0 (because the gate signal is off; this bypass logic is not shown). The comparator circuit and test programs are implemented within the application program that resides in the PES, or a program provided by the vendor within the system software.

The advantages of active diagnostics are obvious. The disadvantages are that they may require more:

- Sensors, final elements, and application programming.
- Components in the SIS system, which can decrease the overall system reliability. This may mean more unwanted (spurious) shutdowns, forcing the process to go through what many feel is the most unsafe processing state—start-up. Obviously, the need for additional diagnostics must be balanced against the problem of spurious shutdowns. Diagnostics should be simple to implement and should minimize unnecessary shutdowns.
- Active diagnostics are accomplished by integrating test software and hardware in a way that does not disrupt the normal process operation. This may be accomplished by various methods, including the:
- Addition of supplemental hardware and software to monitor the PES performance as in a watchdog timer (WDT) (Appendix C).
- Duplication of the monitored area so while one area is being tested the duplicate area is controlling the process.

5.1.2.8. Human/Machine Interface (HMI)

Establishing the proper HMI for safety applications involves many issues (Section 4.6). The single largest impact on the HMI is the amount of operator intervention that will be instituted during a safety shutdown as opposed to having automatic shutdown features. This has a major impact on how the HMI is designed. Other areas of concern are critical alarms and whether their display should be through direct-wired techniques or through multiplexed communications to video display units (VDU). The operator should not have "write" access to the SIS except for predefined functions; however, there should be full "read" access. Where the BPCS HMI is used to configure the SIS logic solver, a hardware write protect feature should be provided in the SIS logic solver to avoid inadvertent changes.

Operator's interfaces should represent the process clearly and simply, avoiding misinterpretation. The more complex the system or the greater the risks involved, the greater the need for faceplate data display organization (e.g., alarms segregated by process function, equipment). This is true whether the operator's interfaces are conventional control panels or VDU graphic display screens.

Human error is involved in about 95% of all accidents. This is often caused by wrong or confusing information, not enough or too much information, confusing controls, and poor design. Complex computer control systems require human engineering to reduce the frequency and impact of operator errors. While some ergonomics is built into a PES and cannot be changed, other features can be programmed or configured. These features should be used to improve the effectiveness of the operator's interface. Manual SIS activation switch(es) (sometimes identified as emergency stop switches) should be considered in the PHA and provided where applicable.

5.1.2.9. Communications

With control systems distributed throughout plants, the result is a high degree of communication (or networking) among various components, subsystems, and systems. Appendix D addresses the relationship of information transmitted over communications links and the suitability of various communications techniques for use in SISs.

A complete understanding of the SIS communication system is required. This will assure that issues relating to communication in SISs are considered. These include (see Appendix F):

- Backplane communications in SISs.
- Identification of common mode failures.
- LAN and data highway use in SISs.
- Communication diagnostics.
- Multiplexing and remote I/O use.
- Use of fiber optics.

5.1.2.10. Equipment

Equipment selected for SIS applications should be constructed from high-performance, industrial grade components and should be user-approved safety

(Section 3.3.2.2). Both spurious and unsafe failure modes, and the failure rates of this equipment should be obtained from the supplier and reviewed by the SIS designers. The equipment should be installed to provide reliable operation. All devices should be products that (1) have proven plant performance in less critical applications and (2) have less than one major engineering design change per year.

It should be recognized that not all equipment from a given manufacturer will provide the required reliability or functionality for SIS application. Therefore, it is important that each application be thoroughly analyzed before any item of equipment is selected to ensure that the required function can be attained.

There are many things to consider when selecting the proper sensors, signal interfaces, communications, logic solvers, and final control elements for an SIS. The reliability of these components will determine the overall reliability of the SIS.

However, there are many more sensors and final control elements in an SIS than there are logic solvers. So sensors and final control elements should be selected to minimize the number of unwanted failures, since their reliability is generally lower than the logic solvers. Availability is even more important in an SIS, because it determines if the SIS will be active when a demand occurs. When there are known or suspected problems with measurement and/or final control devices, redundancy may be necessary even though only a single logic solver is being used. This topic is discussed in more detail in Section 5.4.

It is also important that sensors and final control elements be installed in a manner that does not degrade the overall integrity or performance of the SIS. Field elements that are part of an SIS should be uniquely identified so they cannot be confused with field elements that are part of the BPCS. Some companies paint field elements that are part of an SIS red (or some other distinctive color); some just install red tags to identify these elements.

Attention should be paid to the installation of field devices to make them easier to test while the process continues to operate. In some cases, this may require provisions for bypassing these devices, particularly sensors, to allow for testing. This topic is covered in more detail in Chapter 6.

1. Sensor Considerations. Process variable sensors serving as inputs to SISs should be selected for maximum reliability based on past history and performance. High integrity SIS interlock requirements typically require the use of redundancy and may require diverse sensors to meet performance requirements. Accuracy and repeatability are also important considerations in the selection of these sensors.

Separation between devices that are used as inputs to the SIS from those sensors that are used as inputs to the BPCS may be required. Where the performance of the sensor is questionable (e.g., where the process fluid contains solids that can coat the sensor), diverse sensors should be considered. Redundant and diverse sensors may provide some assurance that at least one of the sensors will be operating properly when a demand occurs on the SIS.

In the past, most sensors for SISs were discrete devices (e.g., flow switches, pressure switches, etc.) that had only two states; on or off. The input to the SIS was either energized or deenergized. There are some problems associated with discrete sensors. The only way to ensure that the sensor will operate when needed is to test the operation of the sensor periodically. A discrete signal can only indicate whether or not switch contacts are open (i.e., that something has happened). It cannot provide information that allows diagnosis of a failing sensor or that supports an analysis about why the problem happened.

Many companies are now using analog sensors for SISs, because it is easier to provide diagnostics with analog sensors. It is easy to build diagnostics into the system (e.g., overrange or underrange detection). Similarly, it is relatively easy to make a comparison between two analog signals; an alarm can be actuated if the difference exceeds some preset limit. An analog measurement can give valuable information about what the value is now and can also be used to compute values. This trend will escalate in the future. There is also evidence that modern transmitters may be more robust and dependable than typical mechanical switches.

Analog transmitters with their output going to current or voltage switches or to analog input cards (where applicable) are preferred over discrete devices (e.g., pressure or temperature switches) as the sensor for integrity level 2 and 3 interlocks (see Section 5.4). This is because of improved repeatability, the presence of a continuous signal that can be used for operator observation to sense component condition, and a significant reduction in the mean time to detect failures.

Blind initiators (i.e. sensors without indicators) of shutdown action are not preferred. Some means of verifying the status of a shutdown initiating variable should be provided. A dual thermocouple—one for control and one for shutdown initiation—is adequate to verify the status of a shutdown initiating variable.

Shutdown switches which are mechanical pickup devices off recorder pens and auxiliary contacts off annunciators are not recommended.

Smart sensors are now available that provide increased accuracy over the previous generations of sensors. Another advantage of smart sensors is the additional diagnostics that can be built into these sensors. A downside to smart sensors is that they incorporate programmable electronic devices that have the same failure problems that have been discussed previously. They need to be treated in the same manner as a PES-based logic solver, because failures of these sensors can have the same detrimental effect on the overall integrity of the SIS. Smart sensors when used in safety applications should be "write protected" (e.g., by altering a direct-wired jumper).

When analyzers are used as inputs to SISs, the length of response time (Figure 3.4) should be considered. Many analyzers (e.g., gas chromatographs) are sampling devices. This means that the output signal cannot be updated in a time shorter than the sampling time. It is not unusual for the update times to be in the tens of minutes.

Sometimes it is possible to use variables as inputs to an SIS that cannot be directly measured. These may be calculated variables or variables that are inferred based on a number of other signals, both analog and discrete. When calculated or inferred variables are used as inputs to an SIS, it is essential that the calculated value be validated before control actions are taken. It is often more difficult to test the operation of the system when calculated and inferred inputs are used because of the complex algorithms used to develop the inferred values.

Since all sensors will fail at some time, sensors should be set up to be fail-safe. It is possible to design fail-safe solid-state devices by using pulsating signals which test each critical element in the circuit (Ref. 5.3). If a gate fails and the pulsating signal is lost, then the device fails safe.

It is common to also configure an alarm for the same measured variable used for shutdown at the opposite end of the scale from the trip point, to monitor the condition of the sensor and to increase the effective system integrity.

There are very few fail-safe alarm switches available. A fail-safe electronic alarm switch has been designed (Ref. 5.1). A commercially available voltage/current alarm switch based on this design is now available (Ref. 5.4). For a low signal alarm, this switch will fail-safe on power failure, on loss of input signals (either low signal input or open circuit input), on an open circuit or short circuit of its set point signal, on excessive drift, or on the failure of any critical component. Often the best that can be done is to purchase devices where the output signal is energized in normal operation. The output will then be deenergized in the alarm condition or if power is lost to the device.

The actual principle behind the sensing device should also be investigated. Appendix E provides an example of fail-safe considerations in sensor design.

Input devices should also be tested to see what happens when something else fails (e.g., a wire comes loose). In some commercially available capacitance-type, single-point level switches, the switch is disabled if the probe wire comes loose. However, this loose probe wire does not generate an alarm or shutdown. This switch is not fail-safe and some additional circuitry is needed.

Switches should be analyzed to understand their failure modes. When those failure modes are unsafe, provide features to make them fail-safe, such as diagnostics, end-of-line detection, etc. The most common failures are either a broken wire, a short circuit, or a contact that fails to operate.

Sensor failure should be detectable. With a single sensor, failure may be difficult to detect. Some detection methods available include:

- Out-of-range analysis
- Time-out of "no-change" period
- Tracking against related sensing
- Compare versus related data

Usually it is assumed that two independent sensors measuring the same variable will not fail at the same time. If, however, the sensors are identical, they may be subject to common-mode failure mechanisms (e.g., environmental, process related) that could cause simultaneous failure of both sensors. This problem can be reduced by using diverse sensors, as long as each sensor provides reliable operation. With two sensors, detection of a disagreement is readily available but there still may be no clear indication of whether an alarm condition actually exists. Three discrete sensors connected for two-out-of-three (2oo3) majority voting can be used to provide an indication of a failed sensor. The failed sensor can be outvoted by the two good sensors and allow the system to continue running (an alarm should be provided to the operator concerning the failed sensor). With three analog sensors, median selection is used. A median selector rejects both the high and the low signals and transmits the third. Median selector redundancy protects against sensor misoperation, while also filtering out transient noise or transients not common to two of the signals.

2. Signal Interface Considerations. Signal interfaces among sensors, logic solvers and final elements require special care to ensure fail-safe operation. Many logic solver output modules have triacs which fail in an indeterminate mode. Therefore, design using triacs should include provision for fail-safe operation. Input devices may have to be redundant to ensure fail-safe design. Diagnostics are typically required (see Section 5.1.2.7).

Failure alarms should be provided on I/O devices to indicate malfunctions to operating personnel.

3. I/O Considerations. The use of remote I/O components is acceptable provided that secure communications between the I/O and the logic solver are provided and any required redundancy is not compromised. A redundant communication link may be required when I/O and logic solver are not physically located in the same room. Another alternative is to utilize distributed I/O techniques that minimize the impact of the loss of any component. External watchdog timer I/O monitoring is recommended (see Appendix C).

4. Logic Solver Considerations. Cost and/or ease of modifications to the shutdown logic should not be the only factor in the selection of the SIS logic solver. The type of logic solver selected should be based primarily on reliability and the functionality requirements of the SIS (see Section 5.3).

The logic solver should be designed so that it is fail-safe when power is lost and when either a single key component or the system itself malfunctions. The SIS should not automatically restart when power is restored, or a malfunction is corrected.

The logic solver software should reside in memory that prevents loss of program on loss of power. Such memory devices include nonvolatile memory and battery-backed-up RAM.

Failure of a logic solver should provide a user selectable option that permits removal of power from all outputs, causing them to go to their fail-safe position.

Logic solvers should have the capability of detecting both failures and abnormal changes of their inputs and outputs and should alarm any detected failures. This self-checking feature should be able to determine that the processor is successfully scanning all I/0 points.

An external watchdog timer feature should be included on all logic solvers that have covert failure modes and are used in SISs (see Appendix C).

When a manual shutdown switch is recommended for initiating shutdown action, the PHA team dictates if it is direct-wired to the SIS outputs or direct-wired to logic solver inputs (typically for sequential shutdown). A properly applied manual switch provides additional safety. When relays are used with the manual shutdown switch, utilize user approved safety relays. Manual shutdown switches serve as a last resort means of removing power from outputs and should not be used for normal shutdowns. However, manual shutdown switches should be integrated into the normal shutdown procedure periodically, so that operators gain the necessary confidence and experience in their use.

PES logic solvers bring a new, powerful tool to SIS layers of protection. The logic solver used as an SIS should have reliability that meets the integrity specification and should be fail-safe. Target reliability is achieved through the:

- Selection of the proper logic solver.
- Proper hardware configuration (e.g., redundancy).
- Elimination of hardware common mode faults.
- Elimination of software common mode faults.
- Selection of the proper application programming techniques.
- Arrangement of each of these features so as to be secure from accidental or undetected changes.
- Selection of independent protection layer diversity and separation criteria so that common-mode faults are reduced to an appropriate level.

To accomplish this, the logic solver is:

- User approved safety (UAS).
- Configured in a single or redundant scheme providing the desired system reliability.
- Programmed in a simple, easily recognizable, modular, self-documenting method with which the plant is familiar.
- Separated from the BPCS through the use of direct-wired information interchange, UAS read-only communications, and the use of dedicated HMIs.

Fail-safe operation is achieved through understanding of the fail-to-danger modes of the:

- Logic solver.
- Application software.
- HMIs (e.g. operator's console).
- Communications.

It is important to understand which fail-to-danger attributes require attention, as well as which implementation of design techniques (i.e. fault tolerance, diagnostics, etc.) minimize the problem.

5. Final Control Elements. Generally, valves and motors are the final control elements in SISs. Valves should be selected for the specific application requirements of the SIS, considering such things as severe service conditions, shutoff classifications required, reliability of the type of valve in the service, actuator requirements (i.e., power for moving the valve and the failure position of the valve), and experience with valves in similar services. The valve design should provide for a high degree of security in terms of being able to perform its action when called upon.

Final control devices should also be fail-safe. When valves are used as the final control device, they are usually set to take either an open or closed position on loss of motive power. Some special situations may call for the valve to remain in its last position on loss of motive power. A valve equipped with a spring-return pneumatic actuator and, for some applications, a pneumatic storage cylinder are required to achieve a predictable failure state. Although each application should be checked to see which way the final device should fail, valves that supply energy to the system normally fail closed (e.g., a steam valve), while valves that remove energy from the system normally fail open (e.g., a cooling water valve).

All valves have the potential for leaking. If a slight leakage can cause an explosive or hazardous condition, then double-block-and-bleed valves should be used (see Figure 5.8). A double-block-and-bleed system consists of two block valves in series with provision to vent the piping between the two valves. If process fluid leaks through the upstream block valve, it will go out

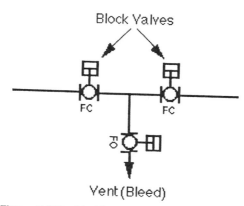

Block Valves

FC   FC

Vent (Bleed)

Figure 5.8 Double-block-and-bleed valves.

the vent line instead of going through the second block valve and into the downstream process equipment. For environmental reasons, the vent from the double block and bleed may sometimes require disposal features in lieu of atmospheric venting, but the system must then be designed so that safety is not compromised. Shutdown or isolation valves should be specified for a shutoff leakage classification based on the application and the appropriate ANSI/FCI 70-2 leakage classification.

For less demanding applications, a double-block-and-bleed valves function is available in a single valve body. A hole is drilled into the bottom of the valve body and up through to the inside of the ball or plug. Any leakage through the upstream seat will drain out the bottom vent.

Double-block-and-bleed valves are commonly used on the natural gas supply to boilers, furnaces, and process heaters. Interlocking of double block and bleed valve arrangements should include monitoring by means of positive feedback devices (e.g., limit switches) and not by timers or rate limiters on actuators. All block and bleed installations should have a periodic full function test to ensure operation that meets SIS availability requirement. The bleed needs to be checked for no blockage and valve opening, and each block valve should be checked individually for low leakage.

The final control element should not introduce a hazard into the system. For example, a ball valve can be overpressured in some applications if the process fluid gets trapped inside the ball when the valve is closed (e.g., this is common for some liquids that vaporize easily, such as ammonia and chlorine). So, for some services, it is important to purchase valves where the interior of the ball is vented into the upstream piping when the valve is closed. Some automatic block valves will only seal in one direction of flow. If these valves are installed backwards in the system, they cannot be relied upon for tight shutoff (Ref. 5.5).

There are times when it is advantageous to use the control valve from the BPCS as part of the final control element for the SIS. This may not be acceptable for all interlock integrity levels. The topic is discussed in more detail in Section 5.4.

Feedback of the position of the final control elements is an important diagnostic tool that should be implemented in some SISs. Whenever the SIS or the operator directs a final control element to a certain state, there should be verification that the final control element actually achieved that state (e.g., through the use of position switches, downstream sensing, etc.). If not, the position feedback signals can be used to alert the operator to the problem or to take automatic corrective action. Field indication of the valve's actual position should also be provided. This indication should be visible from a distance of several feet away from the valve. Special consideration should be given to custom keying of position indication mechanisms on critical valves to ensure correct indication of position at all times.

Solenoid valves on air actuated block valves in the process line are considered part of the final control element. Solenoid valves should be selected for appropriate reliability and availability. Solenoid valves should have an electrical coil minimum temperature rating of Class F. Class H or HT are preferred. Consider implementing redundant solenoid valves for critical applications.

Solenoid valves typically have small orifices that can plug up and impede the flow of the actuating medium. Redundant solenoid valves can be used in these cases to improve the availability of the final control element (see Section 5.4).

Solenoid valves with manual mechanisms that can be operated when the coil is deenergized should not be used in SISs.

Mounting locations of solenoid valves should be chosen to ensure fail-safe operation of the valve. If the valve has a positioner, the solenoid should be installed to vent the actuator and not the positioner.

All piping, tubing, and fittings on control valve actuators used in SIS service should be well designed, properly installed, protected from the weather and freezing, and properly supported. The appropriate material should be selected for the application, although stainless steel is preferred if there is any danger of corrosion. Plastic tubing may be desirable where the valve should deenergize in case of fire.

Quick-exhaust valves may be used to speed up the action of valve actuators where solenoid valves would otherwise restrict flow and slow the action.

Shutdown or isolation valves use pneumatic actuators with spring power to drive them to their safe position, unless provision can be made for an uninterrupted source of air.

There are various methods of using solenoid valves to initiate action of a double-block-and-bleed valve (DB&B) installation. The preferred method is by individual solenoids on each valve. Consideration should be given to the consequences of failure of a single valve, both during normal operation and during the required shutdown action. It may be advisable to limit the flow through the bleed or vent valve in the double-block-and-bleed arrangement by a restricting orifice or by line size to reduce venting of process fluids in case the upstream block valve fails to close.

Most SISs are designed so that the final control elements are energized under normal operation (i.e., deenergize to trip). As a result, loss of power to the final control element can cause nuisance shutdowns and possible subsequent hazardous situations. The integrity of the power source should be ensured.

Another final element is the motor starter. Some Integrity Level 3 interlocks may require an additional automatic power trip in addition to the starter contactor. Consider use of an additional starter or shunt trip breaker for these applications.

5.2 PERFORM SIS REQUIREMENTS ANALYSIS

Design of an SIS begins with the Safety Requirements Specification (Chapter 2, Section 2.3.4), which is prepared by the PHA Team. This document specifies each safety interlock required for risk reduction or control; assigns an Integrity Level classification to each interlock; defines the critical alarm and trip values for each SIS; defines quantitative availability and reliability parameters (when a quantitative validation of the subsequent design is required); identifies IPLs associated with each controlled risk; and specifies any unusual installation or functional test requirement.

The practical application of SIS technology to prevent/mitigate hazardous events results in interlock requirements for all three integrity levels interlocks in typical processes (Section 2.3.4.2). This does not mean that there is a need for three separate SIS systems, one for each integrity level. Where there are a small number of interlocks, a single SIS to handle all three integrity levels may be cost effective. Note that all interlocks implemented in an SIS should meet the logic solver requirements of the highest integrity level interlocks installed. For example, if an SIS is designed to handle three interlocks requiring Integrity Level 3 and five interlocks requiring Integrity Levels 1 and 2, all eight interlocks should adhere to Integrity Level 3 (IL3) logic solver requirements while the sensor, actuator, operation, maintenance, and support integrity requirements are assignable on an individual interlock basis.

Note that combining all IL3 interlocks into a single logic solver with IL1 and IL2 interlocks will result in restricted access to the IL1 and IL2 interlocks. If there are many lower level interlocks, the designer may choose to install another SIS system with a shared logic solver designed for IL1 and IL2, or individual logic solvers for IL1 and IL2.

5.3 SIS TECHNOLOGY SELECTION

A number of technologies can be selected for SISs. These include program-mable electronic systems, direct-wired electrical, electromechanical relays, solid-state logic, solid-state relays, pneumatic and hydraulic systems, fail-safe solid-state logic, and hybrid arrangements. Each of these technologies is detailed in Appendix A.

The SIS selected should be the simplest and most reliable system that meets application requirements. It should meet all of the environmental constraints (e.g., it may have to be suitable for installation in an electrically hazardous area, such as Class 1, Division 2). The technology selected should be capable of attaining the required fail-safe, reliability and availability criteria.

The technology selected for the BPCS influences the technology selection for the SIS. For example, if a PES has been selected as the BPCS technology, it is possible that a PES-based SIS would also be selected, particularly if a high degree of BPCS/SIS communication is needed. Pneumatic BPCSs often will be integrated with relay logic-based SISs using pneumatic trip units (e.g., pressure switches) wired to electromechanical relays that perform SIS logic functions.

An SIS technology selection matrix that will aid in the selection of the correct technology for an SIS is shown in Table 5.2.

Table 5.2 is arranged in the following manner:

- Table 5.2 lists the BPCS technologies commonly used in plants, starting with past technologies on the left (pneumatic) and proceeding to current technologies on the right (PES).
- The table lists, by type and illustration, the typical SIS application complexity encountered. The simple SISs are listed at the top, with the applications increasing in complexity toward the bottom.
- The matrix blocks in the table are filled in with the recommended technology to be used, based on the BPCS technology used and the SIS application complexity. Because the technology selection recommended does not consider all facets of interlock applications [e.g., a large number of safety interlocks may be more economically handled with program-mable electronic systems (PESs) than with electromechanical technology], alternate interlock technologies are provided whenever possible. Table 5.3 provides a prescriptive methodology for selecting the SIS technology.
- Interlock technologies are identified by letter codes (see below).
- Recommended SIS technologies are highlighted by underlines.
- Alternate technologies (without underlines) are provided whenever possible.
- The legend for Table 5.2 follows:

Table 5.2 SIS Technology Selection Matrix

SIS Application Complexity / BPCS Technology	Pneumatic BPCS	Electronic BPCS	PES BPCS
	SIS Technology		
Discrete Sensor ——┤├—— (Load) Discrete Sensor	<u>DW</u> P	<u>DW</u>	<u>DW</u>
Discrete Sequencing (Boolean Logic) S1, S2, Reset, CR, A, (Load)	<u>P/EM</u> P	<u>EM</u> PES FSSL	<u>EM</u> PES FSSL
Discrete Sequencing (with Timing, Counting, Shift Registers, Math, etc.) $\left(\begin{array}{c}\text{Positional}\\0008\end{array}\right) + \left(\begin{array}{c}\text{Positional}\\0072\end{array}\right) + \left(\begin{array}{c}\text{Constant}\\8\end{array}\right) = \left(\begin{array}{c}\text{Movement}\\ \text{Required}\\0079\end{array}\right)$	<u>PES</u>	<u>PES</u>	<u>PES</u>
Analog Input ⊗ — Analog Input Analog Sensor	P PES FSSL	<u>PES</u> FSSL	<u>PES</u> FSSL
Algorithms (PI, Diagnostics) SP $-\,\underset{PV}{\overset{+}{\Sigma}}\,\xrightarrow{e}$ KO (1+1/TrS) → Output	P PES	<u>PES</u>	<u>PES</u>
Model-Based Control Adapter ← Identification Package SP $-\,\underset{PV}{\overset{+}{\Sigma}}\,\xrightarrow{e}$ Control System → Process → Output Measured Variable	<u>PES</u>	<u>PES</u>	<u>PES</u>

Note: Recommended SIS technologies are underlined

DW = Direct-wired

EM = Electromechanical

P = Pneumatic

FSSL= Fail-safe solid-state logic

PES = Programmable electronic system

When applying Table 5.2 consider the following:

Table 5.3   SIS Technology Selection—Prescriptive Methodology
1. List all of the integrity level 3 (IL3) interlocks (from PHA) and their application complexity (from P&ID and other documents).
2. Select the IL3 interlock with the highest degree of application complexity from Step 1.
3. Identify the BPCS technology for the interlock selected in Step 2.
4. Locate the matrix block from Table 5.2 based on the application complexity from Step 2 and the BPCS technology from Step 3. Select a SIS technology from the matrix block from the options shown, based on which technology is most effective for the application.
5. Test each remaining IL3 interlock listed in Step 1 against the solution in Step 4 to determine if: • It should be included in the solution. • The solution should be modified to accommodate all IL3 interlocks (i.e., choosing another technology that is listed under both interlocks). • Another technology should be selected (even if not shown) to attain the most effective solution that satisfies safety requirements. Typically, the most economical way to proceed is to choose one technology solution that will satisfy the safety requirements of multiple interlocks.
6. Repeats Steps 1 through 5 for IL2 and IL1 interlocks.
7. Determine if a separate SIS should be used for each integrity level or if more than one integrity level can be handled in the same SIS.

- Interlock technologies recommended may require unique design adaptations to make them fail-safe.
- All interlock technologies shown should be User-Approved Safety.
- When a technology is not indicated, that technology should not be considered.

5.4 SELECT SIS ARCHITECTURE

This section identifies the elements required to determine the SIS architecture, so that design groups in plants, company engineering groups, and contractor organizations can implement effective, consistent, and safe SISs that meet the requirements of integrity levels 1, 2, and 3.

Note: Section 5.4 assumes the reader has an understanding of all the previous text provided herein.

5.4.1 Integrity Level 1 (IL1)

SISs that are used for implementing integrity level 1 interlocks usually do not require redundancy. Redundancy may be required for the least reliable device

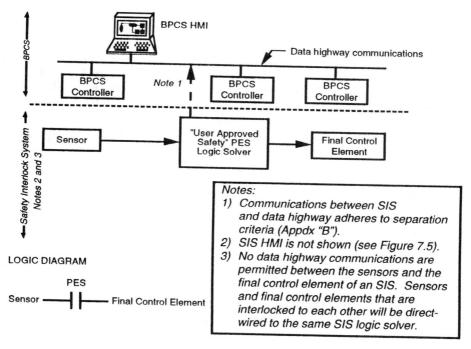

LOGIC DIAGRAM

STRUCTURE

Availability Range: about 0.99 (Reference 5.6)

Figure 5.9 Integrity Level 1 SIS structure and block diagram.

in the SIS. An SIS with single sensor, single SIS logic solver, and a single final control element is shown in block diagram format in Figure 5.9 for a deenergized-to-trip SIS where availability is the prime concern. Note that all components of the SIS should be User-Approved Safety.

Separation between the BPCS and the SIS is necessary (see Section 5.1.2.4). Diversity is primarily needed in hardware, system software, and application programming between the BPCS and SIS. See Section 5.1.2.6 for software application considerations. External watchdog timer is needed for the UAS/ PES (see Appendix C). Passive diagnostics are recommended for the SIS (see 5.1.2.7). The appropriate communications techniques can be selected from

Section 5.1.2.9. Section 5.1.2.10 discusses things to consider when selecting sensors and final control elements.

The system in Figure 5.9 may use the concept of "mirroring" the SIS logic in the BPCS (although not illustrated). This can be accomplished by developing the SIS logic in the BPCS controller, comparing its results versus SIS logic solver results (as a software diagnostic), and using the output of the BPCS controller to deenergize the SIS final control element (s).

Following are typical application considerations for an IL1 system. Specific PHA requirements may alter any or all of these.

5.4.1.1. Sensors for IL1 Interlocks
Discrete sensors are acceptable for IL1 interlocks. Redundancy and diversity in sensors is usually not required. Calculated or inferred measurements are acceptable as a redundant input in IL1 systems.

5.4.1.2. PES Input Modules for IL1 Interlocks
When redundant sensors are used, there is no requirement for the two sensors inputs to be located on separate cards, separate input racks, or separate I/O communicator channels. Use of input modules with a low fail-to-danger mode combined with the use of software signal comparison minimizes the need for additional input module diagnostics.

5.4.1.3. Logic Solver for IL1 Interlocks
A single user approved safety logic solver is acceptable for IL1 interlocks.

5.4.1.4. PES Output Modules for IL1 Interlocks
Output module failure can result in an unsafe condition. Therefore, the output module should be provided with necessary fail-safe diagnostic features to ensure safe operations.

5.4.1.5. Final Control Elements for IL1 Interlocks
Redundant final control elements are usually not required. Valve position feedback switches (i.e., limit switches) may be required if the final control element is a valve. A User-Approved Safety control valve is acceptable as the only final control element, as long as it can meet the leakage characteristics required and is equipped with valve position feedback switches.

Where dual final elements are used (e.g., a control valve from the BPCS is used as a backup to the automatic block valve or where redundant automatic block valves are needed to meet the availability requirements), there is no requirement for the element control signals to be located on separate cards, separate output racks, or on separate I/O communications channels.

5.4.2 *Integrity Level 2 (IL2)*

SISs that are used for implementing IL2 interlocks usually do not require full redundancy. Redundancy may only be required for sensors and/or final control elements and/or logic solvers. An SIS with redundant sensors and logic solvers and a single final control element is shown in block diagram format in Figure 5.10 for a deenergized-to-trip SIS where availability is the prime concern. Note that all components of the SIS should be User-Approved

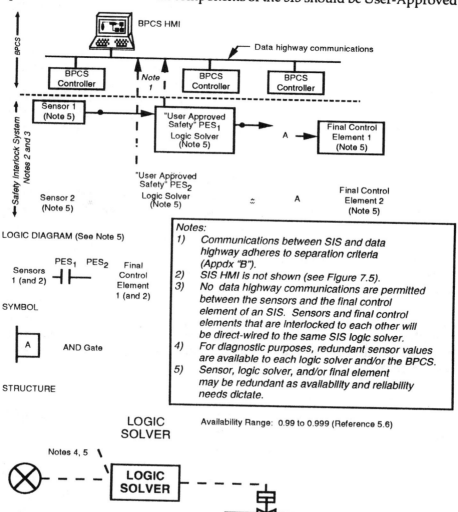

Notes:
1) Communications between SIS and data highway adheres to separation criteria (Appdx "B").
2) SIS HMI is not shown (see Figure 7.5).
3) No data highway communications are permitted between the sensors and the final control element of an SIS. Sensors and final control elements that are interlocked to each other will be direct-wired to the same SIS logic solver.
4) For diagnostic purposes, redundant sensor values are available to each logic solver and/or the BPCS.
5) Sensor, logic solver, and/or final element may be redundant as availability and reliability needs dictate.

Availability Range: 0.99 to 0.999 (Reference 5.6)

Figure 5.10 Integrity Level 2 SIS structure and block diagram.

Safety. As the logic diagram in Figure 5.10 shows, this SIS is configured as a 1oo2 system (Table 5.1).

Separation between the BPCS and the SIS is necessary (see Section 5.1.2.4). Diversity in hardware, system software, and application programming for the SIS logic solvers is recommended (see in Section 5.1.2.5). See Section 5.1.2.6 for software application considerations. Although not shown in Figure 5.10, external watchdog timers are needed for both SISs (see Appendix C). Passive diagnostics are the minimum requirement; active diagnostics are recommended for the logic solver and output (see Section 5.1.2.7). Note that passive diagnostics can be used in conjunction with active diagnostics to improve the availability of the SIS. The appropriate communications techniques can be selected from Section 5.1.2.9 and Appendix D.

The system in Figure 5.10 may use the concept of "mirroring" the SIS logic in the BPCS, although it is not illustrated. Availability can also be improved by having the BPCS continually compare the logic solutions and/or measurements obtained by PES 1 against PES 2. Disagreement typically means that the problem should be quickly corrected or the process should be safely shutdown.

5.4.2.1. Sensors for IL2 Interlocks
Analog sensors are preferred for IL2 interlocks, but discrete sensors are acceptable if some means of verifying the operability of discrete sensors is provided (i.e. formal periodic testing). Discrete sensors typically require more frequent testing than analog sensors. Diversity in sensors should be considered when diversity can reduce the probability of the same failure affecting similar devices without negatively impacting reliability. Calculated or inferred measurements are acceptable as a backup input in IL2 systems, when approved by the PHA team.

Sensor separation should be extended to the actual connection to the process where practical. For example, if two pressure transmitters are used, they should have separate taps into the process (see Figure 5.11). This same principle should be followed for other measurements (e.g., flow, level, and temperature).

5.4.2.2. PES Input Modules for IL2 Interlocks
The two sensor inputs should be located on separate cards, separate input racks, and separate I/O communication channels if practical. Active diagnostics (e.g., mirroring) should be incorporated for the inputs if practical.

5.4.2.3. Logic Solvers for IL2 Interlocks
Various types of SIS technologies may be suitable, such as programmable logic controllers, distributed control system controllers, fail-safe solid state logic, relays, etc. (see Section 5.3). Hybrid schemes may also be desirable (e.g., a PES and a relay logic system).

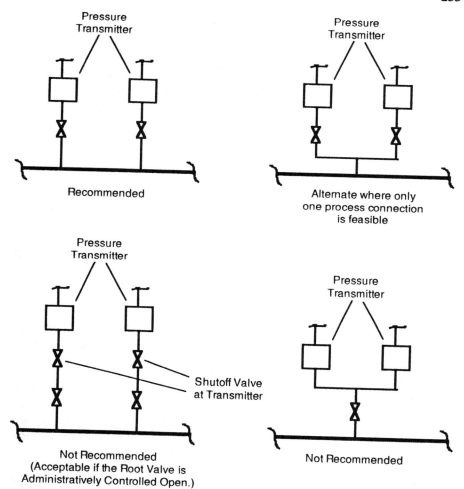

Figure 5.11 Pressure sensor installation connection diversity.

5.4.2.4. PES Output Modules for IL2 Interlocks

Where redundant final control elements are required (e.g., a control valve from the BPCS is used as a backup to the automatic block valve or where redundant automatic block valves are needed to meet the availability requirements), they should be located on separate cards, separate output racks, and on separate I/O communications if practical. The output modules should be designed with diagnostic features to ensure safe operation.

5.4.2.5. Final Control Elements for IL2 Interlocks
Valve position feedback switches (i.e., limit switches) are needed if the final control element is a valve. If a UAS/PHA approved control valve from the BPCS is available, it can be used as a backup to the automatic block valves to improve the availability of the system. A control valve should not be used as the only final control element, unless it is User-Approved Safety and has PHA approval (see Chapter 3).

5.4.3 Integrity Level 3 (IL3)

SISs that are used for implementing IL3 interlocks should be fully redundant, from sensor through SIS logic solver to final control element. This type of SIS is shown in block diagram format in Figure 5.12 for a deenergized-to-trip SIS where availability is the prime concern. This is typical for many batch processes where a short interruption to processing can be tolerated without affecting product quality or safety.

For continuous processes and some batch processes, where availability and reliability are both important, consider the SIS arrangements shown in Figures 5.18 and 5.19. Note that all components of the SIS should be User-Approved Safety. As the logic diagram in Figure 5.12 shows, this SIS is configured as a 1oo2 system (i.e., only 1 out of the 2 sensors, one out of the two logic solver(s), etc. need to switch to cause a shutdown).

Separation between the BPCS and the SISs and between each SIS is necessary (see Section 5.1.2.4). Diversity in hardware, system software, and application programming between SISs is also recommended (see Section 5.1.2.5). See Section 5.1.2.6 for software application considerations. Although not shown in Figure 5.12, external watchdog timers are needed for both PESs (see Appendix C). Note that diagnostics may be required to achieve the desired availability.

Appropriate communications techniques (Ref. Appendix D) can be selected from Table F.4 in Appendix F.

Figure 5.12 may utilize the concept of "mirroring" (not illustrated in Figure 5.12) the SIS logic in the BPCS (see Section 5.1.2.7). Availability can be improved by having the BPCS continuously compare the logic solutions obtained by PES_1 against PES_2, as well as the sensors and final elements' performance comparisons. If the SIS logic can be developed in the BPCS, this can be used in "mirroring" as well. Disagreement in "mirroring" typically means the problem should be quickly corrected or the process should be safely shutdown (see Figure 5.13).

Figure 5.14 shows one way of implementing passive diagnostics in the SIS of Figure 5.12. The analog inputs from both sensors are fed to both PES "A" logic solver (through analog input A/A11 and A/AI2) and PES "B" logic solver (through analog input B/A11 and B/AI2). The output from PES "A"

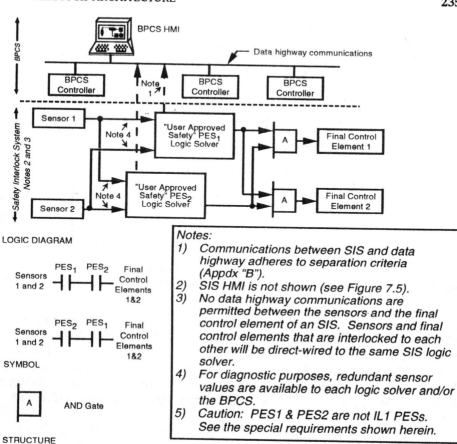

Notes:
1) Communications between SIS and data highway adheres to separation criteria (Appdx "B").
2) SIS HMI is not shown (see Figure 7.5).
3) No data highway communications are permitted between the sensors and the final control element of an SIS. Sensors and final control elements that are interlocked to each other will be direct-wired to the same SIS logic solver.
4) For diagnostic purposes, redundant sensor values are available to each logic solver and/or the BPCS.
5) Caution: PES1 & PES2 are not IL1 PESs. See the special requirements shown herein.

Availability Range: 0.999 to 0.9999 (Reference 5.6)

Figure 5.12 Integrity Level 3 SIS structure and block diagram.

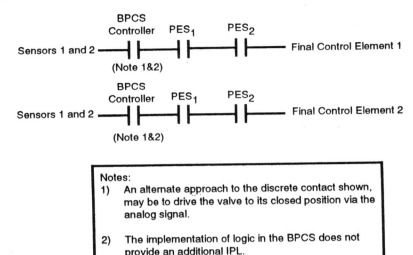

Figure 5.13 Loogic diagram for Integrity Level 3 SIS with BPCS mirroring.

logic solver is connected to the load side of both Final Control Element #1 and Final Control Element #2 through discrete outputs A/DO1 and A/DO2. The output from PES "B" logic solver is connected to the neutral side of both Final Control Element # 1 and Final Control Element #2 through discrete outputs B/DO1 and B/DO2. This ensures that the final control element can be taken to its safe state (deenergized) if there is an output module failure in either PES "A" or PES "B."

Note that no active diagnostics have been shown in this drawing. Passive diagnostics are incorporated on the outputs via feedback from inputs A/DI1, B/DI1, A/DI2, and B/DI2. These are used to verify that the signal being fed to the final control element is the same as the internal output from the PES.

An additional passive diagnostics that would typically be implemented is to use valve position feedback from limit switches on the final control elements and compare those signals with the outputs sent to the final control elements from the PES logic solvers.

5.4.3.1. Sensors for IL3 Interlocks

Sensors used for IL3 interlocks should be analog, rather than discrete switches, because analog process signals can be used to implement active diagnostics on redundant sensors in the SIS controller. Diversity in sensors should be considered when there is concern about common mode failure. Diverse sensors should be carefully selected without negatively impacting reliability. Calculated or inferred measurements should be used with caution as backup inputs in IL3 systems, and should have PHA team approval.

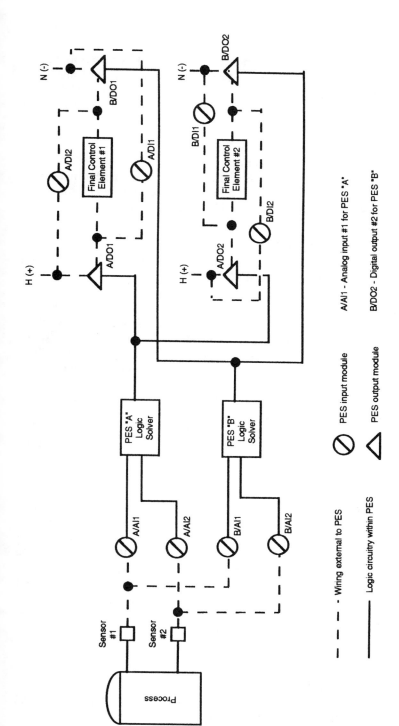

Note3: The use of position switches indicating final control element operation as feedback in lieu of inputs (A/DI1, A/DI2, B/DI1, B/DI2) may be an acceptable alternate. In this case the outputs (B/DO1, B/DO2) can be placed in the line side of the final control elements.

A/AI1 - Analog input #1 for PES "A"

B/DO2 - Digital output #2 for PES "B"

⊘ PES input module

◁ PES output module

– – – · Wiring external to PES

——— Logic circuitry within PES

Figure 5.14. A Passive diagnostic implementation of the SIS from Figure 5.12.

237

Sensor separation should be extended to the actual connection to the process where practical. For example, if two pressure transmitters are used, they should have separate taps into the process (see Figure 5.11). This same principle should be followed for other measurements (e.g., flow, level, and temperature).

5.4.3.2. PES Input Modules for IL3 Interlocks
The two sensor inputs to each logic solver should be located on separate cards, separate input racks, and separate I/O communication channels if practical. Active diagnostics should be incorporated in the inputs, if practical.

5.4.3.3. Logic Solvers for IL3 Interlocks
Although not illustrated in Figure 5.12, active diagnostics in the form of an external WDT are recommended for each PES.

Various types of technology are suitable for IL3 interlock (see Section 5.3). Many consider electromechanical relays as an acceptable logic solver for level 3 interlocks. However, relays require a current/voltage trip to interface with analog sensors to convert the analog signal to discrete. This makes it difficult to perform diagnostics on the analog sensors. This may reduce the system reliability and availability. PESs have a number of advantages over other technologies when the sensor inputs are analog. This includes the ability to do error checking on the inputs (e.g., over range, under range, and comparisons among two or more inputs).

Hybrid systems (e.g., a PES and a relay logic system) may be desirable because they offer the advantages inherent in diversity and the potential for minimized complexity for applications with a few IL3 interlocks. Figure 5.15 illustrates a hybrid system. The hybrid system Figure 5.15 overcomes the analog signal handling deficiencies discussed in the previous paragraph, by performing sensor diagnostics in the BPCS.

An alternative to paralleling input sensors to both PESs is to take sensor 1 into PES1, sensor 2 into PES2, and then to use communications between the two PESs, or to the BPCS, to provide information on the alternate sensor (see Figure 2.5). The arrangement shown in Figure 5.15 is preferred, because the communications link between the two PESs can be a source of reduced availability, if the communication link fails, since the sensor information from one PES is not available (i.e. with communication link failure) to the other PES. Although not shown in Figure 5.15, an external WDT (Appendix C) is recommended the for PES.

5.4.3.4. PES Output Modules for IL3 Interlocks
The redundant outputs should be located on separate cards, separate output racks, and on separate I/O communications, if practical. Active diagnostics

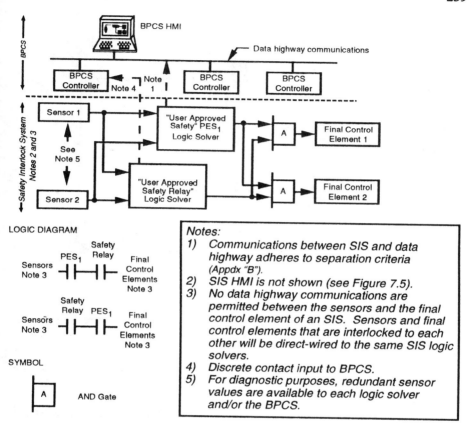

Figure 5.15 Integrity Level 3 hybrid structure and block diagram.

should be used on the output modules (similar to Figure 5.7 except for outputs) if practical.

5.4.3.5. Final Control Elements for IL3 Interlocks

Where the final control elements are valves, separate automatic block valves are preferred. Valve position feedback switches (i.e., limit switches) or other indication of valve position (e.g. confirmation of low flow by flowmeter) are needed. A control valve from the BPCS can be used as one of the automatic block valves if its leakage characteristics are acceptable for the application, it is UAS, and PHA team approval is obtained.

Redundant solenoid valves connected in series (see Figure 5.16) should be used when it is important that a solenoid valve failure not prevent the valve from actuating. In this figure, only one of the solenoid valves must actuate (deenergize) to bleed the air off the pneumatic actuator.

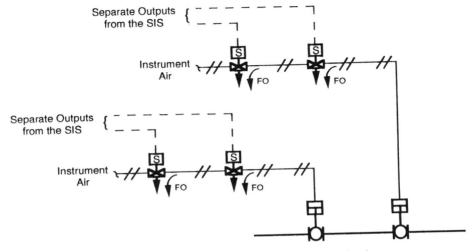

Figure 5.16 Dual solenoid method (series) to implement output redundancy.

The effectiveness of the redundant solenoid scheme in Figure 5.16 can be compromised if the solenoid valve closest to the valve actuates, but the air does not bleed out through the vent port of the solenoid valve, as could happen if the vent port were plugged. Precautions need to be taken when using this scheme to prevent inadvertent pluggage of the solenoid valve vent port (e.g., by a wasp's nest).

The redundant solenoid valve scheme of Figure 5.17, with the solenoid valves in parallel, gets around this problem. Even if one of the solenoid valve vent ports gets plugged, the other solenoid valve is directly connected to the actuator and can bleed the air off the valve. However, if one solenoid valve in this scheme actuates and the other does not, one solenoid valve will be attempting to bleed the air off the actuator and the other solenoid valve will be attempting to put air onto the actuator. The resulting effect on the actuator is indeterminable. Depending on lengths of tubing around the solenoid valves, port sizes in various directions through the solenoid valves, air supply pressures, etc., the valve may either open or close. This redundant solenoid valve arrangement may be acceptable for IL2 SISs but should not be used in IL3 systems.

5.4.4 High Availability/High Reliability SISs

Many applications require that the SIS be both highly available and reliable. Figure 5.18 shows a way to improve the reliability of the SIS (without degrading the availability significantly) using a packaged triple-modular-redundant

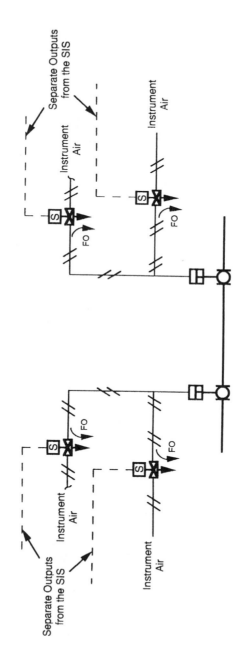

Figure 5.17 Dual solenoid method (parallel) to implement output redundancy.

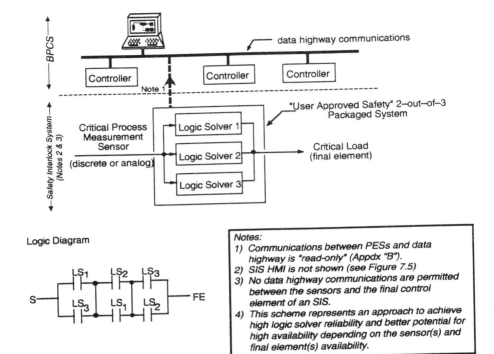

Figure 5.18 High logic solver reliability using TMR (see Note 4).

(TMR) system. TMR systems use 2oo3 voting at various points within the SIS. This means that two out of the three components need to trip before a shutdown is initiated.

Figure 5.18 shows only a single sensor and single final control element for simplicity. In actual practice, duplicate sensors and final control elements are typically used. There may be cases where triplicated sensors and/or final control elements are needed to achieve the necessary reliability level. This system, although suitable as a replacement for a PES for all three integrity levels, is probably only justifiable when infrequent, spurious shutdown of the process caused by faults in the SIS cannot be tolerated.

Figure 5.19 shows a different technique of achieving high reliability, which uses four SIS logic solvers (central processing units or CPUs). As the logic diagram shows, the two PLCs are implemented with redundant CPUs as a "hot backup" system, as are the DCSCs. Hot backup means that one of the CPUs is the primary and one is the backup. If the primary CPU fails, the backup CPU switches in and keeps the system running. The input modules

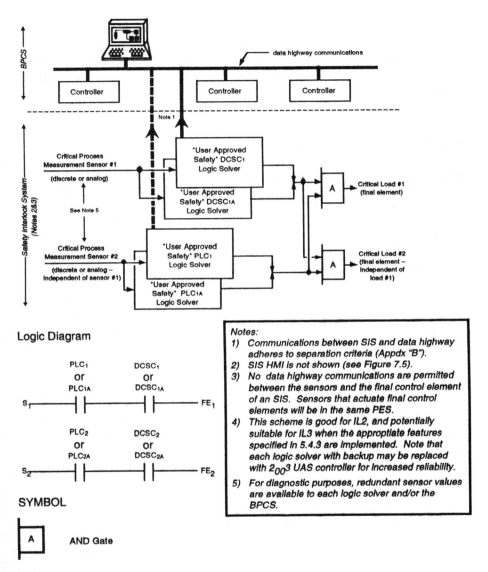

Figure 5.19 High reliability SIS using quad-redundant processors (see Note 4).

and output modules are not redundant. Then these two systems are implemented in a 1oo2 fashion, so that a failure in either of these systems will cause a shutdown. The comments on sensors and final control elements from Figure 5.18 also apply here.

Figure 5.19 is suitable for IL3, when the appropriate features specified in 5.4.3 are implemented. This schematic is suitable for IL3 assuming the sensors and final elements meet availability requirements. Note that each processor with backup may be replaced with a 2oo3 UAS controller for increased reliability.

5.5 SELECT SIS EQUIPMENT

All equipment that is used in a SIS should be User Approved Safety. Chapter 3 provides more detailed information on the stages that control equipment goes through before being classified as User Approved Safety. Appendix F provides considerations and guidance in SIS equipment selection.

5.6 PERFORM SIS DESIGN

Development of a safe process requires a team approach including operations, technical, maintenance, specialists, and consultants (see Chapter 3). The advantage of this approach is the proper identification of the hazards and the potential for reducing or eliminating the hazard within the design of the process, thus minimizing the need for SISs.

Extreme caution should be utilized when not following any of the rigorous PHA methodologies discussed in Chapter 3 and proceeding to an alternate scheme.

The complexities of today's processes, and the business needs requiring constant change during the life of the process, make proper up-front analysis mandatory.

Other resources are sometimes utilized to facilitate development of a safe design. These include:

- Use of standard "prescriptive" approach to ensure that the process is safe. An example of this is reliance on standards developed for repetitive type processes such as the NFPA standards for combustion systems.
- Use of vendor "packaged" equipment to serve as an SIS. An example of this would be prepackaged, "off-the-shelf," ready to install, combustion control equipment available from control system manufacturers.
- Duplication of an existing SIS or portions of an existing SIS from an existing safe and proven process because the new process is the same as the existing process.

Design of an appropriate control system for safety interlocking applications may require both qualitative and quantitative methods to specify the appropriate PES. These methods can:

- Provide SIS mean time to failure (MTTF).
- Provide SIS mean time to repair (MTTR).
- Analyze operating system modes. (e.g., manual, semiautomatic, automatic, etc.)
- Analyze operator interfaces. (e.g., display layout, response time, etc.)
- Analyze process dynamics. (e.g., sensor response time)
- Analyze BPCS and instrumentation. (e.g., technology impact)

For a sampling of qualitative and quantitative methods, see Chapter 3.

An output of the PHA team (see Chapter 3) is a listing of the safety interlocks and their required integrity levels. When equipment is being selected to implement the SIS, the highest integrity level interlocks should be examined first (Table 5.3); then the lower level interlocks can be examined. It is important that all equipment being considered be User Approved Safety (see Section 3.3.2.2).

If the SIS uses PESs or has solid-state devices their failure can vary (i.e. on–off, intermediate value, etc.).

With a relay system, the failure of one relay usually affects only a part of the system. A PES failure could cause a complete control system shutdown. Relays also usually fail in a known mode. PESs can fail in an indeterminant mode. PESs are also prone to internal failures that can affect the scan cycle. It is very important that a PES failure be detected and that the proper shutdown sequence be performed.

Special precautions should also be taken with the solid-state output devices used with logic solvers such as PESs, relays, solid state logic, etc. When triacs fail, they often fail in the ON state. Some logic solver manufacturers have introduced output modules that make these outputs more fail-safe. These typically include built-in diagnostics to detect a triac failure (e.g., shorted triac, open-circuit triac, or half-wave triac failure) and disable the output. Some also include redundancy in the output circuits.

False turn-on of inputs can be caused by magnetic coupling in the field wiring and false turn-on of outputs can be caused by rapid voltage transients during power-up conditions and when recommissioning an output module on line. Utilize inputs and outputs with sufficient signal to noise ratio to provide appropriate noise immunity, to eliminate these problems.

Care should be taken in the design of output power fusing in common output module arrangements to avoid a single electronic component failure that blows a principal fuse and causes the loss of supply power to an entire output array. Special precautions should be taken when using such output devices.

Solid-state output devices in PES applications require particular care against covert failure to the "on" position during operation.

Following are other key considerations when performing the SIS design.

5.6.1 Design Verification and Validation Framework

Verification is done at various stages of the design process. This will usually involve design peer reviews, application program simulations, installation checkout, and trial run.

Verification of safety integrity prior to design is illustrated in Figure 5.3. This type of verification can be either quantitative or qualitative. When quantitative verification is chosen, the reliability and availability of the system as designed is compared against target values that were determined by the PHA team. See Chapter 3 for a more detailed discussion of this technique. Qualitative verification can take many forms. One technique that is shown in Figure 5.4 is by peer review (i.e., have another person review the design who was not involved in the actual design). Comparison against checklists is another technique.

Validation takes place once the SIS has been installed. This involves testing the system to ensure that sensors and final control elements are wired properly and that the application program functions properly with the plant equipment. See Chapter 6 for a more detailed discussion of this technique.

5.6.2 Systems Services

Chapter 4 of this text addresses the proper installation techniques to allow a PES to function safely and reliably as a BPCS. The SIS should adhere to the same criteria established in Chapter 4. This section addresses those requirements involving system services for the SIS (any or all of which may be required for the BPCS as well).

5.6.2.1. Power and Grounding

Reliable energy sources are extremely important to the continuous operation of an SIS. Uninterruptible power sources are often required to achieve the necessary uptime required.

Reliable power supplies should be provided for all SIS components (see Section 4.8). SIS components should not normally be powered from common circuits (same breakers or fuses) as BPCS components. Where processes require that all SIS elements should be powered to operate safely, install input modules to detect power and utilize these inputs in diagnostics. These diagnostics can be implemented in the BPCS.

Special attention should be paid to grounding requirements to avoid ground loop problems. See Section 4.9 for grounding requirements for the BPCS and the SIS.

In addition to that specified in 4.9, provide the following highly recommended features for power and grounding systems for an SIS:

- Design the SIS power distribution and grounding system so that SIS performance criteria (i.e., reliability, safety, failure modes, etc.) is consistent with the SIS integrity requirements.
- Provide clear, concise alphanumeric equipment descriptors, color coding, wire numbering, etc., on the equipment and the documentation.
- Identify with nameplates, signs, etc. all equipment that is part of the SIS. This should be done in a way that provides plant personnel with an immediate awareness and understanding of the equipment functionality.
- Provide references to documentation on the equipment.
- Provide and maintain "AS BUILT" documentation of the SIS power distribution and grounding systems.
- Determine the test frequency required to achieve SIS performance criteria.
- Develop preventive maintenance programs to achieve target reliability and availability.

5.6.2.2. Environment

Installation of I/0 components and control processors used for SISs in outdoor locations should be avoided. Recognize that heat and humidity do adversely impact electronic components. Therefore, all SIS components, except field sensors and final control elements, should be located in areas where temperature control prevents excursions below 40°F and above 100°F. Humidity should not be allowed to become 90% or greater. In addition, adequate ventilation should be provided, dust protection is recommended, and exposure to corrosive fumes must be prevented.

Don't let location be a common mode fault when more than one subsystem is installed in one SIS. This means that the SIS logic solver should be installed in a location where a fire, explosion, etc. could not compromise the safety of the PES installation and prevent the SIS from being available in case of an emergency, particularly for energized-to-trip systems.

5.6.3 Impact on BPCS

Physical separation of the SIS hardware from the BPCS hardware should be provided (i.e., separate racks or cabinets). Application programming can become more complex if the BPCS is used to provide additional diagnostics for the SIS through the concept of "mirroring" (i.e., where all, or a portion of, the SIS logic is duplicated in the BPCS, and the BPCS is then used as a check on the performance of the SIS.) This may also increase the I/O count on the BPCS because of the additional inputs that may be needed to get the process signals into the BPCS and the additional outputs that may be required for SIS functions.

Routine alarms should not be installed within annunciation sequences that provide first-out alarms for SIS. Differentiation of alarm displays for safety interlocks from other alarm displays should also be provided. Safety interlock alarms may be provided with a distinct tone to identify them as critical alarms and not normal operational alarms. Some intelligent alarm systems in PESs may provide this capability.

5.6.4 Impact on HMI

Operator interfaces to SISs should not allow on-line changes to the SIS logic. Only if appropriate security exists should off-line changes to the SIS be allowed from operator interfaces. This should be accomplished through the use of hard-wired write protect security rather than software write protect techniques, unless the PHA team approves the use of software security measures.

There is a need to provide a display of SIS status for the operator on a continuous basis. This can be done in several ways: the choice depends on the particular SIS, the type of control system used for the BPCS; and the requirements for operator interaction with the SIS during normal and abnormal operation, such as complete loss of central operator interface(s).

Acceptable displays of SIS status information include:

- Graphic displays on operator workstations.
- Dedicated graphical displays on separate video display units (VDUs).
- Back lighted annunciator displays.
- Panel mounted graphic displays.
- Panel mounted status lights.

All variables that impact an SIS should be displayed to the operator in a secure format (for security against inadvertent change).

5.6.5 Impact On Communications (Including SIS/BPCS Communications) (see Appendix D)

Data transmission between card chassis located in separate equipment enclosures should use serial buses that may require redundancy and have multiple layers of communication diagnostics and communication failure alarm functions.

The SIS logic solver should be designed to control the communications link between itself and the BPCS. This can be accomplished by having essential parameters within the logic solver available to BPCS via "read only" communication status. A limited set of process status and SIS command signals may be communicated between the Operator Workstation and the SIS logic solver. This occurs often in batch systems where recipe changes may require SIS application program changes.

5.6.6 Alarms

Annunciation of alarmed SIS variables should be of the first out sequence type only when there is value in determining the first indication of an alarm condition. This may be accomplished with a sequence of events recorder, DCS alarm screens, logic programmed into the BPCS, conventional panel mounted annunciators, or equivalent devices. First-out sequences should follow the ISA S18.1 standard. All SIS initiating variables should be provided with a pre-alarm, if appropriate. All functions (not just process variables) that can trip an SIS should be alarmed.

An alternate, or supplementary, approach sometimes implemented is to configure an alarm-cancel function to achieve first-out alarm performance. In this mode, only the first alarm appears on the screen. *Caution: A potentially hazardous situation associated with this approach is that automatically canceled alarms are not displayed as being out-of-limit.*

5.6.7 Integration of SIS/Sequential Interlocks

It is important to have good integration between the SIS logic and any sequential logic. This ensures that an action taken by the sequential logic in the system does not cause any safety problems. This is accomplished by having the resulting actions of sequential logic pass through the SIS logic when sequence operations affect safety functions (see Section 4.10).

5.6.8 Programming

It may be desirable that separate personnel perform the application programming for the BPCS and the application programming for the SIS. This reduces common mode failures that might occur in programming logic functions.

Any changes to the application program for the SIS should go through management of change procedures to ensure the change does not affect the integrity of the SIS (see Chapter 6).

5.6.9 Provisions for Testing

Testing is necessary to find fail-to-danger faults and "as-found" conditions in an SIS. This requires that a periodic test interval be defined and that interlocks be tested at this interval. Any problems that are detected during testing should be promptly corrected and documented to ensure the integrity of the SIS. A complete test of the SIS is necessary, from field sensor to final control element.

Consideration of testing requirements during the design process will ensure that the SIS is set up to allow for these test procedures. Testing provisions may increase the overall complexity of the SIS. Testing is covered in more detail in Chapter 6.

5.6.10 Documentation

It is difficult to overemphasize the importance of adequate documentation for SISs. Without clear, concise (e.g., text sentences describing the SIS change), "as built" documentation, operating personnel will have trouble understanding how the system is supposed to work, maintenance personnel will have difficulty maintaining the system, and engineering personnel will have problems when changes are necessary.

Documentation for SISs should be adequate to completely describe the SIS's function and should be maintained up to date at all times. Some of the documentation that may typically be used for SISs is described in this section.

5.6.10.1. Operational Description

An operational description for the SIS should be text supplemented by binary logic diagrams where needed. This description explains to operating and maintenance personnel how the system is intended to work; it includes the safe and correct methods for operating the equipment. An operational description can prove invaluable to both operations and maintenance personnel. Special emphasis should be placed on areas where a general description would not be adequate. In addition to binary logic diagrams, documentation such as schematics, flow charts, etc., may be required to provide a complete operational description.

5.6.10.2. Schematic Diagrams

This diagram shows, by means of graphic symbols, the electrical and/or instrument functions of a specific circuit arrangement. The schematic diagram facilitates tracing the circuit and its functions without regard to the actual physical size, shape, or location of the component device or ports. Schematic diagrams should show SIS interlocks and their relationship to the rest of the system. Drawing references should be noted that refer the user to SIS interlock logic and/or block diagrams.

The schematic diagram is intended to identify and show the function (logic) of the component devices in the SIS. The diagram serves the twofold purpose of showing complete circuit connections and the manner in which the equipment will function. It can be considered as the "translation" of a written description of operation or logic diagram into electrical symbols.

The schematic is useful during verification and validation to determine and pinpoint malfunctions of the equipment due to wiring errors, faulty electrical components, etc.

The schematic diagram is developed and drawn from information obtained from the operational description, basic data, piping and instrument control diagrams, Operating Department liaison engineers, vendor's data, and meetings with all Design functions.

Description of operation of limit switches, pressure switches, solenoid valves, and other instrument items should be shown on the schematic diagram to clarify operation (e.g. limit switch closes when valve is fully open). If the control is complicated, the sequence of operation should be included with the schematic diagram.

Circuits which function in definite sequence should be arranged to indicate that sequence. The circuits should be arranged in their order of sequence, and be shown from top to bottom of drawing. All devices are to be shown in the deenergized condition (off-the-shelf) unless noted on the drawing. The physical relationship among components normally should be disregarded; however, in some cases, the components can be arranged in the circuit to simplify wiring.

Indication of the source of control energy (e.g., the lighting panel and circuit number, control transformer identification) should be given on the diagram.

5.6.10.3. Binary Logic Diagrams

The binary logic diagram is not concerned with the technology used to implement logic, but stresses logic that the designer should be concerned with to allow implementation of any appropriate technology in a fail-safe manner with the minimal risk of error. A logic diagram should express, without ambiguity, a relationship that exists between an input and output condition or state.

There are two states every binary signal can assume: the 0-state and the 1-state. Logic diagrams should conform to the following conventions:

- The 0-state defines an absence of energy. It represents the unsafe, abnormal, or undesired conditions for original input signals. This is especially critical for PES systems final output signals; it represents the direction of action desired for fail-safe operation. Contact opening should create the 0-state signal. Alarms and interlocks are actuated by the 0-state signal.
- The 1-state defines a presence of energy. It represents the safe, normal, or desired condition for original input signals. For final output signals, it represents whatever is opposite to the fail-safe action or direction. Contact closing should create the 1-state signal. *Note: The above refers to nonprogrammable logic solvers. The PES logic solver may have to be represented opposite to what is noted above to clearly reflect actual operation.*

All logic statements should define the same state, typically the 1-state. The time behavior of contacts should be explicitly noted. Some options are: general statements such as "all contacts are momentary," symbols discriminating momentary and maintained contacts, and timing diagrams included within logic diagrams.

5.6.10.4. Single-Line Diagrams

Single line diagrams (Figure 4.5) illustrate by means of single lines and simplified symbols, the component devices or parts of an electrical circuit or system of circuits. Physical relationships usually are disregarded.

Single-line diagrams are intended to portray major (high voltage distribution) and minor (branch circuit) system layouts, and are analyzed to understand distribution areas with insufficient reliability and unacceptable common mode failure characteristics. For SISs, it is especially critical to show the power supply (UPS) configuration.

Single line diagrams portray how the SIS is connected to the main power feed. Diagram content should allow analysis of how the SIS system would be impacted by a power feeder outage. Power availability, system uptime, stability, and surge-free operation should be considered when designing and developing this diagram. Minor system components and connections such as the branch circuits to SIS logic solver, sensors and final elements, etc., also may be shown.

5.6.11 Plant Support Capabilities

Regardless of the SIS technology chosen, training should be provided to allow plant personnel to operate and maintain the system. The proper implementation of user approved safety equipment requires the plant to have on site spare parts, training and maintenance, all essential to safe SIS performance.

5.6.12 Plant Change Control Capabilities

Change control procedures are important with BPCSs and are an absolute necessity with SISs. There should be strict controls established as to who can make changes to SIS application programs, hardware configurations, software updates, verification, etc. This topic is discussed in more detail in Chapter 6.

5.7 REFERENCES

5.1.   Ida, E. S., Electronic alarm switch uses fail-safe design concepts. *Control Engineering,* March 1983.

5.2. Smith, S. E., "Hardware-Implemented Fault Tolerance Considerations in Designing a Programmable Controller for Protective System Applications." The Eighth Annual Control Engineering Conference, May 23–25, 1989.

5.3. Bryant, J. A., Design of fail-safe control systems. *Power,* January 1976.

5.4. Models ET-1234 & ET-1235 Fail-Safe Voltage/Current Alarms, Rochester Instrument Systems, October 1984.

5.5. AIChE/CCPS. *Guidelines for the Safe Storage and Handling of High Toxic Hazard Materials.* New York: American Institute of Chemical Engineers (1988).

5.6. Bourne, A. J., et al., "Defences against Common-Mode Failures in Redundancy Systems," United Kingdom Atomic Energy Authority, Safety and Reliability Directorate C.21, January 1981.

5.7. Goble, W. M., "Evaluating Control Systems Reliability, Techniques and Applications," Instrument Society of America 1992.

6

ADMINISTRATIVE ACTIONS TO ENSURE CONTROL SYSTEM INTEGRITY

6.0 INTRODUCTION

The administrative actions necessary to ensure BPCS/SIS integrity rely upon a corporate commitment to safety as well as systems designed to ensure sound technical management of safety. The CCPS *Guidelines for the Technical Management of Chemical Process Safety* (Ref. 6.1) provide guidance on this subject.

Chapter 3 discusses the management responsibilities during the development of the BPCS/SIS design and implementation. Corporate management has the prerogative of setting general safety guidelines used in the design process. A Process Hazard Analysis (PHA) team performs evaluations of design integrity at several stages of the design evolution. PHA teams are also used to review major safety-related design changes and to perform safety audits during the operating life of the facility.

The plant operator directs the operation of even the most highly automated chemical processing facilities by routinely selecting product flow paths, stopping and starting pumps, and making changes in operating conditions for quality control. The operator also serves as a monitor of the plant's operation through detection and compensation for failures of processing and control equipment. The Basic Process Control System (BPCS) provides (1) automatic regulation of the process during normal operation and when plant transitions are necessary and (2) plant status information that enables the operator to ensure that the commanded changes are successfully made. The plant operator and the BPCS form an essential protection layer in most chemical plants.

While safe operation of chemical processes can be enhanced through the use of BPCSs and Safety Interlock Systems (SISs), the long-term integrity of these systems still depends on the proper actions of human beings. Engineers design the systems. Operators monitor and direct the operation of the plant. Maintenance personnel maintain the performance of the various components. It is, therefore, important that appropriate management and administrative procedures and controls be in place to ensure that the integrity of BPCSs and SISs is maintained. The implementation and enforcement of these procedures demonstrate management's support and commitment to the safe operation of the plant.

The rest of this chapter addresses procedures and administrative actions that are fundamental in ensuring the integrity of the installed BPCSs and SISs. Minimum procedural requirements are described for:

- Operating during normal and abnormal conditions of the plant.
- Making changes to installed BPCSs.
- Making changes to SISs.
- Maintaining the equipment and software.
- Testing of the systems.
- Training personnel on the use and maintenance of the systems.
- Providing adequate documentation for the systems.
- Maintaining documentation of the systems.
- Auditing the systems.
- Maintaining performance history on the systems.

The importance of having these procedures developed and in place prior to starting operation of the plant is also stressed.

The development of all procedures related to BPCSs and SISs should involve participation of management, engineering, maintenance, operations, safety, and other groups as appropriate.

6.1 COMMUNICATION OF PROCEDURES

It is recommended that procedures critical to plant safety be provided to appropriate personnel in written, easily understood form and be up to date. Written procedures are required to avoid confusion that might occur in verbal communication. (In this chapter, "written" refers to information provided in hard copy [paper] form.) Further, a current paper copy of all plant safety procedures should be on file as backup for any problems that might result from computer system failures.

The written procedures may be in text format, flow charts or schematic diagrams, or graphic representations of the process. Information should be clear and concise and should be available prior to the initial and subsequent plant start-ups. All BPCS/SIS aspects that may impact safe plant operation should be addressed.

Some of the required written procedures, discussed in the following sections, include:

- BPCS Standard Operating Procedures (SOPs).
- Critical operations procedures.
- SIS operating procedures.
- Abnormal condition procedures.
- Batch operation procedures.

- Turnover procedures (the transfer of operation/control from one functional area to another, such as Maintenance to Operations after completion of repair work).
- Facility change procedures.
- Safety review procedures.
- Security procedures.

Each of these procedures, its importance to overall plant safety, and its maintenance will be examined. Maintenance and testing procedures are discussed in Sections 6.2 and 6.3 respectively.

6.1.1 Basic Process Control System (BPCS) Standard Operating Procedures (SOPS)

Procedures are required for operation of BPCSs/SISs. The complexity of the equipment, the flexible modes of operation, and the fact that all functions may not be utilized continuously require written documentation to maintain an adequate knowledge base for proper operation.

These SOPs should address the specific functions of the BPCS/SIS equipment being utilized. Documentation manuals provided by the equipment manufacturers tend to emphasize specifics of the BPCS equipment itself and may not provide the information necessary from an operator's perspective. There will also be application-specific information that only the user can provide. These SOPs may be provided through help screens on the BPCS displays but should also be in hard copy documentation manuals that are readily accessible.

These procedures should stress the importance of what is to be done and training in the use of the procedures should emphasize the reasons why this is considered necessary. An understanding of the importance of these procedures will reduce "deviations" that could adversely impact plant safety.

Operators should be provided with specific information concerning the BPCS's impact on operational safety. These procedures may include information such as:

- Documentation of any special control loops (ratio controls, cascade controls, etc.) with emphasis on maintaining them in the automatic mode during normal operation.
- Specific values for any tuning parameters that should not be changed arbitrarily.
- Range limits, if any, for set point values.
- Descriptions of any process interlocks or override controls that may be provided.
- Descriptions of review requirements prior to any control program modifications.

- Procedures for notification of other personnel when abnormal conditions are present.

All SOPs relating to BPCS safety should be reviewed with operations personnel periodically. This will allow the "normal" changes that take place during routine operation to be incorporated into the procedures and uncover any undocumented changes that might have taken place. This review also serves as a training tool to reinforce the operators' and support personnel's knowledge of plant operation.

6.1.2 Critical Operations Procedures

Some process operations may be more critical to safety than others. These may include:

- Unit start-ups and shutdowns.
- Initial start-ups versus routine start-ups.
- Start-ups after maintenance.
- Switching from normal operating equipment to backup units.
- Starting or stopping equipment.
- Changing of personnel at shift change time.
- Preparing equipment for maintenance.
- Performing maintenance on operating equipment.
- Verifying the accuracy of signals to an SIS.
- Switching from one operating mode to another (e.g., MANUAL to AUTOMATIC and vice versa).
- Switching from one operating state to another (e.g., START-UP to NORMAL, or HOLD to NORMAL).
- Making significant operating rate changes in the unit.
- Converting operation to an alternate product.
- Performing on-line tests on control or safety equipment.
- Performing on-line maintenance on a BPCS or an SIS.
- Response to any pre-alarms on trip signals.

When these situations occur, additional instructions on the safety implications of the control functions may be necessary. This should be done through written procedures that identify them as critical operations and emphasize the safety of the operation. They should be included in the SOPs but with some means of highlighting them as critical operations. For example, they may be printed on different colored paper, placed in a separate section of the SOP manual, or whatever means proves to be most satisfactory for the particular situation.

BPCSs can play very important roles in maintaining plant safety during these critical periods in the plant's operation.

6.1.3 Safety Interlock System (SIS) Operating Procedures

Since SISs provide an independent protection layer for the processing unit, it is recommended that procedures be in place to prevent actions that could compromise this protection. SOP's may have some instructions related to the safety interlocks, but there may be special procedures describing items that are specific to an SIS. Areas that should be covered by these procedures include:

- Bypassing criteria for SISs.
- On-line calibration of SISs.
- Response to SIS Alarms.

Procedures for operation of an SIS should emphasize the importance of the interlocks and how they provide the desired safety function. Training should be provided for *all* personnel to ensure adequate understanding of, and the reasons for, these interlocks.

6.1.3.1 Bypassing Criteria for SISs

Bypassing of SIS functions during operation of the plant can have significant safety implications. It is recommended that each facility establish a policy concerning this action. The policy should be applied uniformly throughout the facility to ensure that movement of personnel among units does not result in different interpretations of the policy.

A policy is recommended that does not permit master bypasses (action that bypasses all trip initiators simultaneously) on SISs. This will prevent having an SIS out of service when it could be required to perform its function.

Bypasses on individual SIS trip initiators may be acceptable, provided that written procedures are established defining how they may be used, for example, describing monitoring of specific process conditions while the bypass is being used.

The accepted bypassing method should be determined ahead of time, and special operating procedures should be established for the period of time the SIS trip initiator will be out of service.

The bypass procedures should include a cautionary note identifying improper ways that safety interlocks could be bypassed through some seemingly unrelated actions, such as:

- Elevating or suppressing zeros on transmitters.
- Adjusting spans on the transmitters.
- Adjusting transmitter purge rates.
- Providing hand jacks on automatic valves.
- Installing bypasses around automatic valves.
- Installing filters to reduce noise on instrument signals.
- Leaving span gas open to analyzers.
- Defeating limit switches.

It is recommended that all procedures related to the bypassing of SIS functions be approved in writing by appropriate plant management prior to use of the bypass. These procedures may include decisions that require thorough analysis, testing, documentation, and communication to appropriate personnel before they are implemented. It may also be appropriate to establish a time limit for which a bypass may be in place and a means by which to prevent return to normal operation until the bypass has been removed.

Bypass procedures should also require special tagging on all SIS trip initiators that are in a bypass mode during operation of the plant. The tags should be visible and may include items such as :

- Identifies the function bypassed.
- When bypass initiated.
- Who approved the bypass.
- Personnel authorized to remove the tag.

The tags should not be removed until the system is returned to a normal operating mode. The time removed and the individual removing the tag should be noted in the operations logbook.

6.1.3.2 On-line Calibration of SIS Inputs

During the normal operation of a unit, the need may arise for verifying the calibration or status of an input to the SIS. Procedures should be developed prior to initial operation of the unit, detailing the steps to be followed during this verification process. Consideration should be given to the following points:

- When verification can be accomplished, (i.e., time of day, weekends, holidays, etc.) without disrupting normal operations.
- The time allotted to perform the calibration.
- Whether or not the calibration can be accomplished while complying with bypass procedures.
- How the calibration is to be done.
- Who can authorize the work.

6.1.3.3 Response to SIS Alarms

Operating SOPs should cover normal response to process alarm conditions. However, SIS-related pre-alarms may require a different response. SIS pre-alarms may be an indication of impending SIS action if preventive action cannot be taken immediately. SIS pre-alarm situations should be defined and procedures and training developed ahead of time to prevent confusion when these alarms occur.

6.1.4 Abnormal Condition Procedures

The same philosophy that applies to general operating procedures relating to abnormal operations in chemical processing plants should apply to the BPCS/SIS during these situations. Operators may need specific guidance on actions to be taken when addressing such situations as:

- Indicated loss of instrument power or air.
- Indicated loss of key utilities to the plant, such as cooling water, steam, and electricity.
- Loss of a feedstock or some other key ingredient.
- A fire or chemical release especially with potential impact on the community.
- Failure of a pump seal.
- Pipe breakage.
- A phoned-in threat, or the occurrence of other acts of terrorism.
- Severe weather conditions.
- Blank CRT screens during normal operation.
- An indication that the SIS has failed in a "stall" position (i.e., the program sequence has stopped in an unknown condition).

Whether or not the BPCS/SIS is affected directly by the abnormal condition may not be the critical concern. The fact that out-of-the-ordinary conditions exist requires operators to respond in a different manner, and the BPCS and SIS are a key part of their response.

All identifiable abnormal situations should be defined ahead of time and appropriate response procedures should be developed. These procedures should be written and included in the SOPs.

Providing both the procedure and the training ahead of time may prevent a serious incident when an abnormal situation occurs.

6.1.5 Batch Operation Procedures

The frequent start-ups and shutdowns, which are normal operation in a batch process, require special attention to operating procedures (Ref. 6.1.1).

6.1.5.1 Batch-Process Operations

Many operations in a batch process are typically performed manually, for example, addition of materials, sample extraction, and vessel discharge. The process steps may be manually executed by operators following written instructions. These written procedures describe in detail raw material additions, valve sequencing instructions, set point changes, and instructions for recording process variables, but the interpretation of these instructions may vary from operator to operator. If the batch operation has features that are safety critical, it may be necessary to require confirmation in the form of date, time, and operator identification for each step in the operation.

In a complex batch process, sharing of equipment may become a coordination problem. If the operators are paying attention to other simultaneously executing processes, they may be distracted from important observations.

These varying modes of operations may cause variations in the yield and quality of the product. They may also cause problems with materials and equipment planning in addition to impacting productivity and safety. If an operator makes a mistake in the proper execution of one of these steps, and the mistake is not recognized in time for corrective action to be taken, off-specification product will likely result, or a potentially hazardous condition could be initiated.

Batch operations require strict adherence to the sequence of addition and the quantities of various raw materials, catalysts, inhibitors, solvents, or other ingredients to ensure final product quality, consistency, and safety. The large quantities of two-state devices, such as automatic block valves, pumps, and motors contribute to the difficulty and complexity of the BPCS application and can lead to problems. Potential problems include:

- The processing steps not being performed in correct sequence, at the correct time, or for the correct time duration.
- Wrong material added.
- Incorrect quantity of material added.

The implementation of ISO 9002 or similar quality control programs will address these issues and provide guidance to develop appropriate procedures. Automation of batch processes has the potential to minimize the problems noted in this section. The use of Statistical Process Control (SPC) or Statistical Quality Control (SQC) techniques may provide a good means for monitoring both the product quality and the adherence to procedures.

6.1.5.2 Multiproduct Plants

If multiple products are produced with the same equipment, additional procedures and controls may be required to prevent contamination among batches that could lead to unsafe conditions. Batch production of multiple products requires recipes for each product, or grade of product. It is extremely important that these recipes be kept up to date. Administrative procedures should define who has the responsibility for the recipes, who is allowed to change a recipe, who can select and edit a recipe to make a batch of product, etc.

6.1.5.3 Recipe Management

If the BPCS is Programmable Electronic System (PES)-based, it may include a recipe management program. A recipe management program allows the operator to select a master or control recipe; make the necessary changes, based on laboratory data and equipment characteristics, to produce a "working recipe"; and then download the working recipe to the batch controller. This recipe is then used to produce the batch. The working recipe is a control recipe, operating in real time in the BPCS that is making the batch of product. Even at this level, there may still be a need for the operator to be able to make changes to the recipe as situations occur throughout the batch operation.

The contents of each recipe should be defined. For each product or grade, a list of required recipe parameters and their values should be created. These parameters are individually named by the user and are accessed throughout the system by that name.

The operational characteristics will determine what sort of operator actions will be allowed within the recipe management system. Will the operator be allowed to modify recipe data to create a working recipe? Are there limitations on what data can be entered? When multiple process units are used to make the same products, should the system automatically scale recipes to different unit capacities? These types of questions will determine the structure of operator interaction.

After the operating characteristics have been defined, the next step is to actually enter the data into the recipe management system. This can be done in either an on-line operating mode or an off-line engineering mode.

1. Recipe Selection, Editing, and Downloading. Typically, the operator would simply select the control recipe to be used from a recipe summary display. If necessary, the operator then makes necessary changes to the recipe to meet the needs of the particular batch. If multiple destinations (i.e., multiple reactors) for the recipes are possible, the operator then selects the appropriate destination for downloading. Some security checks are necessary at this point (Ref. 6.2):

- The system should compare the selected recipe with a list of allowable control programs and inhibit final downloading until the operator activates the appropriate control programs.
- The system should make sure that the selected recipe is compatible with the selected process equipment.

In some systems this security check may be the responsibility of the operator, foreman, supervisor, etc. Procedures may be required to handle these necessary checks in the batch control logic.

2. Ongoing Revision and Maintenance. Maintenance of the recipe database includes the following functions (Ref. 6.3):

- Protecting the recipes from unauthorized changes (e.g., with a key lock or password system).

- Copying the recipes to other systems as needed.
- Maintaining updated copies of any edited recipes.
- Determining allowable destinations for recipes (e.g., which recipes can be used with which equipment).

3. Documentation/Verification. Documentation of the recipes (from basic recipes to working recipes) is needed by both Engineering and Operations groups in an electronic source file and a paper copy.

Verification involves ensuring that the recipe still exists as designed and produces statistically repeatable results.

6.1.5.4 Human Interaction
For automatic and semiautomatic operation, troubleshooting, and problem correction, a wide range of human interface possibilities may be available (Ref. 6.3). Operators may be required to:

- Pause in a sequence.
- Bypass a phase.
- Insert a phase.
- Override linking conditions.
- Terminate a phase.

It is important that, in addition to automatic operation, the BPCS provide a form of operation that allows initiation of individual operations or phases from the BPCS. Administrative procedures are recommended to define who has the authority to take these kinds of actions and to make sure that all such actions are documented, either automatically or manually.

6.1.6 Turnover Procedures

When responsibility for plant facilities are transferred from one functional group to another, there is potential for compromising safety that might not otherwise be present. This is primarily due to change of personnel and the potential for loss of critical information. Procedures should be established describing those criteria that should be covered in the turnover operation. These might include such things as:

- Special tagging of equipment to identify responsible functional group.
- Special requirements when checking out equipment.
- Special requirements for equipment start-up (e.g., any run-in requirements, lower speeds for a period of time).
- Special conditioning (e.g., like seals in valves, pumps).
- Establishing initial conditions for control equipment.

The smooth transition of information among functional groups to prevent compromising safety can best be accomplished with the establishment of checklists, with a sign-off procedure for both the group transferring the facility responsibility and the one receiving the responsibility. A list of some items that might be included in a checklist for transferring a BPCS/SIS from maintenance to operations after major maintenance work might include:

- Hardware components, whether different or exact replacements, used in maintenance activities have been verified as compatible with current system configuration.
- Field sensor calibrations have been verified against a master list of ranges for field instruments.
- Wiring and communication links from field to control room equipment have been verified and tested to ensure correct operation.
- Failure directions of all final elements worked on during the maintenance activities have been verified as correct.
- Operator displays modified or developed during the maintenance activities have been tested with operator actions.
- Software operating system upgrades made during maintenance have been tested to ensure safety controls still function as designed.
- New control applications have been tested to verify correct function.
- Operators have been trained on any new features and control applications implemented during maintenance.

This is not an exhaustive list, but one can get the flavor of the type of questions that should be addressed.

Another area where turnover procedures are very important is at shift change time. Transfer of plant equipment and BPCS/SIS status from those leaving the plant to those coming into the plant can be critical to plant safety. The lack of information or the transfer of incomplete information could potentially compromise safety. Procedures may consider written documentation (e.g. checklists) of the status of critical equipment and the BPCS/SIS between the two shifts involved and verbal review of the information at shift change time. This will ensure that the necessary information has been transferred in both a timely and understandable manner.

6.1.7 Facility Change Procedures

Modifications can have significant impact on operating plant safety. Processes frequently undergo changes to improve efficiency and productivity, conserve energy, reduce waste materials, etc. These changes may be in the process itself or in the BPCS/SIS on the process and may create problems unless an adequate safety review of the proposed changes is performed (e.g., Ref. 6.4).

This can be especially true for changes relating to PES-based BPCSs/SISs. They allow changes to be implemented relatively quickly, without hardware additions in many instances, and ideas can be tried out before final decisions are made. Because of this ease of implementation, key safety issues that could lead to potentially hazardous situations could be overlooked.

This concern also applies when software upgrades are made by equipment vendors. A revision to system software may impact a BPCS/SIS in an unexpected manner. It is therefore important that software revisions and updates be controlled. Examples when changes are required include:

- Maintain spare parts for system (supplier does not support the revision level you have)
- Revision corrects defect in the software version you are using.

The concept known in the chemical industry as "Management of Change" addresses these issues. Details of this concept are outlined in the CCPS book "Guidelines for Technical Management of Chemical Process Safety", published in 1989 (Ref. 6.1). The concept of managing change is also addressed in the OSHA regulation 29 CFR 1910.119, "Process Safety Management of Highly Hazardous Chemicals" (Ref. 6.5). Both of these documents stress the importance of establishing a plant-wide review of all proposed changes before they are implemented. This should be accomplished by a team of technical and operating personnel familiar with the process and plant in question. A formal approval procedure should be a part of this policy, requiring written approval before any changes are made that impact safety. The approval process should include responsible individuals from engineering, process, operations, maintenance, and safety.

The review should be documented, with the logic for arriving at the agreed upon decision clearly stated. The policy should require a comparison with previous control logic to ensure that subtle safety issues are not overlooked. The proposed change should have a written description, detailing what is to be done, why, and the initiator's evaluation of the safety impact. A checklist of concerns that should be addressed may provide a convenient method of facilitating the review. It is recommended that all plant personnel be aware of this policy and the need for its existence.

The BPCS and SIS require different levels of review. The specific considerations for each system are discussed below.

6.1.7.1 Changes to the BPCS

Proposed changes to the BPCS that involve items with only minor safety consequences like alarm limits, display features, and the like can be done with minimal review. (Some formality, for example, checklists to follow for these changes, may be required.) This generally can be done by the person responsible for the maintenance of the system, with operations approval. There are, however, some changes that may require additional review and approval. These might include the following changes:

- Control strategy.
- Adding/deleting control loops.
- Adding/deleting cascade or ratio functions.
- Control algorithm changes.
- Process interlocks.
- Range changes.
- Final-control element failure position.
- Type of primary sensor used.
- Vendor-supplied operating systems, including software revisions and updates.
- Any third-party software.
- Any affect on backup control capability.

After a proposed change is reviewed, approved, and implemented, and before it is released for normal operations, it should be tested. This testing may be done on-line, under supervision of the operations and technical groups; or it may be done using an off-line simulation program. There may also be on-line simulation testing capability within the BPCS. The testing should verify the operation of the change for all anticipated operating scenarios including start-up and shutdown of the unit. This information should be used in the training of operators on the new control procedures.

Minimum essential steps that a BPCS change procedure would include are:

- Change the drawing(s).
- Review and agree to the change to the drawing.
- Have one person change the configuration per the drawing.
- Have an independent person verify the change made correctly.
- Verify system operation.

6.1.7.2 Changes to the SIS

Changes to an SIS require all of the same considerations as changes to the BPCS as well as those that relate directly to the SIS safety function. A formal procedure should be in place to ensure that the safety function provided by the SIS is not compromised. Additional considerations that should be reviewed include the following changes:

- Trip initiator values.
- Logic within the control processor.
- Any valve actions.
- Any sequencing modifications.
- Type of hardware used for inputs, outputs, or other components.
- Addition/deletion of any trip initiators.

- Addition/deletion of any final-control elements.
- Type of control processor or I/O components themselves.

The review process should include formal, written approval before implementation. Testing of the changes is recommended prior to placing the modified system in operation. The testing required may include a full functional test as described in this chapter.

Documentation of the changes, approval, and testing are all very important for SISs. A modified system should never be placed in operation unless acceptable documentation has been completed and appropriate personnel have received instructions and training on the operation of the revised system.

6.1.8 Safety Review Procedures

Recommendations for process safety reviews related to the process itself currently exist. (See Ref. 6.4.) These address the overall process safety philosophy and cover such things as when a formal review should take place, of what the review should consist, who should perform the review, etc. The BPCS/SIS is an integral part of those reviews, but there may be instances where the BPCS/SIS themselves require separate and independent reviews. The makeup of the group performing such a review should include those individuals familiar with the detailed workings of the BPCS and SIS as well as those normally involved in process safety reviews. There may also be a need to include the manufacturer of the equipment, or their representative, in the review process to ensure all considerations relating to the operation of the system are covered.

The complexity of some BPCS/SIS makes them difficult to analyze in a simple manner. This may not be evident in the safety review technique being used for the process. There is a danger of looking only at the endpoints of the controls, (i.e., the variable being measured and the variable being controlled), and evaluating the effects of changes or modifications to these variables on the process safety. This can sometimes obscure side effects that may occur or intermediate conditions that could result due to the control logic being used. This can be especially true of complex control strategy embedded in a PES. Software changes intended for a single control loop may, through implementation, impact other loops. A simple thing like an address change of a variable could have a serious impact on process safety if not handled properly.

Adequate review procedures should be established to prevent such potential occurrences during software changes.

In new plants, this review should be a part of the overall process safety review but may require a more detailed procedure such as a fault tree analysis of key safety system control functions. Considerations that should be addressed include:

- Verification of wiring to input and output devices in the BPCS.
- Analysis of control logic as it may impact other loops.
- The consequences if the BPCS halts or otherwise stops working.
- Status of any redundant (backup) system in terms of how it is updated to current operating conditions.
- Methods of making hardware changes or replacements for faulty equipment and how they might impact safe operation.
- The policy concerning facility changes discussed previously.

This is not an exhaustive list, but one can see the need for adequate review covering all the identifiable situations that can occur as well as some that may be highly speculative.

It is important that the reasons and considerations used to arrive at SIS-related decisions be documented to ensure the correct design. The reviews should be made by qualified process control, process, and hardware- and software-knowledgeable individuals with input from operations, maintenance, and safety personnel where appropriate. It may be desirable to involve personnel from other company locations in these safety reviews, if applicable, to provide fresh, unbiased input. In some instances, where the safety impact could extend beyond the plant boundaries, it may be desirable to involve personnel from outside the company.

6.1.9 Security Procedures

The use of PES-based BPCSs and SISs also introduces security concerns. Although control logic changes may be clear to one operator, if not communicated to all operating personnel, a potentially hazardous condition could result. Therefore, security methods and procedures are recommended to maintain the control logic integrity of the systems. Security procedures should be implemented in the following areas:

- Engineering activities that modify, add, or delete control logic.
- Maintenance activities that diagnose, replace, or repair components or systems.
- Operations activities relating to changes in informational alarms, setpoints of control loops, data reporting, and documentation of events.
- Vendor activities related to the upgrading of vendor- supplied operating systems and hardware modifications.

These procedures should address: (1) who is authorized to perform these activities, (2) what method will be used to prevent unauthorized access to the system, and (3) how authorized or approved activities will be implemented. This may require multiple levels of security. Changes made by the operator or maintenance technician that have no adverse impact on process safety may not require special access to the system. Changes impacting control logic may require access through a key or password available only to supervisory personnel. Changes impacting safety should require an additional level of authorization, with either separate key access or additional levels of password protection.

6.2 MAINTENANCE FACILITIES PLANNING

Regardless of how much effort is put into the design, selection, installation, and normal operation of the BPCS and the SIS, proper maintenance of the system is necessary to ensure safety. Procedures should be in place addressing:

- What on-line testing is required, how it will be done, at what frequency, and how it impacts process safety.
- How repairs will be handled during normal working hours, at nights and on weekends and holidays.
- What equipment is required for maintenance and testing, where will it be available, and who will have responsibility for its maintenance.
- What testing will be done during unit down times, scheduled or otherwise.
- Who will perform the various maintenance functions.

Maintenance procedures should also address documentation issues such as, "as found" conditions, repairs made, modifications and replacements made, and who performed the work. In modern BPCSs, the verification of communication among components should also be covered by maintenance procedures.

6.2.1 Maintenance and Engineering Workstations

The Human/Machine Interface (HMI) interfaces (Fig. 2.1), used to configure BPCSs and perform certain maintenance functions, are generally separate from the operator interfaces used in normal plant operation; for SISs, it is recommended they be different devices. Maintenance and Engineering interfaces typically have capabilities over and above those of operator interfaces, and may be referred to as maintenance or engineering workstations. Actions taken through these stations should be controlled to prevent compromising the control or safety functions. Access to BPCS/SIS through these stations should be limited to those individuals who are familiar with the process, the technical aspects of the control logic, and its impact on plant safety. They should also be familiar with the functional operation of the BPCS/SIS and the steps required to effect changes both during operation and during plant downtime.

A security technique that limits access to engineering workstations should be in place. This may be accomplished by a key lock arrangement on the station, but may also include software password protection to further ensure proper authorization. The use of passwords by themselves may provide a level of security, but it should be recognized that they may be easily compromised.

Those authorized to change tuning constants on control loops or perform changes to the BPCS configuration should be so designated. These functions may be controlled by a separate level of security, that is, a separate key that does not enable the engineering keyboard of the workstation. Users not authorized to make changes could negate a critical safety function of the BPCS without realizing they had done so. Communication of a written policy on engineering workstation security should take place prior to plant start-up.

The interfaces associated with SISs require a higher level security technique. A single individual should be authorized to make changes to the controls and function of the safety interlocks. At least one other knowledgeable individual should be able to perform these functions, but they should only do so when the primary responsible individual is not available. It is recommended that proper documentation of any changes be included in the SIS documentation package.

Provision may be required for a separate HMI to the SIS. (Figure 2.1 in Chapter 2). The access to the safety interlocks should be limited to this separate interface, and then only by those authorized. The maintenance/engineering interface to the SIS should not be left active during normal operations of the plant.

It may be desirable that the individual working on the SIS be different than the one performing programming, configuration, or modification work on the BPCS. This could prevent one individual from making a common-mode error in both systems without realizing they had done so. The use of different individuals also provides some additional "eyes" on the process and protective systems for increased assurance of safe system design.

Any access to the SIS from the BPCS should adhere to the separation criteria defined in Appendix B. The operator may see and determine the status of any inputs, outputs, or pending control actions, but should not be able to change any of them directly.

6.2.2 Use of Vendor's Remote Diagnostic Tools

Some PES-based equipment may be connected to a remote site diagnostic center via a modem and a telephone line. This could allow the vendor's home office or central repair facility to run specialized diagnostic programs, analyze a faulty system, and perhaps, determine the necessary maintenance actions to restore it to normal operation. While this may appear to be a convenient feature for expediting repairs, it has the potential for an adverse impact on the safety aspects of the BPCS or SIS. A thorough understanding of the diagnostic procedures to be performed through the remote connection and what affect, if any, they might have on the system operation is very important.

Before this technique is performed, a system backup should be done to ensure that a copy of the system configuration containing the latest information will be available.

Following are some additional considerations that apply to the BPCSs and the SISs.

6.2.2.1 BPCS Considerations
Ensure that any remote diagnostics are clearly descernible between diagnostics of the BPCS-embedded software and user installed application programs. A written procedure stating that diagnostics of the embedded software can be performed only when the process is shutdown, should be in place prior to implementation of this technique.

Actions to ensure that during a remote diagnostic operation, a downloading of a new or revised vendor-supplied operating system does not occur is also a concern. Such an occurrence could have an impact on a key, safety-related system without anyone's knowledge.

6.2.2.2 SIS Considerations
Due to the potential impact on safety, remote diagnostics applied directly to the SIS should not be used. An acceptable alternate method would be to copy the internal diagnostics of the SIS and send it to the vendor for analysis. A written procedure should be provided covering this aspect of the operation prior to initial start-up.

6.2.3 Spare Parts

A key consideration in planning for maintenance of a BPCS or SIS is the provision for spare parts. Components will fail and most likely at the most inopportune time. When this occurs, having the required replacement parts available and in working order will not only ensure rapid return to normal operation but also will provide a measure of safety; the shorter the time without the BPCS/SIS in operation, the safer the operation.

Because spare parts are on hand does not necessarily guarantee that they are operational. The conditions under which the parts are stored may result in inoperable components when they are needed. Advance provision is recommended to ensure an adequate supply of functioning spares for any safety-related control equipment. Areas addressed to achieve these requirements may include:

- An adequate number of appropriate spare components.
- A storage facility that is free from dust, excessive humidity, and the potential for mechanical damage to the components.
- For microprocessor components, provision for preventing damage due to magnetic field exposure.
- For CMOS circuitry, provision for preventing damage due to static electricity.
- Provision for the verification of operational status of critical components when received.
- Determination if spare components require storage in powered up condition, to ensure smooth transition from standby to normal operation, and provision to allow this, if applicable.
- Verification of operating system revision level for microprocessor-based spare parts prior to their need.
- Provision for security of the spare parts to prevent their use in other than intended applications, as well as protection from malicious damage.

In some instances it may be desirable to store spare parts off-site, at a vendor warehouse, for example. When this is done, the accessibility of parts during off-hours should be considered along with provision for ensuring the working condition of the spares.

6.2.4 BPCS Preventive Maintenance Program

In addition to normal preventive maintenance recommended by the BPCS vendors, systems that have a safety related impact may require additional procedures to ensure continuing safe operation. When formulating these procedures, ensuring that attention given to safety-related systems is appropriate to its importance in the overall safety is a must, that is, the more safety-critical, the greater degree of attention required. This may include establishment of formal record keeping systems that maintain a watch on the system reliability to detect any potential problem areas. Statistical Quality Control (SQC) charts on key components may prove useful in detecting early signs of degradation.

Preventive maintenance procedures that may prove beneficial include:

- Running off-line diagnostic programs that check both hardware and software functionality.
- Simulation programs designed to test specific safety-related control functions.
- Scheduled replacement of components known to have limited shelf life.
- Scheduled cleaning of components to remove buildup of dust or other foreign materials that could cause a breakdown during normal operation.

Perhaps just as important as what maintenance is done is the determination of who performs the maintenance. Only those persons who are fully competent with both the architectural and functional makeup of the system should perform the maintenance. They should also understand the reasons why maintenance procedures are important. This would be especially true for the use of diagnostic routines and simulation programs. Maintenance personnel should also understand their duties in the context of PES technology. The need for care in making and breaking connections on printed circuit cards to prevent damage to the connectors, and the potential problems due to static electricity, should be communicated. They should also be aware of the need for ensuring that replacement components are fully compatible with those being replaced. In some instances, this will require additional training of maintenance personnel.

Timing of any preventive maintenance should be based on the vendor's recommendations, analysis of any problem areas that have been defined, and scheduled periods of other maintenance required on the plant. Account should also be taken of any assumptions made in the safety integrity assessment on the plant to ensure that such assumptions are not violated by the maintenance procedures established for the system.

Carrying out preventive maintenance on the BPCS should not present any hazards to the personnel performing the operations or the process being controlled. Normal safety procedures should always be followed. Total reliance should not be placed on the indications given by the BPCS instrumentation when determining whether the plant is safe. An independent verification should always be made before personnel attempt any actions. Control signals should not be used to maintain equipment in a safe condition during maintenance. For example, keeping a valve closed with a control signal to provide isolation is not recommended. Independent primary control devices should always be used to isolate sources of any potential hazard.

6.2.5 Program to Maintain BPCS Modules and Software at Supported Level

A key aspect of PES-based BPCSs relates to the potential for having "look alike" equipment that does not function in a like manner. This is due to the programmable characteristics of microprocessors. A common hardware component can be programmed to perform many different functions.

A potential problem in this area is the constantly changing technology that vendors are utilizing to "upgrade" their systems. These upgrades may not function in an identical manner to the ones applied to your plant and could, in fact, contaminate safety features designed into your system. Upgrading any embedded operating system to a later revision should not be done without thorough validation against the existing operating system to verify that it does not introduce any problems.

Ensuring that these problems do not occur may require that a specific software revision level be specified for a safety-related system; and that arrangements are made with the vendor for the long-term support of this revision, even if later revisions of the software are released. It may even be necessary to have the vendor assign a totally separate identifying number to that revision level, or specific hardware modules for that matter, and agree to maintain support for a specified period of time. A means of testing any components for safety-related systems, prior to their need, is also recommended to ensure the availability of supported components.

It will be necessary at some period in time to upgrade the operating system to a new revision level. When this occurs, special provisions are recommended for validation testing of the upgraded system to verify that it meets the functional requirements specifications and the safety integrity requirements specifications for the system. Sufficient time should be allowed to ensure that this validation can take place prior to the required operational date. This time allowance should include sufficient time to correct any "bugs" that might arise.

Special precautions should also be made relating to board level repairs of system components. There are two concerns that require attention: (1) Adequacy of the in-house repair capability and (2) Status of factory repaired boards for identical capability to original versions. Before any in-house board level repairs are attempted, the capability of the personnel and equipment in effecting the repairs should be assessed and determined to be adequate. When boards are returned after repairs, verification of the board's performance against the original requirements should be obtained. Failure to obtain acceptable results in either instance could have adverse impact on the safety of the plant.

6.2.6 Contract Maintenance

The trend toward the use of contract personnel for maintenance of BPCSs/SISs rather than in-house maintenance personnel, presents some additional considerations relative to the BPCS and SIS. There are at least three methods by which this is implemented: (1) contracting with the supplier of the BPCS or SIS for the maintenance, (2) contracting with a third-party service organization for the maintenance, or (3) contracting with individuals to perform the maintenance. There may be variations of each of these methods that are also possible. While there may be cost advantages of one method versus another, the user should ensure that the service being contracted will provide the level of support required to maintain the systems not only in an operable condition but in a timely manner as well.

Considerations that should apply to the selection of a contract maintenance method or supplier include:

- Capability of the personnel relative to the system to be maintained.
- Availability of personnel in a timely manner.
- Level of support required from the external source. For example, will only hardware maintenance be required or will software maintenance also be required.
- Will there be capability for maintaining the process control technology with an external service organization?
- Will the contract personnel be committed to meeting your needs?
- Stability of the contract organization for long-term support.
- Are you willing to transfer critical safety system know how to an external source?
- Confidentiality and security concerns.

There may be other concerns that should also be included in the evaluation of the most satisfactory method for providing the required maintenance. Those listed have been highlighted to precipitate thinking prior to action.

6.2.7 Other Maintenance Considerations

After maintenance has been performed, checks are recommended to confirm that equipment is in a fit state to place in operation. Checking of the control program and data stored in memory will be necessary if any work has been performed that could have caused corruption (e.g., inadvertent application of test voltages, test probes causing short circuits, temporary removal of electromagnetic screening, or inadequate protection against electrostatic discharges). If any corruption of program or stored data has taken place, it is essential that it be revealed. Programs in PES-based equipment should incorporate error detection routines (e.g., check sums that operate at start-up and periodically during program execution).

Procedures are recommended to ensure that any bypasses required for the maintenance work are removed prior to start-up of the equipment. This can be a potential problem whether the unit is shutdown during the maintenance or is on-line. It may be desirable to maintain a list of all bypasses installed, with a sign-off required for removal to ensure they are all removed after the work has been completed.

6.3 TESTING BPCS HARDWARE AND SOFTWARE

6.3.1 Need for Testing

It is important that a BPCS perform properly when initially installed in order to instill confidence in the operating personnel involved.

Validating that a BPCS performs the required functions in a safe manner can best be accomplished through testing. Testing should validate both the functional requirements specifications, and the safety integrity requirements specifications as well as the application programming that customized the system to the process being controlled. There are several types of testing that may be appropriate: (1) factory acceptance testing of systems prior to shipment, (2) off-line testing of installed systems, and (3) on-line testing of operating systems.

6.3.1.1 Factory Acceptance Test (FAT)
Before a BPCS/SIS is shipped from staging to final plant-site, a factory acceptance test (FAT) may be appropriate. This will ensure that no surprises will be found upon installation and that the system will perform the specified functions. This could be a formal test, witnessed by user personnel, and follow a mutually agreed upon procedure. A typical FAT procedure is shown in Appendix H as a guide in conducting this function if required.

6.3.1.2 Off-Line Testing
Off-line testing refers to those tests that are conducted while a system is not performing its intended control function, that is, the system is off line. This testing can be very detailed and complete since the plant operation is not affected by the testing. Off-line tests may consist of the following:

- Verification of all inputs to a system for proper termination assignments, functionality, ranges, etc.
- Verification of application programming of the system through functional or simulation tests.
- Verification of proper operation of all outputs from the system.
- Diagnostic tests on system hardware and software.
- Verification of the accuracy of all custom graphic displays with associated data.
- Verification of controller cycle time.
- Validation of the current operating program against a master (verified) program.

Off-line testing should be performed on all new systems prior to placing them in operation (this could be accomplished with the factory acceptance test (FAT) but field devices would not be included) and at any times when modifications have been made to the system during a plant outage. The personnel performing the testing should be knowledgeable in both the hardware and software operation of the system as well as the functional requirements for the process being controlled. Documentation of this testing should be included in the system documentation package.

Attention should be given to ensuring that any "add-ins" used in the testing are removed before the system is returned to normal operation.

6.3.1.3 On-Line Testing

There will be occasions when it will be necessary to perform tests on an operating system, that is, on-line tests. This type of testing requires special safety considerations since any unexplained or inappropriate actions that the test might precipitate could result in a potential hazardous event. Plans should be developed and approved, prior to any testing, describing the purpose of the test, the test procedure, the persons performing the test, the expected results of the tests, and any special precautions that may be required during the test to ensure safe operation of the plant. If on-line simulations are to be a part of this testing, they should have been previously validated off-line to ensure suitability for the system and the operation being tested. Typically on-line tests include:

- Verification of output status from the BPCS.
- Verification of any on-line control function changes.
- Verification of range changes of any variables.
- Verification of a simulation program that was previously validated off-line.
- Verification of input status to the BPCS.

Only those persons who have proper knowledge of the system and the process should be allowed to perform any on-line tests on operating plants.

6.3.2 Time Constraints of Testing

Testing in any form takes time. Most of the testing required of BPCSs or SISs, can only take place at the end of the project life, just before start-up of the plant. Therefore, there is always pressure on those responsible for these systems to "hurry up" or to take short cuts to complete the work faster. Where there is any concern for safety in the plant, it is recommended that NO short cuts be allowed, and the FULL time necessary to complete the required testing be allowed. This may require adding time in project schedules for this to take place and recognition that it will likely take longer than projected to perform the testing. Management commitment to accomplishing this testing is recommended.

6.4 SIS AND ALARM TEST PROCEDURES

Validation of complete SIS operations and their related alarm systems is a critical function prior to the start-up of the plant, during normal operation, and after any modifications have been made. In PES-based equipment, software programs should be validated against the functional requirements to ensure plant safety. There are at least three areas that testing of SISs should cover:

- Testing of the SIS hardware (including all interconnections of the components).
- Testing of the written software (application program).
- Testing of the process operation performed by the system (functional test).

6.4.1 Hardware Testing

Testing of the SIS hardware should include verifying that the vendor-supplied internal diagnostic program detects hardware failures. This may be accomplished by fault injection testing (i.e., creating failures by disconnecting components, shorting inputs or outputs, cutting power to components, etc.).

Diagnostics typically repeat simple operations, such as writing a fixed pattern in memory and reading it back. These operations are most meaningful when repeated a large number of times in order to identify marginal components that cause intermittent failures.

Some types of devices, typically electromagnetic ones such as disks and tape drive, cannot be expected to operate completely error-free. Manufacturers implement both hardware techniques (retry) and software capabilities (error detection and correction code) to detect and accommodate these occasional errors.

For these reasons, it is important that the hardware diagnostic procedures be thoroughly documented as to the number of passes required and the "normal" number of soft or recoverable errors (pass/fail criteria). It is equally important that the procedures be carefully followed and the results documented.

The hardware test should also verify the correct physical and soft (communications link) connections of all inputs and outputs associated with the SIS. This includes the primary sensors, I/O interface devices, and the final-control elements. Validation of the functional operation of all primary sensors and final-control devices should also be part of the hardware testing.

6.4.2 Software Testing

Testing of the written software (application program) should include a review of the program logic by someone not directly associated with the program development. Simulation of the program using either the actual system or other PES-based equipment may also be necessary to ensure an error-free program.

6.4.3 Functional Testing

Testing the correct process operation performed by the SIS is generally referred to as the functional test. This provides validation that the programmed logic controls the action of the final-control elements as specified by the functional requirements.

It is recommended that the functional test be designed to validate each function of the SIS logic and the interactions of the various components. This will allow any problem areas that might exist to be uncovered and corrected prior to placing the SIS in operation.

The functional testing should be done by a team with technical, operations, and maintenance capability using a written procedure describing each step to be performed during the test. The written test procedure should be specific to each SIS.

There should also be formal documents for recording the results of the testing. These documents should provide for sign-off by the team performing the test as a reference for any questions that might arise concerning the system operation.

The SIS functional test may require validation of:

- The operation and range of all input devices, including primary sensors and input modules (i.e. the field sensor and all connecting wiring).
- The logic operation associated with each input device.
- The logic associated with combined inputs where appropriate.
- The trip initiating values (set points) of all inputs or the contact position of all switch inputs.
- The alarm functions that may be included.
- The operating sequence of the logic program.
- The function of all outputs to final-control elements.
- The correct action of the final-control elements.
- The first-out alarms, if appropriate.
- Any variable or output status indications that might be provided for operator monitoring (e.g., printed messages).
- Any computational functions performed.
- That the manual trip provided for bypassing the SIS logic program works to bring the system to its "fail-safe" condition.
- The software version in the processor is the correct version.
- System action on loss of power, both instrument and utility.
- Operation in presence of RFI and EMI. (This may be covered in the FAT.)
- Proper operation through the range of its environmental rating (e.g., temperature and humidity). (This may be covered in the FAT.)

A separate part of the functional test should validate the "fail-safe" position of all inputs, outputs, and final-control elements that are a part of the SIS. Checklist-format documents are preferred for recording status found during testing, and any corrections that may be required should be noted on these documents. An example of the information that might be included in a checklist for this purpose is shown in Figure 6.1.

Person(s) performing the functional validations should initial and date each step they verify on the checklist. Validation is interpreted to mean observation of the correct function, value, display, etc. and indicating this on the checklist.

It is theoretically possible to develop an automated test procedure utilizing another microprocessor in performing the functional test, but this is not recommended. Such a procedure would depend on the development of software programs that would have to be checked and validated. It may be

Field Transmitters:

Tag Number	Range	Validation of Displays		Logic
		SIS Display(s)	BPCS Display(s)	Function
_____	_____	_____	_____	_____
_____	_____	_____	_____	_____

Pre-shutdown alarm Set-point:

Tag Number	Set-point (Inc/Dec)	Validation	Display(s) SIS BPCS PRINTER
_____	_____	_____	___ ___ _____
_____	_____	_____	___ ___ _____

Trip Set-point:

Tag Number	Set-point (Inc/Dec)	Valida- tion	Logic Function	Display(s) SIS BPCS PRINTER
_____	_____	_____	_____	___ ___ _____
_____	_____	_____	_____	___ ___ _____

Final Elements:

Tag Number	Operating Position	Trip Position	Fail-safe Position	Feedback Indication
_____	_____	_____	_____	_____
_____	_____	_____	_____	_____

Figure 6.1 Typical information for functional test checklists.

more appropriate to validate the actual SIS than to validate another system program to check the SIS.

If more than one SIS is installed on the same process unit, the testing procedures should ensure that each one is tested independently of the other. They should also be tested at different times, and not simultaneously.

There may also be a concern related to testing SISs where portions of a single channel system (entire SIS logic included in a single processor) require testing while the processing unit continues to operate. This would require that the SIS either be out of service or partially bypassed during the testing. Special precautions should be taken if testing of this nature is attempted to ensure that adequate protection is constantly in place for the unit. Concerns that may require special attention include:

- How testing of a portion of the system can be accomplished safely and without potential for inadvertent changes to remainder of system.
- Means of bypassing only the logic being tested.
- Existence of monitoring of key variables being tested by some other techniques, direct or inferred, during the testing.
- Operating conditions that might need to be adjusted for the testing to take place safely.

It should again be emphasized that each SIS is an independent system that will require its own testing procedure. There may be some synergy among parts of other systems, but each SIS should have its own, written and approved, functional test procedure.

6.5 TESTING FREQUENCY REQUIREMENTS

6.5.1 BPCS Testing

The BPCS is an active system where problems, should they occur, will be quickly brought to the attention of the operator. This reduces the need for formal testing of the BPCS but there will still be times when additional testing will be required. These are:

- After upgrading the vendor-supplied operating system.
- After major work that changes system communications network configuration.
- After major upgrades of hardware components.

6.5.2 SIS Testing

The frequency of testing required for SISs is dependent on the safety protection the SIS provides. Caution, this frequency of testing should never be less than that required by the safety integrity level availability requirements. If the safety function is minimal, the testing interval can allow longer periods between tests. If the safety function is critical, the testing may have to be done more frequently.

Another factor that can impact the testing frequency is the finding of faults or failures of any system components during a test. The number of failures may dictate more frequent testing or the lack of failures could allow longer intervals between tests. There should be balance, however, between the time taken for testing and the estimated time the equipment will be out of service due to failures. In no instance should the frequency of testing be less than that included in the process hazards analysis performed on the plant.

This testing should be performed prior to initial operation of the SISs for all new installations. It should be repeated for all modifications prior to their initial operation. It should be repeated, in total, after all major turnarounds where work has been done that might impact any SIS components. For critical safety systems, an annual repeat of this test may be in order, even if known changes have not occurred. If the frequency of testing of SIS components has been factored into the overall plant risk assessment, the test frequency used in the risk evaluation should be used.

The testing after minor maintenance or minor modifications to an SIS may not require the same level of testing that would be required for initial validation or after major modifications. Procedures should establish whether or not the SIS is still capable of meeting the safety requirements specifications by appropriate testing. Some sound engineering judgment will obviously be required in this area.

The internal diagnostics which are part of the vendor-supplied system, should not run at a frequency that could have an adverse impact on the sequencing time of the CPU, as it might affect plant safety. It should, however, operate at a frequency no less than that time included in the calculated availability for the equipment.

6.5.2.1 Application Software Testing
This testing or reviewing of the program logic should be completed prior to installation of the system. It should be repeated after any changes are made to control logic or at any time an operating system upgrade is performed.

6.5.2.2 SIS Functional Testing
This test should be performed prior to placing any SIS in operation for the first time. It should be repeated any time changes have been made to SIS logic or when physical changes have been made to arrangement of inputs or outputs. For minimal-risk safety systems, it should be repeated no less than every other year or during the scheduled major turnarounds, whichever is more frequent. For high-risk safety systems, it should be repeated at least annually or at times of major maintenance, whichever is more frequent.

6.6 INSTALLED TEST CONNECTIONS AND NECESSARY BYPASSES

Provisions should be made for performing the required testing during the design phase of the project. If on-line testing is to be required, test points or other means should be provided to eliminate the need for removing and replacing wires during the testing. The need for any bypasses required to perform testing should also be addressed during the design phase, with the ultimate goal of eliminating bypasses wherever possible. Criteria for the bypassing of SIS interlocks, detailed in this chapter, should be followed.

One means for providing the test points would be to include additional terminal connections, with the capability for attaching test equipment, on the inputs and outputs of the SIS equipment. It may be necessary to provide removable plug-type jumpers in this arrangement to allow the necessary tests to be performed. Should such an arrangement be utilized, a system of numbering, or otherwise identifying, the jumpers should be employed. A written log should be maintained of their removal and replacement in proper location to provide for normal functioning of the SIS. Only those with written approval should be allowed to move these jumpers, whether the plant is in operation or not. This system should allow for testing of a single variable or input, one at a time, except for those instances where multiple variables or inputs must function together to provide the required trip signal. This is especially important for operating systems, to ensure that the remainder of the system is still functionally available.

When bypass switches, or some other bypassing method, are required, there should be a written procedure that prevents having more than one signal bypassed at the same time. All instances of bypassing should be documented; and the return-to-normal position should be a requirement, prior to signing off that any work has been completed. Where the SIS system allows, changes in positions of all bypass switches should be automatically logged by the system. This is just as important for off-line as for on-line testing. Only those bypasses that are truly required for maintenance or testing should be allowed in the system.

As stated earlier, no master bypass switches should be allowed.

6.7 PLANT OPERATIONS TRAINING WITH INSTALLED CONTROLS

If BPCSs and SISs are to perform properly, it is recommended that those who design, install, operate, and maintain the systems be properly trained. A plant management commitment to provide the training is recommended. Since this requires no small commitment of resources, adequate time and money should be allocated prior to the need. Those performing the design configuration may require training in the methods and procedures necessary to accomplish their work. Those who will maintain the equipment may require training in the necessary routine, breakdown and preventive maintenance techniques. Those who will operate the process using this equipment may require training in how to perform the functions required of an operator quickly and efficiently. All of these personnel must be able to perform their functions without compromising a safety function or creating a potentially unsafe condition.

A continuing training effort is just as important as the initial training on the installed system. Provision should be made for providing scheduled training updates for both operations and maintenance personnel to ensure they do not become "rusty" on the use and operation of the BPCS and the SIS. This should cover not only changes made to the system but also functions that may not be used on a regular basis.

A concept currently being considered in the chemical processing industry is that of certification of operators and maintenance personnel. In such a program, specified skills and abilities would be determined for each job or classification level; the individuals at these levels, or aspiring to them, would be tested to determine their adequacy. This would not be a one-time occurrence, but a scheduled and continuing procedure to ensure that those operating and maintaining the BPCS and SIS equipment are, in fact, qualified to do so. A continuing training program will be a necessary part of such a certification program.

OSHA regulation 24 CFR 1910.119 (Ref. 6.5) also addresses the requirements of training for personnel who have an impact on process safety. This regulation should be consulted in the development of any training programs to ensure those working with BPCSs and SISs meet the requirements.

One method that may be useful in training involves simulation of the process and controls either off-line or on-line. A discussion of this technique is included in Section 6.10

There are four basic operational conditions where specific training in the use of the BPCS and SIS equipment is necessary:

- Normal conditions.
- Start-up conditions.
- Shutdown conditions.
- Abnormal conditions.

6.7.1 Normal Conditions

During normal operation of the plant, there are functions that the operator may be required to perform on a regular basis to maintain control of the plant. These include monitoring of:

- Control loop status, automatic or manual.
- Set points.
- The status of any interlocks or safety-related systems.

Training should address the means of accomplishing each of these tasks and also include things to look for that might indicate deviate conditions. This may include:

- Responding to routine alarm messages.
- Identifying input variables that might be false or out of normal bounds.
- Determining inferred flows from control valve positions.
- Comparing the indicated pressure and temperature profiles to those expected.

Training should be specific to the operating plant and process, whenever possible, but generic training in use of equipment, diagnostic procedures, and the like should also be included. Again, it should be stressed that when changes are made, follow-up with training to cover the changes is recommended. All operating procedures relating to process control safety should be reviewed with operations personnel on a regularly scheduled basis.

6.7.2 Start-up Conditions

For many hazardous processes, the most dangerous time is during start-up or shutdown. During training, emphasis should be placed on the transitional nature of these times and the increased potential for problems. Strict adherence to predefined procedures, operational sequences, use of start-up permissives, and perhaps, different variable-controlled levels should be covered in detail. Special requirements such as control logic initialization and alarm suppression during start-up should also be addressed. The use of training simulators should be considered whenever practical. Areas where strict adherence to procedures/sequence of operation is required for safe operation should be defined and emphasis on why this is necessary should be communicated to operations personnel prior to initial start-up and on a regular basis afterwards to prevent any "forgetting" that might take place due to lack of use.

6.7.3 Shutdown Conditions

In any training program, special emphasis should be given to the shutting down of plants or equipment. This may be especially true when the purpose of shutting down is for the performance of maintenance work. In addition to the normal process-related concerns, emphasis should include the preparation of equipment for direct personnel contact. There may also be special considerations related to the BPCS and SIS themselves, such as updating to a new software release, testing of any system components, functional testing of any SISs, or any other work that can only be done when the plant is not in operation. If system reinitialization is required prior to restart, it should also be covered.

6.7.4 Abnormal Conditions

There may be conditions that require advance training relating to abnormal situations in the plant. This might include potential emergency situations arising during the operation of the plant; preparation for on-line maintenance; operation without a backup system, such as reserve power, for some time period; or other similar situations. Operating personnel should be trained in the proper responses to these potential situations. The use of simulation training may prove helpful in this instance also.

6.8 DOCUMENTATION OF THE BPCS AND SIS

The presence of documentation that accurately reflects the current state of BPCSs and SISs is important. An adequate understanding of what each does is accomplished only when adequate documentation describing the intended function is provided.

In today's constantly changing environment, we are always looking for ways to improve but we also need to know where we have been. This can be accomplished by maintaining up-to-date documentation describing the basic system, its functions, and previous alterations (e.g., P&ID showing revision) to the system.

Adequate documentation should be maintained, so that should a loss occur through some undesired incident, transfer of knowledgeable personnel, or some other means, adequate information exists to recreate it. There are legal requirements relating to documentation in Occupational Safety and Health Authority (OSHA) regulations (24 CFR 1910.119) and in Environmental Protection Agency (EPA) regulations that may require consideration in establishing documentation requirements.

It is suggested that documentation for process control systems be maintained not only in software and on some storage medium such as floppy discs, hard discs, or magnetic tape, but as hard copy (paper) as well. This may include:

- Functional requirements specifications for all system components.
- Configuration data for all control and indicating loops, including schematic drawings where applicable.

- Up-to-date process and instrumentation diagrams (P&IDs) for the plant.
- Details of all BPCS and SIS logic.
- Details of any special control strategies.
- Location of all inputs and outputs to the system by termination assignment locations.
- Details of all displays that are used for normal, critical, and emergency operations.
- Details of any routine or demand reports that are part of the normal operational philosophy of the plant.

Attention should also be given to scheduled updating of this documentation after significant changes. Copies should be stored in safe locations, but not all copies in the same location. Special precautions may be required to prevent multiple copies of documentation that are not in agreement with each other. The number and location of master file copies should be designated and communicated to those individuals involved. A time frame for updating hard copies of documentation after changes should also be defined. For a normal operating unit, after it has passed the start-up break-in period, a monthly or quarterly update may be desirable.

Special attention should be given to documentation provided with vendor supplied package equipment or system integrator supplied packages. Purchase orders for these systems should outline the requirements for documentation. Anything less than you, as a user, would provide if you did the job, should be unacceptable from these suppliers.

Documentation of software based application programs is also important. Not only is the original documentation describing the system function critical, but the maintenance of backup for this software is also critical. Backup should not be taken to mean an archived copy of the system. If changes have taken place, they should be included in the backup copies. The backup should allow full recovery from loss of system function due to corruption or failure of equipment without long down times.

Documentation of SISs is even more important than the BPCS documentation. Here the system is designed to provide protection from events that have been identified as potentially causing serious safety concerns. Procedures should be in place to ensure that the documentation for the SIS is always in a current state. The requirements for this documentation are similar to those for basic process controls. In addition, a description of the safety system logic, which includes the following, should be provided:

- Variables that are being monitored to determine the potential for unsafe conditions.
- Reasons these variables are important for safety.
- Values that are critical for safe or potentially unsafe conditions.
- Actions required to prevent excursions into the unsafe regions.

A description of the logic performed by the system, as well as a copy of the program itself, should be maintained in a safe location. This may be presented in any format understandable by those using the equipment, but typically is provided in the form of ladder diagrams, boolean logic, or in some cases a special programming language specific to the equipment being used. A written description of how the SIS is intended to operate may be provided.

In many instances, detailed wiring diagrams of the safety system may be required to impart the total understanding required.

It may also be desirable to have a simplified logic diagram, which may be in the form of a flow schematic, that shows the relationships between the monitored variables and the safe actions taken in a manner that even the least experienced operator can understand.

The number of copies of the documentation maintained for both BPCSs and SISs, should be tightly controlled and dated, or else copies that do not reflect the latest philosophy may be confused with current operation. A minimum number of copies should be defined as acceptable, and a location for their maintenance should be specified. A practical minimum on the number of copies of SIS documentation might be three (a master copy in engineering, an operations copy, and a maintenance copy). One person should also be identified with the responsibility of updating all official copies when any changes are made. It is recommended that a time limit be set for updating the documentation after a change. This should be the shortest practical time to avoid confusion should a problem occur with the system.

A log of all changes should be maintained for the SIS indicating the change made, the reason for the change, the date performed and the individuals approving and performing the change. This will allow anyone reviewing the system to evaluate its current status.

It may also be desirable to maintain a log of all SIS problems requiring repairs or replacement of components. This could provide early indications of potential chronic problems with hardware components.

6.9 AUDIT PROGRAM TO KEEP EMPHASIS ON BPCS/SIS MAINTENANCE AND DOCUMENTATION ACTIVITIES

Maintenance of the technology embodied in the BPCSs and the SISs will require that procedures be in place to monitor the health of these systems. Periodic validation of their function should take place. This may be accomplished by scheduled audits of the system, the administrative procedures, the documentation, and the understanding of the operation of the system by operations personnel.

For the BPCS, there is a continuing audit of the operation by those monitoring the plant's performance, and this should identify any problem areas that might creep into the system. There may be, however, some critical loops or control functions that should be checked during each plant turnaround to verify their function and operation is as designed.

For the SIS, a scheduled review of the system operation should be incorporated into normal practice. As a minimum this should include:

- Review of all changes made since the last review and verification of documentation status.
- Review of all problems with equipment or logic associated with the SIS since the last review to ascertain if potential problems are developing that might degrade the system's function in the future.
- Functional check of the system operation during annual, or other turnarounds.
- Verification that all official copies of the documentation are in agreement.
- Review of operations personnel's understanding of the system's function and operation.
- Review of any proposed changes for compliance with the design intent of the system.

The audit program may be conducted by plant personnel, including representatives of process, engineering, maintenance, operations, and safety. It may also be desirable to include at least one individual from outside the plant, either a company employee from another location or an outside consultant, to ensure the review is not biased by the views of those closest to the plant or process.

Another area that can be helpful in maintaining safety in a process relates to incident investigation. Uncovering the root causes of abnormal events, whether they result in an injury or not, can be very beneficial. Findings can be used to "fine tune" procedures, replace unreliable equipment, evaluate protection layers on the process, evaluate effectiveness of these protection layers, and determine the capability of SISs. The analysis of abnormal incidents can also be used for training personnel in techniques to avoid reoccurrences. This tool should be used in the overall process safety management program for the plant.

6.10 SIMULATION

Computers are useful tools for both process design and control system design. Computer modelling generally refers to the use of computers to perform various process calculations at steady-state (or equilibrium) conditions. Computer models may include material, energy, and momentum balances, along with phase–equilibrium relationships, chemical-reaction-rate expressions, physical property correlations, equations of state, and equipment correlations (e.g., head/capacity curves for compressors and pumps). Modelling is used to specify process and control equipment for steady-state conditions (e.g., the line and valve sizes required for design operating rates).

Computer simulation (sometimes called dynamic modelling), on the other hand, refers to the use of computers to mimic the time-varying behavior of a process by solving the dynamic forms of the balance equations (i.e., ordinary and partial differential equations with time as an independent variable). Thus simulation finds direct application in process dynamics and control. More specifically, real-time simulation refers to the use of a simulation running on a computer that is connected to external hardware (and is sometimes called hardware-in-the-loop simulation). For example, in the case where a computer is simulating a distillation column, the external hardware may be a BPCS specifically configured for the control of that distillation tower.

Real-time simulation can be used to:

- Design processes for dynamic operation.
- Develop process control strategies.
- Evaluate normal and emergency operating procedures.
- Train process operators on actual BPCS consoles.

Real-time computer simulation can also be used to improve the safety and performance of process controls by providing a means for thoroughly testing control equipment prior to its actual use in the plant. The factory acceptance tests should be performed with realistic simulations to confirm that the equipment meets the specifications of the project and that the equipment performs adequately under conditions of load. Such computer simulations should be verified against mathematical or conceptual models and validated against actual plant data whenever possible.

The scope of a simulation may be an entire processing unit (a so-called flow-sheet simulation) or one particular component of a plant. Simple simulations can be carried out on most BPCSs and even on single-loop controllers operating in simulation mode. Rigorous real-time simulation for control, however, usually requires specialized, high-speed computers with real-time I/O capability, as well as carefully chosen numerical integration methods, to ensure:

- That the simulation computer maintains synchronization with the external hardware (i.e., that time in the computer does not speed up or slow down).
- That the external input signals to the computer are not required for the numerical integration calculations prior to their actually being read (e.g., converted and updated in the case of analog signals) from the external hardware.
- That the numerical integration step sizes are sufficiently small for numerical accuracy and stability and also small compared with the control interval (or sample time) of the external (digital) control hardware.

Real-time simulation generally precludes the use of variable-step and implicit (iterative) numerical integration methods. See, for example, the general-survey articles, Refs. 6.6–6.11. Also, the monthly journal *Simulation* (published by Simulation Councils, Inc.) periodically offers directories of special-purpose simulation hardware and languages.

6.11 PROCESS CONTROL ORGANIZATION AND STAFFING

The application of current technology in controlling chemical processes requires skill and capabilities that were not necessary when pneumatic control systems and electromechanical relays were the only methods available. It is, therefore, very important that a facility consider this when making a decision to install modern BPCSs and SISs.

Some factors that should be considered include:

- The capability requirements of the process control personnel.
- The number of support personnel required.
- Managed constancy (ensuring that movement of personnel does not reduce support capability).
- A continuum of support.

6.11.1 *Capability Requirements for Process Control Personnel*

Process control personnel need to have adequate knowledge of both the process being controlled and process control technology. In addition, they need to have a minimum working knowledge of the following:

- The basic operating system of the BPCS and SIS selected for use.
- The programming or configuring capability of the system selected.
- An understanding of process dynamics and the associated mathematical techniques for achieving control.
- Knowledge of fundamental measurement techniques.
- The application of SIS logic.
- An understanding of the operational requirements of the system selected from both an operator's perspective and the process control engineer's perspective.
- An understanding of the communication between the BPCS and the peripherals that may be used.
- An understanding of the facility's requirements or standards relating to the use of both BPCS and SIS systems.
- An understanding of the underlying safety concepts relating to the BPCS/SIS, as described in this book.

6.11.2 Staffing

Providing adequate staffing for process control functions is recommended if safe, reliable, and efficient systems are the goal. Competent personnel should be assigned at an early stage in the project; the number of individuals required should be determined by the project's magnitude. Considerations in planning staffing requirements may include the following:

- Having one individual responsible for both the BPCS and the SIS may not be desirable.
- The actual time required for system configuration will always be greater than estimated by a factor of at least two.
- A backup person should be included on all projects to allow for uncontrollable eventualities.
- Depending on contract personnel does not totally eliminate the need for company personnel on the project.
- The training of personnel in the performance of the required engineering will take a significant amount of time, which cannot be recovered during the compressed project schedule.

The organizational structure within which the process control function is placed should also be evaluated. A close working relationship between the process control function and both the process and instrumentation engineering functions is recommended. They should work together as a team and communicate with each other. And further, all three should communicate with and work with operations. An organization that allows flexibility and is not rigid in reporting relationships will allow a more efficient mode of operation for achieving the desired results from process control projects.

6.11.3 Impact of Personnel Movement (Promotions, etc.)

The maintenance of support for BPCSs and SISs can be adversely impacted when qualified personnel are promoted to new assignments or moved to new projects. Not only should consideration be given to the need for providing qualified support to fill in behind the one leaving, but the impact of any potential delay in having the qualified support in place should be evaluated. Management of change should apply in the area of personnel movement just as it does to changes in the process or equipment. It is recommended that equal attention be given to the impact of safety of the process due to a personnel change as is given to the "growth" (professional development) needs of the person being moved or promoted.

6.12 REFERENCES

6.1 AIChE/CCPS. *Guidelines for Technical Management of Chemical Process Safety.* New York: American Institute of Chemical Engineers (1989).

6.2 Smith, E. M. and E. Whitmer, "Recipe Management in a Distributed Process Control System," ISA 1987, Paper #87-1171.

6.3 Kletz, T. A., *What Went Wrong?* Gulf Publishing Company, 1985.

6.4. AIChE/CCPS. *Guidelines for Hazard Evaluation Procedures, 2nd edition with worked examples.* New York: American Institute of Chemical Engineers (1992).

6.5. Occupational Safety and Health Administration. Regulation on Process Safety Management of Highly Hazardous Chemicals, OSHA 29 CFR 1910.119 (1992).

6.6 Hunter, D., E. Johnson, W. Short, and R. Zanetti. Accelerating interest in simulation. *Chemical Engineering,* 96:30–33, 1989.

6.7 Amsden, D. M. Process simulation. *Chemical Processing,* 51:74–80.

6.8 Biegler, L. T. Chemical process simulation. *Chemical Engineering Progress,* 85:50–61, 1989.

6.9 Egol, L. Simulation and process control. *Chemical Engineering,* 96:161–164, 1989.

6.10 Palas, R. F. Dynamic simulation in system analysis and design. *Chemical Engineering Progress,* 85:21–24, 1989.

6.11 Pritchard, K. Applying simulation to the control industry. *Control Engineering,* 36:70–72, 1989.

7

AN EXAMPLE: THESE GUIDELINES APPLIED TO THE SAFE AUTOMATION OF A BATCH POLYMERIZATION REACTOR

7.0 INTRODUCTION

This chapter selects a subsystem of an existing process and applies to it the design philosophy, procedures, techniques, and verification methodology discussed in the first six chapters of this book. The SIS selection methods chosen for use in this example illustrate both qualitative and semiquantitative approaches, that are within the broader spectrum of qualitative and quantitative methods. Each company should have policies and guidelines for selection of appropriate SIS integrity classification methods.

Because of the amount of detail that is required to achieve a high-integrity, safely automated design, the example used in this chapter necessarily includes a number of simplifications, but is presented to show the application and discussion of the principles described earlier. Further, the specific design choices do not reflect practices that are part of a particular company's standards, but are representative of good practices. It certainly does not represent a complete design for a polymerization process.

The process, described in Section 7.1, is the polymerization of vinyl chloride monomer (VCM),

$$CH_2=CHCl$$

to make polyvinyl chloride (PVC),

$$[-CH_2-CHCl-]_n$$

The example is based on a well-known process scheme and involves a hazardous reactant, VCM, which is flammable and has toxic combustion products, as well as being a known carcinogen. The process also illustrates a larger-scale batch operation that operates in a semicontinuous manner during an approximately 10-hour period while the polymerization progresses. A simplified description of the process steps is also provided. In Section 7.2, hazards are identified; in Section 7.3 the process design strategy is described, considering some of the design philosophy issues presented in Chapter 2. Section 7.4 addresses development of the process control system design strategy. Then, in Section 7.5, a BPCS is designed following the guidelines in

Chapter 4. Section 7.6 reviews the risk assessment and control evaluations following the techniques covered in Chapter 3. Section 7.7 treats the SIS design and validation as described in Chapter 5; Section 7.8 covers issues relating to installation and testing based on material in Chapter 6. Finally, in Section 7.9, administrative procedures needed to maintain operational integrity are outlined, as discussed in Chapter 6.

Although the material in this chapter is based on the design of a new process unit, in Section 7.10 the differences are contrasted between the approach taken and the approach that might be necessary if automation were being added to an existing process unit, designed using older process control approaches.

7.1 PROJECT DEFINITION

7.1.1 Conceptual Planning

Once a business decision is made to consider producing a certain product, in this example, polyvinyl chloride, the initial project team is assembled. This team will start by evaluating potential process routes to identify a technology that will satisfy production needs while meeting responsibilities for health, safety, and protection of the environment. In the very early stages of process evaluation and project definition, a process hazards analysis team starts to interact closely with the designers. For projects handling hazardous materials, the team will include not only process design engineers but also health and safety specialists. The team will often need to have access to other specialists— such as chemists, operating personnel, consultants or engineering contractors with experience with the same or similar processes, process licensors, etc. In this example, a well-proven process is available as a starting point. Therefore, we will proceed with the business decision to produce this product, and concentrate on the aspects of the design process that influence or directly involve the design of the process control systems and safety interlock systems. More detailed information on related aspects of the design process can be found in CCPS *Guidelines for Hazard Evaluation Procedures* (Ref 7.1), *Guidelines for Chemical Process Quantitative Risk Analysis* (Ref 7.2), *Guidelines for Safe Storage and Handling of High Toxic Hazard Material* (Ref 7.3), *Guidelines for Vapor Release Mitigation* (Ref 7.4) and *Guidelines for the Technical Management of Chemical Process Safety* (Ref 7.5).

7.1.2 Simplified Process Description

The manufacture of PVC from the monomer is relatively straightforward. The heart of the process is the reactor vessel in which the polymerization takes place over a period of about ten hours, while the reactor contents are agitated mechanically and the heat of reaction is removed by the circulation of cooling water through the reactor jacket. Because the process involves the charging of a batch to the reactor, process systems are designed with multiple reactor units in parallel, so that the process can operate on a semicontinuous basis. For simplicity, this example will focus on one of the units, recognizing that a real production facility will typically have several parallel units operating in sequence.

Figure 7.1 is a simplified process flow diagram for a typical PVC manufacturing facility. If the reactor vessel has been opened for maintenance after the last batch was processed and dumped, it must first be evacuated to remove any residual air (oxygen) in the vapor space, to minimize the oxidation reaction of monomer which produces HCl and may lead to stress corrosion damage to the reactor vessel as well as to poor product quality. Otherwise, the first step is to treat the reactor vessel with antifoulant solution to prevent polymerization on the reactor walls. This is followed by charging the vessel with demineralized water and surfactants.

Then the liquid vinyl chloride monomer (VCM) charge is added at its vapor pressure (about 56 psig at 70°F).

The reaction initiator is a liquid peroxide that is dissolved in a solvent. Since it is fairly active, it is stored at cold temperatures in a special bunker. Small

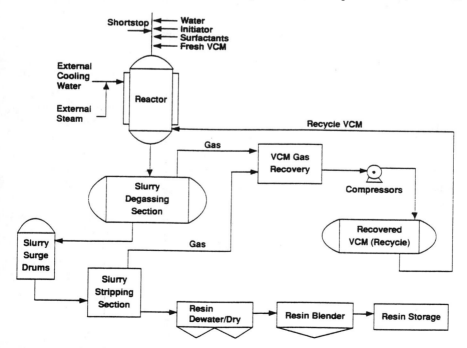

Figure 7.1 Simplified flow diagram: the PVC Process

quantities are removed for daily use in the process and are kept in a freezer. It is first introduced into a small charge pot associated with the reactor to assure that only the correct quantity is added.

After the reaction initiator is introduced, steam-heated water is applied to the reactor jacket to raise the temperature to about 130 to 140°F (depending on the batch recipe for the particular grade of product), where the reaction will proceed at a satisfactory rate. Agitation is necessary to suspend the VCM in the water (control particle size), improve heat transfer throughout the batch, and produce a uniform product. Since the reaction is exothermic, cooling water is then circulated through the vessel jacket to control the reactor temperature. Reactor conditions are controlled carefully during the approximately ten hours required for completion of the polymerization.

The reaction is said to be completed when the reactor pressure decreases, signalling that most of the monomer has reacted. Reacted polymer is dumped from the reactor and sent to downstream process units for residual VCM recovery, stripping, dewatering and drying.

7.1.3 Preliminary Design

At this point in the design, scoping and site evaluation are completed. The first step is to understand the regulatory and corporate policies and constraints on the project, based on the potential hazards. In an initial policy meeting involving the plant manager, the project leader and a representative of the safety/health/environmental activity, general guidance for the project is established. Any special local requirements are reviewed, applicable regulations are identified, and general risk guidelines are established. Shortly thereafter, a formal Process Hazards Analysis (PHA) Team is assembled, made up of design and operating specialists familiar with the technology, an electrical/instrumentation/control expert, a safety/environmental/health risk expert and a risk assessment specialist. After receiving general guidance from the initial policy meeting, the PHA team begins systematic identification of potential hazards. This information is fed back into the process design in a way that attempts to eliminate or reduce the hazard by a process change or by sizing and locating process components appropriately. A basic approach is to minimize hazardous inventories as much as is consistent with sound operation. This is an example of an improvement towards a more inherently safe design. Once these inventories are set, then decisions can be updated on safe separation of components. Similarly, the layout may be modified based on the considerations given to reducing the potential for significant accidents. At this stage, more analyses of the BPCS and the system design philosophy can lead to improvements.

Throughout the design development, there are frequent discussions among the PHA team members. Different organizations may handle these interactions in different ways, but they all have a common objective: to develop and implement a reliable and safe design that meets, or improves upon, all relevant regulatory requirements, company standards and safety criteria.

7.2 HAZARD IDENTIFICATION

The hazard identification process should have started during the business decision analysis. It is one of the most important functions of the PHA team and is ongoing until the process is turned over to plant operations, and becomes subject to operational safety review and audit programs.

7.2.1 Preliminary Hazard Evaluation

The first step in any process development planning, after general safety guidelines have been established in an initial policy meeting, is to identify the broad parameters of the production process, to define safety and environmental hazards (or hazardous events), and to seek opportunities for making the process inherently safer. To do this, information is required about the physical and hazardous properties of all the feedstocks, intermediates, products and wastes involved in possible alternative processes. For this example, where a specific polymer is being made from its monomer, there is little choice about the basic reactant. The available alternative processes vary the polymerization medium—solution, suspension or emulsion. The significant properties of VCM are summarized in Table 7.1. However, the reaction conditions and the initiator (plus any additives) need to be carefully chosen to assure that the reaction rate can be safely controlled to prevent runaway reactions, while producing adequate quality and yield. The selected technology involves polymerization in water, but does require small quantities of a relatively dangerous liquid initiator. The hazards associated with the initiator also need careful attention, but are not included in this simplified example.

7.2.2 Accident History

Next, the potential hazards are identified. In this example, the primary hazards are associated with the flammability and toxicity of combustion products from VCM. In actual plant design, personnel exposure and environmental ambient VCM limits would also be major considerations, but, for simplicity, these are not covered in this example. As a first step, it is useful to review the past history of accidents associated with similar operations. One source of information, which is used for illustrative purposes, is the compilation of accidents in the chemical industry from 1951 to 1973, prepared by the Manufacturing Chemists Association (Ref 7.6):

Table 7.1 Some Physical Properties of Vinyl Chloride

Formula: $CH_2=CHCl$

Synonyms: vinyl chloride monomer (VCM)
  chloroethylene
  chloroethene
  vinyl chloride (VCl)

Shipped as compressed liquefied gas; Reid Vap. press. = 75 psia
Gas, colorless, sweet odor; Mol. wt. = 62.5; Sp. Grav. (vap) = 2.16
Normal boiling point = 7.1°F; Sp. Gr. (liq$_{NBP}$) = 0.97; Floats and boils on water
Critical T = 317°F; Critical P = 775 psia; Melting point = –245°F

Heat of Vaporization = 160 Btu/lb; Heat of Combustion = 8136 Btu/lb
Heat of Polymerization = –729 Btu/lb; Normally stable at ambient conditions; polymerizes in presence of air, sunlight, moisture, heat, or free radical initiators unless stabilized by inhibitors.

FIRE HAZARDS:
Flammable Limits in Air: 3.6–33%
Flash Point: –108°F (o.c.);   Autoignition T = 882°F
Spills flash, boil and produce heavier-than-air gas cloud that may be ignited with flashback.
Poisonous gases (HCl, CO, etc.) produced in fire
May explode if ignited in confined space
External fire exposure to container may result in BLEVE

HEALTH HAZARDS:
Irritating vapor to eyes, nose and throat
If inhaled, causes dizziness, difficult breathing, and may cause serious adverse effects, even death.
Excessive exposure may cause lung, liver and kidney effects. Human carcinogen, listed by OSHA, IARC and NTP.
Threshold Limit Value: 5 ppm
OSHA PEL: 1 ppm TWA, 5 ppm excursion limit average over any period not exceeding 15 minutes.
Odor Threshold: 260 ppm
Liquid contact may cause frostbite

WATER POLLUTION:
Limit in process water: 10 ppm
Limit in water discharged offsite: 1 ppm

AIR EMISSIONS:
Limit in process discharge to atmosphere: 10 ppm (local standard)
Limit for annual concentration at plant boundary: 0.2 µg/m^3 VCM in air

RESPONSE TO DISCHARGE:
Issue Warning—High Flammability, remove ignition sources, ventilate
Stop flow
Evacuate area, allow entry only with proper protective gear
Let large fires burn; extinguish small fires with dry chemical or CO_2
Cool exposed containers with water
Prevent entry into sewer systems to avoid potential explosions

In the case of VCM, we find reference to an accident (#816) in 1961 in a PVC plant in Japan, in which four lives were lost and ten people were injured. This accident was due to discharging a batch from the wrong reactor vessel, so that unpolymerized monomer was released into a room containing the parallel reactors. VCM vapor was presumably ignited by a spark from electric machines or by static electricity and the building housing the reactors exploded.

In another accident (#1132), a worker mistakenly opened a manhole cover on a reactor that was in service, releasing a large quantity of vinyl chloride that ignited and caused a flash fire resulting in the death of the maintenance person and two laborers.

Another accident (#1932) involved charging a reactor with 250 gallons of VCM with the bottom valve of the reactor open. Although a serious hazard was created by this release, ignition did not occur and no one was injured. Other incidents are noted, such as one in which an explosion occurred during maintenance work on a vinyl chloride pump (due to a polyperoxide contaminant that was present as a result of three simultaneous, abnormal situations). VCM was also released from a scrubber in a VCM production plant due to maintenance problems with a plugged valve during periodic recharging. Ignition of VCM resulted in one death and several injuries.

There also have been VCM releases and fires associated with transportation. In 1971, a derailment of 16 cars near Houston led to the escape of VCM from a 48,000-gallon rail tanker with immediate ignition. After 45 minutes of exposure to the fire, a second rail car of VCM ruptured violently producing a large fireball (BLEVE, see Section 7.2.4), killing a fireman and injuring 37 other people. Large sections of a tank car were found about 400 feet from the derailment site after the explosion.

There are probably numerous minor incidents for every major accident reported. These may have cost impacts or cause some small environmental impact, but are too minor to be noted in the published incident lists, even though the more likely causes of minor equipment failures or small releases will be known to those familiar with plant operations and maintenance. Nevertheless, attention must be paid to the potential for small releases since these may be partial pathways to major accidents. Particularly with a highly flammable pressurized material, ignited small releases may cause larger failures if they heat other system components. Thus, integrity of a VCM system needs to be at a high level.

7.2.3 Preliminary Process Design Safety Considerations

For this example, the desired production rate of PVC is 200 million pounds per year (90,000 Tonnes/year) or about 23,000 lb/hr, based on known reaction kinetics, at a reaction temperature of about 140°F, the corresponding cycle time is about 8 hours. In setting the reactor inventory, judgment is usually used with some awareness of the fact that hazard magnitude for catastrophic vessel failure is related to the amount of hazardous material. At one extreme, a single reactor might be used, with a production batch of 180,000 lb of PVC in a 40% slurry mixture, requiring a reactor sized for about 50,000 gallons capacity. This would be unwise, since no redundancy is present and there is a very large, flammable, high pressure inventory. Also since capacity is not distributed, production batches would be large, infrequent, and would require downstream equipment sized for large inventories. Furthermore, this reactor would require addition of a large quantity of the dangerous initiator solution—a large enough inventory to raise serious safety concerns.

At the other extreme, a large number, say ten, small reactors, each designed for a 18,000 lb production batch (about 5000 gallons), might be used. In the first extreme, inventories are large; in the second, batches are small, switching operations are much more frequent, and there are many more interconnecting lines, valves, and complexities. Tradeoffs would be considered, based on operational needs, availability of equipment, and cost, as well as safety.

Refinement of such analyses leads to selection of the number of reactors in parallel and the size of the reactor unit. At this point, it is well to provide for any potential for future expansion in capacity. In this example, it is decided to install three parallel reactors, each with a 17,000 gallon capacity. The 5 gallons of initiator solution required per batch is a manageable quantity for safe handling. The maximum inventory of VCM in a reactor is estimated to be 60,000 lb.

Reaction temperature is selected to achieve a desired molecular weight, which is end-use driven. Proper reactor cooling water temperature control for stable reactor operation is required to prevent runaway reaction. Stable control of the polymerization reaction temperature requires a low temperature difference between the cooling water and the reaction temperature (see Ref.7.7). For this example, the tempered cooling water supply temperature is high enough to provide a low temperature difference, versus the 140°F reaction temperature, for safe operation.

7.2.4 Process Hazard Identification

The **major acute hazards** associated with VCM release are fire and explosion, with generation of toxic combustion products. These types of hazard include the following:

Jet fires: A leak from a pressurized system ignites and forms a burning jet that might impinge on other equipment and cause damage. (In rough terms, jet length is about 150 times the jet orifice diameter—a jet from a 2-in. hole could produce a burning jet about 30 feet long.)

Flash fires: A pressurized liquid release flashes producing flammable vapor that travels to an ignition source. Upon ignition, the flame travels back through the flammable vapor cloud. (The flammable plume in this case can be substantially larger than the flame jet.)

Pool fires: Residual liquid from a flashing release forms a pool which may ignite and burn with a flame height that is two or three times the width of the pool.

BLEVEs (Boiling Liquid Expanding Vapor Explosions): A pressurized tank of VCM or associated piping exposed to an external fire may fail due to metallurgical weakening. Such failure may result in a catastrophic tank failure, a fireball and the potential for rocketing fragments. Relief valve overpressure protection will not prevent a BLEVE,

Explosions: Leakage of flammable gas into a confined space with subsequent ignition may lead to explosions or detonations with substantial overpressures.

Hydraulic Failure: Overfilling of a tank with subsequent liquid expansion through heating may lead to collapse of any vapor space and rapid pressurization. Sudden tank failure may ensue.

Stress Corrosion Failure: Air (oxygen) in the system may increase the presence of chloride ions and may lead to loss of metallurgical integrity.

Toxic Combustion Products: The combustion products of VCM include phosgene, hydrogen chloride and carbon monoxide along with other toxics. (These will be present in the aftermath of a fire, particularly if the fire is within a confined space).

Runaway Polymerization Reaction: VCM polymerization has the potential to rupture the reactor, releasing the VCM with major damage possible.

In addition, VCM also presents *chronic exposure hazards,* being a *known human carcinogen* and is a regulated substance with regard to personnel exposures to its vapors, having an OSHA PEL (personnel exposure limit—time weighted average) of 1 ppm in air. Further, federal and local regulations limit its discharge levels from process vents and plant water treatment systems. There are also stringent limits set on the amount of residual VCM that may be present in the PVC product.

There are some lesser short-term hazards involved with inhalation of VCM vapor and the potential for autorefrigeration of flashing fluid. Personnel require protection both from inhalation and possible freeze burns.

At this point, scoping hazard zone estimates are made to indicate the magnitude of major potential accidents. A 60,000-lb release of VCM could produce a flammable vapor cloud equivalent to a cubic volume that is about 400 feet on a side. Because VCM is a heavy gas and may contain aerosols from flashing, a major vapor cloud is much more likely to be pancake-shaped, but still might have a flammable footprint of 1000–1500 feet in diameter. This indicates that the maximum accident involving a single reactor might have offsite impacts, and could fill a substantial confined volume with flammable gas. In terms of the assessment criteria discussed in Chapter 2, this impact should be considered to be at least "serious," and probably "extensive," depending on specific data considerations. To be conservative, the PHA team considers it to be in the "extensive" impact category. (*Note:* The bulk storage of VCM on site is not considered in this limited example. There is likely to be a bulk storage tank somewhere that is sized to hold operating inventory. That tank may have a capacity in the range of 50,000 gallons to accommodate pipeline upsets or the periodic unloading of rail cars. Such an inventory may be capable of offsite impact and its location and design, as well as those for the associated transfer facilities, require detailed attention to safety.)

The PHA team has decided, because of a northern site location that weather protection against freezing is required and that the reactors should be located inside a building. If freezing were not a problem at the site, it would be preferable to locate equipment outdoors to avoid the potential for accumulation of hazardous gases indoors, should a leak occur. The building will be designed for extensive monitoring for leaks and for potential explosion conditions (in terms of active ventilation, blowout panels, and appropriate electrical design). Gas detectors throughout the enclosure will be tied to an alarm system to provide early warning of leaks. Any process vents will be collected into a header system that is tied into a properly sized VCM recovery disposal system.

Moreover, the process will be operated to minimize the need for operating personnel to be in the vicinity of the equipment. A few manual operations are still necessary; the most sensitive of these is the charging of the small quantities of initiator prior to the start of a new batch cycle. Operators will need sound training in this critical operation, even though hazards are reduced by the small quantities of materials involved.

There also is a requirement for an extensive leak and fire detection and protection system as well as the need to provide personnel with positive-pressure, self-contained breathing apparatus in the event of leak detection.

At this point, the PHA team needs to be fully satisfied that the site and preliminary layout will provide adequate spacing for present and future operations both to tolerate the impacts of potential major accidents in the unit and also to protect the unit from impacts initiated elsewhere in adjacent facilities.

7.3 PROCESS DESIGN STRATEGY

7.3.1 Process Definition

The process design then progresses into a more detailed phase. A plot plan, process flow diagram and piping and instrumentation drawings (P&IDs) are developed. Figure 7.2 presents a simplified P&ID for a single reactor unit in the process.

The full P&ID set would show the full process from the pipeline receipt or rail unloading facilities, to the bulk storage of reactants, to parallel reactors, to the plant utility systems, to waste treatment systems, to relief valve collection and venting systems, to VCM recovery and storage systems, and to downstream systems for stripping and drying the PVC product. Since this is a well-established technology, the designer will draw on past designs and experience as a guide.

Basic instrumentation required for process control and monitoring will be incorporated in the design at this point. However, this will be done on a preliminary basis, since a more comprehensive development of the control system will be considered later.

The details of the design require definition of the basic operating procedures and maintenance strategies for the facility. The operational steps for the process are outlined below:

Pre-evacuation of air: If the reactors have been opened for maintenance, oxygen must be removed from the system for quality and metallurgical integrity reasons. This is done using steam ejectors to pull a vacuum.

Reactor preparation: The empty reactor is high pressure water rinsed, leak tested if the hatch has been opened, and treated with antifoulant.

Demineralized water charging: A controlled charge of water is added. An overcharge might lead to a hydraulic overfill; an undercharge may cause quality problems and potential runaway reaction. Any surfactants or other additives are introduced.

VCM charging: An accurate charge of VCM is added to the reactor.

Reactor heatup: The initiator is added from the charge pot to the batch, and steam is added to the cooling water circulating through the reactor jacket until the batch is at a temperature where the reaction will proceed (about 10°F below the steady-state reaction temperature).

Reaction: The steam system is isolated and cooling water is circulated through the reactor jacket to control temperature by removing the heat of polymerization while the reaction progresses.

Termination: When the reactor pressure starts to decrease because most of the VCM present has been consumed by the polymerization, the batch will be dumped.

Reactor Discharge: The reactor contents are dumped under pressure to a downstream holding facility where the system is degassed for subsequent stripping and drying. To prevent resin settling in the reactor, the agitator operates during the dumping procedure. Unreacted VCM is recovered for reuse.

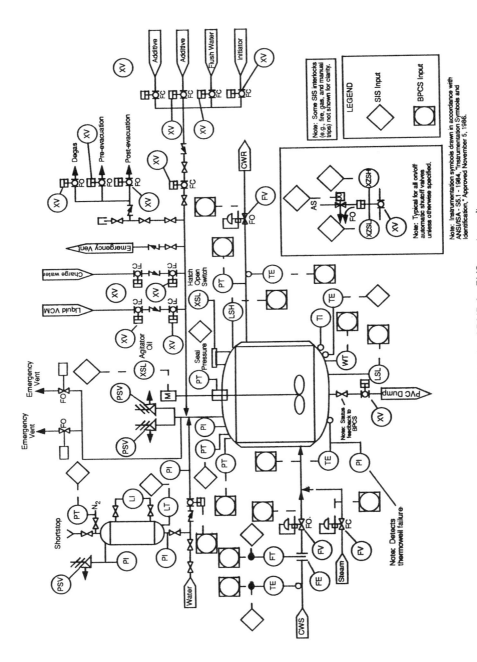

Figure 7.2 Example of P&ID for PVC reactor unit.

306

There are two additional process systems which are provided for an emergency situation. In the event of an uncontrolled reaction or the potential for such an event, the polymerization can be stopped rapidly by **addition of a Shortstop chemical** (chain stopping agent) to the batch. However, agitation of the batch is necessary for good distribution of the Shortstop to rapidly terminate the polymerization. If the agitator has failed, the Shortstop must be added within a minute or two, to allow mixing before the liquid swirl in the reactor dissipates. As a back-up, the reactor contents can be mixed by "burping" the reactor—dropping pressure to generate rising bubbles within the bulk liquid mass.

The second emergency system is an automatic depressurization system. In the event of an uncontrolled reaction, the reaction can be safely limited by depressurizing the reactor to the vent system. The heat of vaporization of the boiling reaction mass safely removes heat from the reactor.

At this time, the specific components are also selected and specifications are set for design conditions, materials, performance, etc.

7.3.2 Preliminary Hazard Assessment

Once the design is developed in some detail, the PHA team subjects the design to a preliminary hazard assessment. The PHA team will use some systematic technique, such as a preliminary HAZOP or FMEA (See Chapter 3), to review the design and the proposed control strategy. From the hazard identification process and from the past accident history, it seems that the example process reactor has the potential for "minor" through "extensive" severity events as defined in Figure 2.4, the Process Risk Ranking Model. In addition, design integrity must consider the need to meet the strict containment requirements to prevent emissions of VCM that might endanger worker safety and health. The results of this hazard review need to be carefully documented, with particular regard to event sequences that might lead to uncontrolled releases.

In this example, a HAZOP was done by the PHA team for the preliminary hazards assessment. Based on the HAZOP results, a team of appropriate PHA process and instrumentation experts prepared a list of accident events for development of an initial safety interlock functional design.

Table 7.2 is a partial list of accident events and the associated prevention strategy used to propose interlock strategy and actions to help further identify or create additional independent layers of protection.

This interlock team proposed the following safety interlock preventive strategy:

- **A.** To deal with the runaway reaction scenarios where the agitator remains on, items 1, 4, 5, and 7 in Table 7.2, the team proposed the following sequence:

Table 7.2 Partial Summary of Preliminary Hazard Assessment Information for Development of Safety Interlock Strategy

#	INITIATING EVENT	PROCESS UPSET	PROCESS VARIABLES AFFECTED	PREVENTIVE STRATEGY
1	Cooling water control fails	Loss of cooling leading to runaway reaction	·Low C.W. Flow ·High Reactor Temp. ·High Reactor Pressure	·Add Shortstop ·Emergency Cooling H_2O Flow (SIS) ·Depressurize Reactor (SIS) ·Pressure Safety Valves (IPL)
2	Agitator Motor Drive Fails	Reduced cooling, temperature non-uniformity leads to runaway reaction	·Low Agitator Motor Amperage ·High Reactor Temperature ·High Reactor Pressure	·Add Shortstop and Burp reactor to stop runaway. ·Depressurize Reactor (SIS) ·Pressure Safety Valves (IPL)
3	Area wide loss of normal electrical power (UPS Instrumentation power remains)	Loss of agitation leading to runaway reaction	·Agitation Motor off ·Low Coolant Flow ·High Reactor Pressure ·High Reactor Temperature	·Add Shortstop and Burp reactor to stop runaway. ·Depressurize Reactor (SIS) ·Pressure Safety Valves (IPL)
4	Cooling water pumps stop, pump power failure	Loss of cooling leading to runaway reaction	·Low C.W. Flow ·High Reactor Temp. ·High Reactor Pressure	·Steam Drives on Pumps ·Add Shortstop ·Depressurize Reactor (SIS) ·Pressure Safety Valves (IPL)
5	Batch recipe error; two charges or initiator are used	High initiator concentration causes runaway reaction	·High Reactor Temp. ·High Reactor Pressure	·Add Shortstop ·Depressurize Reactor (SIS) ·Pressure Safety Valves (IPL)
6	Control system failure overfills reactor	Reactor becomes liquid full as the temperature increases, possible hydraulic reactor damage and VCM release.	·High charge level ·High charge weight ·High Reactor Pressure	·Compare Hi-level & weight with recipe (SIS) ·Depressurize Reactor (SIS) ·Pressure Safety Valves (IPL)
7	Temperature control failure causes overheating during steam heat-up step	High temperature leads to runaway reaction.	·High Reactor Pressure ·High Reactor Temp.	·Add Shortstop ·Emergency Cooling H_2O Flow (SIS) ·Depressurize Reactor (SIS) ·Pressure Safety Valves (IPL)
8	Reactor agitator seal fails	Seal failure can lead to dangerous VCM fume release	·High pressure in reactor seal ·Fume detection in reaction area	·Additional ventilation around reactor seal ·Depressure reactor on high seal pressure (SIS)

—At a "high" temperature or pressure condition, activate the emergency full rate cooling water flow SIS, and alert the operator by alarm.

—If the reactor temperature or pressure continue to increase, sufficient time is available for the operator to remotely add Shortstop.

—If neither of these methods stop the runaway, a "high–high" temperature or pressure SIS will open the emergency depressure vent valves to safely control the runaway.

- B. For runaways that occur because the agitator is not working—items 2 and 3 of Table 7.2—protection is needed in addition to the proposal given in A above:

  —Loss of agitation (low amps) will be indicated to the operator by an alarm, and after adding the Shortstop, "burping" is required to mix the Shortstop into the reaction mass.

  —As in Proposal A above the emergency depressure SIS is a backup to control the runaway.

- C. Low or no cooling water flow upsets are controlled by the protection in Proposal A above. In addition the emergency full cooling water flow SIS is initiated by a "low–low" cooling water flow signal. If low cooling water flow was caused by power loss to the water pumps, the operator is alerted by the low flow alarm to turn on the steam turbine water pump drive.

- D. Overcharging the reactor with water or VCM, can cause overfilling and possible reactor hydraulic overpressure damage. This upset is avoided by preventing the batch heat up if the weigh cells or the reactor level exceed the "high" limit for that batch addition step in the BPCS. Backup is provided by the "high–high" reactor pressure SIS that activates the emergency depressure vent valves.

- E. Failure of the reactor agitator seal causes dangerous releases of VCM. To protect against this it is proposed to activate the emergency depressure SIS for high pressure in the agitator seal.

- F. Because the Shortstop system is so important in controlling all runaway reactions, interlocks to assure Shortstop availability are also proposed by the team. The interlocks do not allow VCM charging to reactor if the Shortstop tank level is low, or if the nitrogen pad pressure on this tank is low.

7.4 SIS INTEGRITY LEVEL SELECTION

Using the proposed interlock list, a PHA team meeting is held to classify the integrity levels for the SISs. In this section, two methods will be used, the "risk-based" method and SIS selection as part of a HAZOP.

7.4.1 The "Risk-Based" Method Description

Table 7.3 presents an example of an assessment format. It documents the assessments involved in developing protective layers and estimating their risk control effectiveness. It captures and documents the layers of protection that exist. In addition, it aids in identifying ways to further reduce risk, if necessary. In the first column, each impact event is listed and the corresponding estimate of its severity is given in Column 2. Column 3 lists direct or indirect initiating causes that might lead to a failure sequence causing the event in Column 1 and Column 4 lists the estimated frequency of the initiating event. The next group of columns represent the available layers of protection. Ideally, the individual layers are selected to represent systems that are separate and diverse in nature. A well-chosen process chemistry and conservative design conditions are the start. Next, the BPCS is designed to maintain the process within safe limits. Further, a series of warnings or alarms are provided to alert a well-trained operator to take corrective action if a failure is evident.

In very qualitative terms, list each protection layer in column 5, if it offers at least one order of magnitude reduction in failure sequence likelihood. This should be a conservative assumption of protection layer effectiveness. If a protection layer includes more than one independent and diverse protective system, additional orders of magnitude reduction in likelihood of failure may be achieved.

In this methodology, a failure in any of the layers is assumed to remove that entire protective layer (unless it contains multiple independent systems) from functional status. Care and judgment need to be exercised in making likelihood estimates, but the procedure is not very different from the thought processes that are exercised qualitatively in HAZOP or FMEA evaluations. The basis for this procedure is the assumption that the individual protection layers are independent of each other and are not susceptible to common mode failures. With the implementation of PES-based controls, the potential for common mode failures must be reviewed.

Column 6 identifies physical protection layers—pressure safety valves, dikes, etc.—act to reduce the likelihood of more severe hazardous events, but may lead to a separate hazardous event of lower impact. For example, a pressure safety valve may reduce the likelihood of a storage tank rupture, but may cause a lesser release of hazardous material, if it operates to protect the storage tank. The pressure safety valve discharge is then considered as a separate hazard event in the system evaluation.

In Table 7.3, the presence of protective layers is described, and each layer listed is conservatively estimated to reduce the initiating likelihood by about an order of magnitude. An event likelihood can then be estimated in approximate terms. The PHA team might decide to make the evaluation more quantitative by using actual failure rate estimates for protection layers. An

Table 7.3 A Framework for System Risk Evaluation/Documentation and SIS Integrity Level Selection

Note: Severity Level: E - extensive, S - serious, M - minor (see Fig. 2.4)
Likelihood values are events per year, other numerical values are probabilities.

#	1	2	3	4	5			6	7	8	9	10	11	12
	IMPACT EVENT	SEVERITY LEVEL	INITIATING CAUSE	INITIATION LIKELIHOOD	PROTECTION LAYERS			ADDITIONAL MITIGATION	INTERMEDIATE EVENT LIKELIHOOD	NUMBER OF IPLS	SIS NEEDED?	SIS INTEGRITY LEVEL	MITIGATED EVENT LIKELIHOOD	NOTES
					PROCESS DESIGN	BPCS	ALARMS ETC.							

estimate of the "intermediate event likelihood" is entered in Column 7. If this event likelihood is appropriately controlled for the consequence involved, no further action is needed.

If not, additional risk control measures need evaluation. The first step in identifying additional mitigation measures to reduce likelihoods is to look for additional protection opportunities in the innermost layers shown in Figure 2.2. Only after these options have been utilized to the extent practical, should the PHA team suggest that the designers add an SIS. The PHA team then incorporates any additional protection identified in this mitigation analysis and generates a revised "intermediate event likelihood" value in column 7. At this point, if the intermediate event likelihood is not appropriate, use of an SIS is indicated. Utilize Figure 7.3 (duplicate of Figure 2.6) which shows the relationship of SIS interlocks to other IPLs and then select the integrity level based on severity and initiating event likelihood.

An IPL is expected to reduce the identified risk likelihood by a factor of 100. The degree of risk control provided by an SIS interlock can be increased as indicated in Figure 2.5 by the addition of redundant SIS elements, the application of structured design practices, and an extensive use of system validation methods. However, some industry experience suggests that an SIS alone is not satisfactory to protect against events of extensive severity. Therefore, it is necessary to identify all IPLs planned and to record these. Table 7.3, Column 8, records the number of non-SIS IPLs identified. Column 9 indicates whether an SIS is needed; Column 10 identifies the SIS integrity level as determined from Figure 7.3. This information is the basis for the "Safety Requirements Specification," which is given to the SIS design team (see Section 5.1.1).

The SIS Integrity Level describes the availability of the SIS, and the corresponding reduction in hazardous event likelihood provided. An Integrity Level 1 SIS provides about two orders of magnitude of likelihood reduction; an Integrity Level 2 SIS, about three orders of magnitude; and Integrity Level 3 SIS, about four orders of magnitude. Column 11 in Table 7.3 shows the likelihood of the particular hazardous event after the SIS IPL is added. The mitigated event likelihood in Column 11 should be considered by the PHA team in determining if the risk meets the corporate guideline. Obviously, the availability targets for Integrity Level 3 safety interlock systems are extremely stringent and the design practices to achieve these high levels are extensive and costly.

7.4.2 "Risk-Based" Method Applied to Example

The PHA team did the safety interlock system integrity level selection during the project process hazard analysis activities. This risk-based method , as described in the previous section, is applied to the VCM reactor example. For

SAFETY INTERLOCK INTEGRITY LEVEL

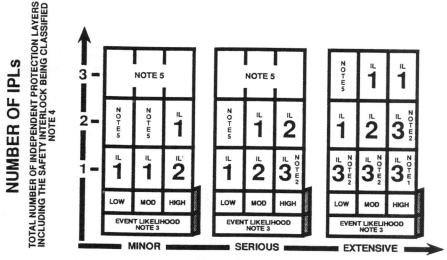

NOTES

1. One Level 3 Safety Interlock does not provide sufficient risk reduction at this risk level. Additional PHA modifications are required.
2. One Level 3 Safety Interlock may not provide sufficient risk reduction at this risk level. Additional PHA review is required.
3. Event Likelihood - Likelihood that the hazardous event occurs without any of the IPLs in service (i.e., the frequency of demand).
4. Event Likelihood and Total Number of IPLs are defined as part of the PHA team work.
5. SIS IPL is probably not needed.

Figure 7.3 Linkage of process risk to SIS integrity classifications.

the events listed in Table 7.2 the risk-based method was used to select SIS integrity level for several of the reactor SISs.

Table 7.4 documents this process and a step-by-step description follows.

Event 1: *Cooling Water Fails*
This upset initiates a runaway reaction that can catastrophically rupture the reactor, with impact judged to be extreme. Several failures in the control system could cause this upset, so an initiating event likelihood of "high" was estimated (10^{-1} per year). Protection per Table 7.2 was the Shortstop addition, conservatively evaluated at only 90% effective, and the pressure safety valves

Table 7.4 VCM Reactor Example: Risk-Based SIS Integrity Level

Note: Severity Level E—extensive; S—Severe: M—minor (see Fig. 2.4)

Likelihood values are events per year; other numerical values are probablilties. SS = Shortstop.

#	1	2	3	4	5		
	Impact Event	Severity Level	Initiating Cause	Initiation Likelihood	PROTECTION LAYERS		
					Process Design	BPCS	Alarms etc.
1	Reactor Rupture	E	Coolant H_2O control fails	10^{-1}	No	No	TAH/PAH/add SS 10^{-1}*
2	Reactor Rupture	E	Agitator motor drive fails	10^{-1}	No	No	Low amp alarm add SS & Burp 10^{-1}*
3	Reactor Rupture	E	Area wide loss of normal electrical power	10^{-1}	No	No	Alarm on UPS add SS & Burp 10^{-1}*
4	Reactor Rupture	E	Cooling water pump power failure	10^{-1}	No	No	Low CW flow alarm, add SS 10^{-2}**
5	Reactor Rupture	E	Double charge of initiator	10^{-2}	No	No	TAH/PAH/add SS 10^{-1}*
6	Reactor Damage Hydraulic Overpress.	S	Overfill reactor, control System Failure	10^{-1}	No	No	Level and weigh cell alarms 10^{-1}
7	Reactor Rupture	E	Temp. Control fails add excess steam	10^{-1}	No	No	TAH/PAH/add SS 10^{-1}*
8	Release of VCM	M	Agitator seal fails	10^{-1}	Spot ventilation at reactor seal 10^{-1}	No	No

*Shortstop addition effectiveness conservatively evaluated at 10^{-1}.
**Low cooling water flow alarm turn on steam turbine (10^{-1}) and Shortstop addition (10^{-1})

6	7	8	9	10	11	12
Additional Mitigation	Intermediate Event Likelihood	Number of IPLs	SIS Needed?	SIS Integrity Level	Mitigated Event Likelihood	Notes
PSVs 10^{-2}	10^{-4}	1 (PSVs)	yes	3 (Depress.) 10^{-3}	10^{-7}	Credit not taken for Emergency Cooling H_2O SIS
PSVs 10^{-2}	10^{-4}	1 (PSVs)	yes	3 (Depress) 10^{-3}	10^{-7}	
PSVs 10^{-2}	10^{-4}	1 (PSVs)	yes	3 (Depress) 10^{-3}	10^{-7}	
PSVs 10^{-2}	10^{-5}	1 (PSVs)	yes	3 (Depress.) 10^{-3}	10^{-8}	
PSVs 10^{-2}	10^{-5}	1 (PSVs)	yes	2 (Depress.) 10^{-2}	10^{-7}	Emergency cooling H_2O inadquate, cannot stop this upset.
PSVs 10^{-2}	10^{-4}	1 (PSVs)	yes	1 (Depress.) 10^{-2}	10^{-6}	
PSVs 10^{-2}	10^{-4}	1 (PSVs)	yes	3 (Depress. (10^{-1}) & emergency cooling water(10^{-1})) 10^{-4}	10^{-8}	Emergency CW SIS provides a conservative safety improvement and helps control the batch.
Low occupancy in reactor area 10^{-1}	10^{-3}	0	yes	2 (Depress.) 10^{-2}	10^{-5}	

(PSVs), 99% effective (10^{-2}). This led to an intermediate event likelihood of a 10^{-4} per year. Per the conservative assumptions used in this example, only the PSVs qualified as an IPL. The PHA team reviewed all the process safety risk issues and decided that an SIS was appropriate. Using Figure 7.3 for this event, which has extensive severity, "high" likelihood and one non-SIS IPL, an SIS integrity level of 3 is indicated for the depressure SIS. Credit was not taken for the emergency cooling water interlock because this event could have been initiated by a cooling water control valve mechanical failure. This mechanical failure would impair the availability of the cooling water interlock.

Event 2:   *Agitator Motor Drive Fails*
This upset initiates a similar runaway reaction as event #1, except that reactor "Burping" is required for stopping the runaway reaction by adding Shortstop. With no agitator, emergency full flow rate cooling water is insufficient to stop the runway reaction, so the depressure interlock function is the only effective SIS for this event, and it requires integrity level 3.

Event 3:   *Area Wide Loss of Normal Electrical Power*
While the upset has obvious differences to Event 2, the SIS integrity level selection comes to a similar result as Event 2.

Event 4:   *Cooling Water Pump Power Failure*
The upset in Event 4 is similar to the upset in Event 1, except for one significant difference. Operator intervention can stop this runaway by starting the steam turbine driven water pumps, or adding Shortstop. While this operator action was judged to be very effective, 0.99 availability, the PHA team decided additional interlock protection was appropriate. Figure 7.3 led to integrity level 3 for the depressure SIS, since the operator intervention was conservatively judged not to qualify as an IPL.

Event 5:   *Double Charge of Initiator*
This upset leads to a very energetic runaway reaction with a high rate of heat generation, even though the cooling water is operating. Both the PSVs, and the depressure SIS are designed to safely control this runaway, as is the Shortstop addition. Because of design and procedure safety features, this upset requires a very unlikely combination of failures. Therefore, initiating event likelihood of "moderate" was selected. The "moderate" likelihood, one non-SIS IPL, and extensive severity, led to an integrity level 2 SIS. Even though an integrity level 2 SIS is indicated, this SIS should be an integrity level 3. Integrity level 3 is needed because this depressure SIS is used for other higher risk runaway reaction events (e.g., Events 2 and 3).

Event 6: *Overfill Reactor Caused by Control System Failure*
The impact of this upset would be to hydraulically overpressure the reactor, causing a blown flange gasket or similar release. With the large numbers of batches per year, the likelihood is judged to be "high". The team judged the effectiveness of the BPCS level and weigh cell alarms to be 90% effective (10^{-1}). The PHA team judged this release to be moderate severity. The PHA team considered using high level and high weigh cell signals as a safety interlock. This would have a very bad effect on reliability and these interlocks were judged not to be necessary. The pressure safety valves and the depressure SIS will be effective in dealing with this upset. With a severity level of serious, a "high" likelihood initiating event, and one non-SIS IPL, Figure 7.3 indicated an integrity level 1 was appropriate for the depressure SIS. However, since this interlock has dual logic solvers, dual final elements, but only one sensor it achieves integrity level 2 availability.

Event 7: *Temperature Control Failure during Heatup Step—Overheats Batch*
This event leads to a runaway reaction similar to Event 1. The event impact and protection aspects are similar to Event 1, except the emergency cooling water is fully functional to stop the runaway reaction.
    Effective interlock functions for this upset are both the emergency cooling water and the depressure SIS. These interlock functions are identified in Column 10, Table 7.4 and an integrity level of 3 is indicated per Figure 7.3. The PHA team decided that the emergency full cooling water function should be included in the total SIS. Full flow rate cooling water interlock function is achieved by using the high reaction temperature and pressure inputs to trip a solenoid valve that fully opens the cooling water control valve. The combined depressure and cooling water SIS was conservatively estimated to have availabilities of 0.9999 (10^{-4}).

Event 8: *Agitator Seal Fails*
The special seal design used for this reactor restricts VCM releases to small flow rates if the seal fails. The spot ventilation provided will be sufficient to minimize this fire and explosion hazard. The PHA team judged the severity as moderate and decided that spot ventilation was 90% effective (10^{-1}). Low occupancy in the reactor area further reduces the chances of injury by a factor of 10 (10^{-1}). There are no IPLs, moderate severity, and high likelihood, so using Figure 7.3, an integrity level of 2 is appropriate.

7.4.3 SIS Integrity Level Selection as Part of a Hazop

The basic HAZOP methodology is briefly described in Section 3.1.1.6, and appropriate CCPS guidelines referenced in Chapter 3. These CCPS guidelines describe the implementation details and provide examples. Assuming the reader understands the HAZOP method, this section describes how SIS integrity level selection is included in HAZOP and PHA activities.

As previously discussed, the most efficient way to develop SIS function design proposals is in a meeting prior to a PHA meeting. An appropriate group of process, technical, and instrument experts can develop proposals for SIS functions, as summarized in Section 7.3.2 and Table 7.2.

Before starting the HAZOP, corporate and applicable regulatory process safety and risk reduction guidelines are reviewed by the PHA team to help assure consistent risk related decisions during the HAZOP. SIS integrity level decisions are risk reduction decisions that need to be considered in conjunction with other risk reduction process safety decisions. With this focus for consensus on risk reduction guidance, the PHA team starts into the HAZOP.

During the HAZOP, SIS functions are encountered in two HAZOP activities. When identifying protective items for deviation events, and when developing the "Design Intention" for lines and equipment that include SISs. In developing and documenting "Design Intentions," the SIS function is briefly described, and the integrity level is not addressed.

During the HAZOP, SIS integrity level selection occurs as part of the risk considerations for specific deviation events. Integrity level selection for an SIS takes place when the following steps are completed for an event involving an SIS:

1. A specific deviation event has been identified.
2. The severity of the hazardous impact has been estimated.
3. Applicable protection and mitigation layers have been identified and their effectiveness evaluated, including the SIS function effectiveness (i.e. if it functions correctly, will it always prevent the hazardous impact).
4. Risk is evaluated by considering the likelihood of the deviation event occurring, the effectiveness of protection and mitigation, and the hazardous impact severity.
5. The risk evaluation is compared with applicable guidelines.

At this point, an SIS integrity level can be proposed, such that the risk evaluation will be consistent with the applicable guidelines. Now, an overall consideration of risk reduction methods takes place. First, risk reduction improvements for other protection and mitigation layers need to be considered as effective ways to comply with risk guidelines. After the PHA team has considered all the options and different ways to meet the risk guidelines, including SIS integrity level, the integrity level selection is made. This selection decision should involve consensus of the PHA team that the integrity level, along with the other protection and mitigation layers, will appropriately address process risk.

In many cases, a given SIS will be addressed as protection for several deviation events in a HAZOP of a process. Some of these deviation events may involve low risk, where a low SIS integrity level selection is appropriate. Other deviation events for this SIS may need a higher SIS integrity level for compliance with risk guidelines. Integrity level for this SIS should be selected to meet the highest risk, or highest integrity level, indicated by the HAZOP. Additionally, before the PHA team considers the HAZOP complete, they should review all the deviation events that put a demand on an SIS. If any of the SISs have a large number of demands from many deviation events, the PHA team may consider a higher integrity level SIS selection, and/or change that SIS design to improve its availability, consistent with the higher risk that results from a higher demand frequency.

7.4.4 HAZOP Method Applied to the Example

The starting point for this SIS is similar to the Risk-Based method, and the necessary process information includes:

- Reasonably complete process and instrumentation drawings (P&IDs).
- Initial or proposed interlock strategy.
- Basic process information.

This information was utilized to do a HAZOP for the VCM reactor. This section will illustrate how SIS integrity levels can be selected during this HAZOP. For purposes of illustrating this SIS integrity level selection method, only a portion of the HAZOP activity and documentation will be described, and the example is limited to only the upset events described in Table 7.2. This approach is similar to that used in illustrating the risk-based method.

In the following description, each of the events in Table 7.2 is addressed in the order shown in the table. Event 1 is shown below as documented in the HAZOP meeting. This cooling water control system failure was identified as the deviation upset for the key word "NONE" and the process deviation event is defined as "no flow of cooling water." The deviation event is defined as "cooling water control fails" resulting in an impact event of "runaway reaction leads to reactor rupture." The severity is shown (i.e., extensive) and mitigation and protection techniques are listed. The HAZOP team defines action taken (i.e. IL3 SIS) and describes how the decision was reached.

Event #	Guide Word	Deviation	Deviation Event	Impact Event	Severity	Protection, Mitigation	Action
1	None	No flow of cooling water (CW)	CW control fails	Runaway reaction leads to reactor rupture	E	• Shortstop • PSVs • Emergency cooling water • Depressure (SIS)	IL-3 SIS

Reason for Event #1 IL-3 SIS Decision:
The PHA team evaluated applicable protection and mitigation approaches. A brief review of these approaches follows:

- Provide the operator with the ability to open the Shortstop addition valve from the control room. This should be very effective since there is sufficient time between high reaction temperature and /or pressure alarm indication to reactor runaway to allow the operator to make this decision.
- While the PHA team feels the PSVs qualify as an IPL, there is potential difficulty with polymer pluggage, which could limit their protection effectiveness.
- The emergency cooling water would provide protection if the portion of cooling water control that failed had no effect on the cooling water valve operation. However if the cooling water valve could not go full open, this interlock would be inadequate.

The extensive consequence and a high likelihood initiating event led to the need for extremely effective protection. Although the PHA team thought an integrity level 2 for depressure SIS might be sufficient, they selected integrity level 3 for the depressure SIS to meet their safety guidelines. These guidelines encouraged risk reduction for continuous improvement of safety performance.

Event #	Guide Word	Deviation	Deviation Event	Impact Event	Severity	Protection, Mitigation	Action
2	None	No agitation	Agitator drive has local power failure	Runaway reaction leads to reactor rupture	E	• Shortstop • PSVs • Depressure (SIS)	IL-3 SIS

Reason for Event #2 IL-3 SIS Decision:
This runaway reaction is similar to event 1, and reinforced the PHA team judgment that integrity level 3 was appropriate for the depressure SIS.

Event #	Guide Word	Deviation	Deviation Event	Impact Event	Severity	Protection, Mitigation	Action
3	None	No agitation	Areawide loss of electrical power	Runaway reaction leads to reactor rupture	E	• Shortstop • PSVs • Depressure (SIS) (Fail-safe design)	IL-3 SIS

Reason for Event #3 IL-3 SIS Decision:
Even though this area wide power failure has obvious differences from event
2 (i.e. in event 2 the depressure SIS logic solver must be operational to perform
its function, while in event 3 the depressure SIS operation is based on fail-safe
design), the integrity level 3 depressure SIS is appropriate.

Event #	Guide Word	Deviation	Deviation Event	Impact Event	Severity	Protection, Mitigation	Action
4	None	No cooling water (CW)	Local power failure to CW pumps	Runaway reaction leads to reactor rupture	E	• Steam turbine drive for CW pumps • Shortstop • PSVs • Depressure (SIS)	IL-3 SIS

Reason for Event #4 IL-3 SIS Decision:
Although the operator can stop this runaway by Shortstop addition or turning
on the steam turbine drive, the integrity level 3 SIS is appropriate.

Event #	Guide Word	Deviation	Deviation Event	Impact Event	Severity	Protection, Mitigation	Action
5	More	High initiator concentration	Double charge of initiator	Runaway reaction leads to reactor rupture	E	• Shortstop • PSVs • Depressure (SIS)	IL-3 SIS

Reason for Event #5 IL-3 SIS Decision:
This upset leads to a very energetic runaway reaction with a high rate of heat
generation, even though the cooling water is operating. Because of design and
procedure safety features, a double initiator charge requires an unlikely
combination of failures. As a result the PHA team considered likelihood of
this event as moderate. But, the extensive severity level of this event results in
the selection of an integrity level 3 depressure SIS.

Event #	Guide Word	Deviation	Deviation Event	Impact Event	Severity	Protection, Mitigation	Action
6	More	High level	Control system failure	Reactor damage, hydraulic overpressure	S	• Level & weigh cell (BPCS) • PSVs • Depressure (SIS)	IL-2 SIS

Reason for Event #6 IL-2 SIS Decision:
The impact of this upset would be to hydraulically overpressure the reactor, causing a blown flange gasket or similar release. With the large numbers of batches per year, the likelihood is judged to be "high." The team judged the effectiveness of the BPCS level and weigh cell alarms to be reasonably good.

The PHA team judged this release to be moderate severity. The PHA team considered using high level and high weigh cell signals as a safety interlock. This would have a very bad effect on reliability, and these interlocks were judged not to be necessary. The pressure safety valves and the depressure SIS will be effective in dealing with this upset. Even though this event invloved only serious severity, the PHA team selected an integrity level 2 for the depressure SIS.

Event #	Guide Word	Deviation	Deviation Event	Impact Event	Severity	Protection, Mitigation	Action
7	More	High temperature	Steam valve fails during heatup step	Reactor damage, hydraulic overpressure	E	• Shortstop • PSVs • Emergency cooling water • Depressure (SIS)	IL-3 SIS

Reason for Event #7 IL-3 SIS Decision:
This event leads to a runaway reaction similar to Event 1. The event impact and protection aspects are similar to Event 1, except the emergency cooling water is fully functional to stop the runaway reaction. The team integrated the depressure and emergency cooling water protection schemes into an integrity level 3 depressure SIS to address this extensive severity high-likelihood event.

Event #	Guide Word	Deviation	Deviation Event	Impact Event	Severity	Protection, Mitigation	Action
8	Other than	Flow other than to the reactor	Reactor agitator seal fails	Release of flammable VCM into the reactor area	M	• Seal design restricts flow release • Spot ventilation • Low occupancy • Seal pressure initiates depressure (SIS) • Seal pressure gauge monitor	IL-2 SIS

Preliminary Reason for Event #8 IL-2 SIS Decision:
The special seal design used for this reactor restricts VCM releases to small flow rates if the seal fails. The spot ventilation provided will be sufficient to mimize this fire and explosion hazard. Even though this event is considered to be of minor severity, the team selected an integrity level 2 for the depressure SIS as appropriate.

7.5 DESIGN OF THE BPCS

The design considerations for a BPCS are presented in detail in Chapter 4. This section gives a general overview of the BPCS design for the VCM polymerization example.

7.5.1 Control System Functional Specification

Technology selection. A PES is chosen as the BPCS for several reasons, including:

- The facility is large and has multiple large reactors.
- The control room is remote from the actual reactor area.
- Critical batch valves (on–off) will have valve position feedback switches.
- Operational capability from the control room will enhance operator safety by limiting the time for personnel in the reactor area.
- Electronic inputs to a supervisory control computer will be useful in batch recipe and sequencing control.
- A process computer will have a model of operating parameters to compare against sensor outputs for process performance.

Operating Stations. The operators in the control room will have access to several BPCS console screens which allow display of process status, control variables, alarm status trending, etc. At each reactor in the operating area, a local panel will display readings from field mounted instruments and other process information. Critical sensors are supplemented with local-readout, diverse instrumentation.

Installation Considerations. All sensors will be direct-wired with 4–20 mA analog signals, with dc low-level signals, or with 120 V discrete signals. The PHA team has recommended that the reactor room be designed to Class 1, Division 1 standards—a conservative choice since a flammable atmosphere does not normally exist in the enclosed reactor area. All low level signal wiring will be twisted pairs, shielded in conduit with appropriately sealed fittings in accordance with the requirements of the electrical hazard classification. Termination boxes will be properly shielded. Power and low signal wiring will be separated. Grounding will be properly designed. In addition, a reliable, process-wide communication system will be provided so that an operator in the reactor area will always be in good contact with the control room.

7.5.2 Sensor Selections

The control variables have been identified as part of the process design and their ranges have been specified, along with criteria for alarm limits. The actual choice of appropriate sensors requires knowledge of ranges, operating environments, available devices, and requirements for reliability and availability.

Reactor Level. We choose a noncontact nuclear level instrument, since it can be mounted external to the reactor. This instrument is used primarily for monitoring reactor fill and emptying rates. An internal instrument would be less desirable since it would have to operate inside an agitated fluid mass and withstand periodic very high pressure water washes of the vessel. Eliminating a vessel penetration is also desirable since the material being handled is a carcinogen.

A diverse technology is chosen for the high level sensor—a probe which can detect the level transition from gas to liquid.

Temperatures. The main batch temperature will be measured by metal sheathed matched RTDs inside a bayonet-type well that is welded at its penetrations through the water jacket. A seal fitting is provided on the well to facilitate replacement of the instrument. The well also is fitted with a pressure indicator to detect leakage through the well. A pressure indication will warn of leakage to protect workers performing on-line instrument maintenance. Because of the well, this sensor will have a long lag time, but because of the relatively slow temperature transients, it will have adequate response for its use in controlling water jacket flow.

Three other temperature sensors are provided on the vessel. For diversity and ruggedness, bimetallic local indicators, installed in wells, are chosen.

Thermocouple temperature sensors are used on the water circuits.

Pressure. The primary pressure measurement is from a pressure transmitter with a diaphragm seal. The secondary sensor is a diaphragm-sealed Bourdon-type gauge.

Flow. For charging VCM, two turbine meters are selected for installation in series. These devices have the needed accuracy and rangeability, and outputs can be readily integrated to give total volumetric flow. They also are easily proven. Diversity is sacrificed because there is not a satisfactory alternative sensor. (The VCM pressure varies with ambient temperature and is near saturation conditions. Other types of flow meters are likely to have larger local pressure losses which may cause cavitation or flashing or do not have adequate accuracy).

The flow measurements are backed up by instruments monitoring VCM charge tank level as well as by weigh cells on the reactor vessel.

On the charge water, orifice meters are chosen since the supply pressure is constant and there is no problem with flashing. Orifice meters are also used on the cooling water circuit.

Weight. Load cells are provided for each reactor vessel to provide an indirect measure on the level status and to give a rough check on fill rate.

Agitator Motor Current. A current sensor is provided for each reactor vessel to indicate the presence of agitation.

7.5.3. Final Elements

Valves are selected based, primarily, on allowing minimal leakage to the environment and, secondarily, on avoiding polymer buildup. Each line to the reactor therefore contains two air-operated valves in series. For the inlet and outlet lines, metal-seated, tight shut-off, butterfly valves or full port ball valves are selected. These valves are selected to be full line size and are designed to minimize polymer buildup. Each valve is also fitted with position switches. Therefore, two correct commands to the valves with status verification feedback are required to open or close a line to the reactor.

Modulating control valves are chosen for the water system to satisfy rangeability and control characteristic requirements.

The agitator requires high reliability so a direct-drive configuration is selected. Speed can be varied using variable frequency drives. The seal pressure is monitored and alarmed to provide early detection of any leakage. The current draw of the motor is monitored along with the status of electrical power to the motor. High current draw is an indicator of problems with the batch—lumps or imbalance. The electric power system design incorporates dual feeders with automatic transfer in the event of a power interruption. The motor has remote restart capability.

7.5.4 Controller Design

The BPCS consists of a distributed control system (DCS) and a model-based supervisory control computer. Because process transients are relatively slow, normal DCS scan times are satisfactory for the application. The process supervisory control computer monitors process states against a recipe-specific model and alarms abnormal status. In the event of computer malfunction, process control reverts to VDU-based operator control for switching of valves for batch steps and PID operation for regulatory control.

For most situations, it is usually best to buy a well-tested, "user approved" DCS system complete with an integrated supervisory control computer from a highly qualified vendor. Another option is to select a DCS and a separate supervisory control computer and take on the responsibility for integrating them. In either case, a functional specification is prepared to give the vendor all the critical design features that are needed—for example, redundant data highways, power supplies and I/Os, as well as a watchdog timer (WDT) or an equivalent checking capability. For this example, it was assumed that the in-house applications group would develop the supervisory software. They will design the screens, develop the program applications model, design the diagnostics, select the alarms, determine the event logging information sequences, constraint control and the alarm management system.

They will also develop a batch recipe control system, with interlocks from the diagnostics, and recipe-specific process models for use in process optimization. In addition, the in-house instrumentation/control specialists will also do configuration programming for the PES.

Issues of security of the system need to be addressed, design verification plans developed, and preliminary operating, testing, and maintenance procedures outlined. An important aspect of the BPCS design process is the careful documentation of each element in the design and evaluation of system integrity. Careful and clear assignment, configuration, and documentation of BPCS application software is particularly important.

7.6 RISK ASSESSMENT AND CONTROL

Once the basic design, equipment, operational and software functions are developed, risk assessment and evaluation techniques are employed to evaluate the integrity of the design. The design should meet all the applicable codes and standards adopted or used by regulatory groups having jurisdiction as well as other requirements set by the facility owner. The next evaluation step is to conduct a detailed design PHA utilizing a systematic technique such as FMEA (Section 3.1.1.7) or HAZOP (Section 3.1.1.6). In these techniques, the P&IDs and other process design information are used to check the failure modes of each process component to identify any potential for unsafe conditions.

Table 7.5 is an example of a HAZOP sheet for part of this system. The output of the HAZOP (or FMEA) is a list of action items that need to be resolved before the design is finalized. Administrative project procedures are needed to assure that all the items identified in the HAZOP are successfully resolved and implemented. Further, as design changes occur, it is important that they be reviewed carefully with respect to their impact on the process safety and results of the HAZOP or FMEA.

The automated control system may require some additional validation using techniques to ensure that independent systems are sufficiently separate and diverse. Fail safe features and maintainability of equipment and software need verification. Diagnostic systems need careful review to ensure that any actions they initiate leave the system in a safe status.

Table 7.5 Sample HAZOP Sheet from PHA Team Review

Line/Vessel: R-1 Reactor (reaction mode)   Date: 0/0/00   Page: 123

Scenario #	Guide word	Deviation	Cause	Consequences	S*	P**	Protection	Action
r-145	More	Pressure	Loss of coolant	Vent VCM; Fail R-1	E	M	PSVs PAH, TAH Emergency vent Shortstop	OK
r-146	More	High reaction rate	Excess initiator	Runaway reaction Fail R-1	E	M	PAH, TAH Shortstop Charge drums Emergency vent PSVs	OK
r-147	More	High level	Overfill	Fail R-1 seals	M	L	Charge drums PAH Weigh cell Flowmeters Emergency vent LAH, PSVs	OK
r-148	More	Temperature	Temperature control failure	Runaway reaction Fail R-1	E	M	TAH Shortstop Emergency vent PSVs	Oper. actions add SIS

*S—severity (Extensive, Serious, Minor)   Team members: PHA Team—XXXX
**P—probability (High, Moderate, Low)

Particular attention needs to be paid to the role of human factors in the operation (Section 3.1.3). This is because the process is normally operated remotely, because it requires sequential operations, and because human error has played a major role in the past major release accidents.

Following a HAZOP or FMEA or equivalent review, the safety features listed in Table 7.2 can be evaluated using methods similar to those presented in Section 7.4: This evaluation results in the specification of the SIS requirements.

Sometimes, however, designers proceed with the design of a SIS based on their overall knowledge of the process and design philosophy. This is done with the knowledge that system design integrity will be validated in full after the SIS design is completed.

7.7 DESIGN AND VALIDATION OF THE SIS

Chapter 5 presents the elements in the design of an SIS in considerable detail. This section presents an overview of the SIS design process for our VCM polymerization example.

The following list describes some of the safety interlocks identified in Section 7.4. They are:

SIS DESIGN			
ITEM	SENSORS	FINAL ELEMENTS	IL
A	High–high Pressure High–high Temperature	Depressure	3
B	Low amps	Depressure	2
C	Low cooling water flow	Full cooling water flow	1
D	High–high pressure from overfilling	Depressure	2
E	High seal presssure	Depressure	2
F	Shortstop low level and low presure	No VCM	1
G	Manual P.B.	Shortstop valve	—

Based on the small number of interlocks and the low complexity, most designers would normally select electromechanical relays for the interlocks. However, to illustrate the principles of Chapter 5, PES and electromechanical technology is selected for this example. See Figure 7.4 for a block diagram of the system developed.

The SIS controller(s) logic is developed to meet integrity level 3. Once the highest integrity level interlock is developed, the design of lower integrity level interlocks can be incorporated into the same logic-solver system. When combining integrity level 2 and 1 interlocks with integrity level 3, all shared interlock components (e.g., the PES controllers and block valves) must be treated as integrity level 3 interlocks. A direct-wired, operator-initiated manual shutdown (interlock G, from push button stations) is also required as an independent protection layer and is provided using an electromechanical relay (R1).

Figure 7.4 illustrates how the eight safety interlocks are integrated with the BPCS. The SIS sensors for interlocks "A" are user approved safety (UAS), separate, and diverse. To meet integrity level 3 availability, two sensors and PESs are required and are connected to two final elements, which are provided with valve-position feedback switches (not shown in Figure 7.4). Redundancy, to improve reliability, is shown for SIS PESs and the BPCS, but may not be

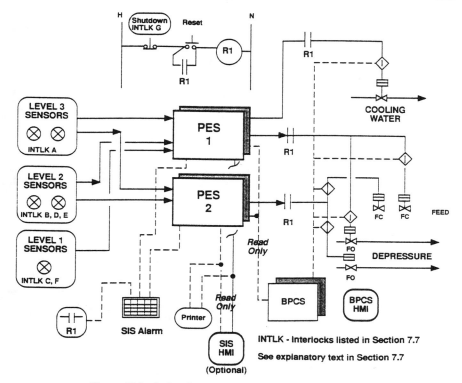

Figure 7.4 Safety interlock system for the VCM reactor

required for all applications (i.e., batch). Note that Figure 7.4 shows two SIS HMIs. Most applications require the SIS alarm and printer. The SIS HMI is optional.

The SIS is designed with batch sequencing interlocks that prevent human errors in selecting the wrong position for critical valves, based on system status indications (reactor charged or empty) and valve and vessel access cover limit switch indications. Information interchange between the BPCS and SIS is required. Accordingly, a "read only" communication link is provided between the systems.

In functional terms, the SIS alarms and causes automatic depressurization of the reactor, which drops reactor pressure and temperature, slowing the reaction. The operator may elect to manually activate addition of Shortstop to kill the reaction.

The SIS also includes inputs from area gas and fire detectors. Upon detection, an alarm is sounded and recorded. Reactor inlet valves are shut and the reactors are automatically depressurized. The operator may take additional actions like the addition of Shortstop if necessary.

The manual addition of Shortstop meets safety requirements because of the relatively long response time available for action. It also prevents the killing of good product batch due to an upset that can be controlled by limited depressurization.

The SIS logic and key parameters are mirrored in the BPCS to give the operator additional information. As discussed in Chapter 4, a hierarchical control system design is required for this batch operation.

Subtle common mode failure possibilities are often present in SISs even when independent and diverse system elements are selected. Consequently, a separate review of all control-related protection layers is suggested after the SIS design is complete. The review is conducted systematically, starting with the highest integrity level interlocks.

After the SIS design is completed, a documentation package is prepared for each interlock, including complete descriptions of system components, manufacturers, model numbers, software versions, programmers and clear descriptions of application programs and configurations, diagnostics, testing procedures, maintenance, etc. This information is reviewed by the PHA team to ensure that the actual design requirements have been met and that the SIS has appropriate availability, separation, diversity, fail-safe performance, diagnostics, testing procedures, and other important attributes.

7.8 INSTALLATION AND TESTING

This section deals with the factors to consider when installing a PES-based SIS–PES systems are vulnerable to many factors, including poor grounding, static electricity, voltage variations, lightning, airborne particulates, electrical noise (radiated and common mode), and environmental conditions (heat, humidity, etc.). Chapters 4, 5, and 6 address these issues. The bottom line is that the installation of a BPCS and SIS should be of the highest caliber with clear documentation updated to reflect actual operating conditions.

The installation instructions provided by the BPCS and SIS vendors, need to be understood, and integrated with local and national codes utilized in the design of the installation. Any deviations should be defined and reviewed for acceptance with the BPCS or SIS vendors.

This same modification, review, and acceptance structure will also be followed with the application software development. The application software will be simple, well-documented, partitioned into subsystems compatible with the process subsystems, and use generic logic functions clearly understood by users—KEEP IT SIMPLE. The SIS will not include application programming that requires frequent maintenance (HMI graphics, etc.). The SIS will have, as a separate subsystem, any application programming related to "read-only" communication links to the BPCS. A system with memory mapping is being used, so space will be provided between each subsystem to allow for expansion/growth and to avoid inadvertent changes to the wrong subsystem. (*Note:* Should memory mapping not be available, consider using "dummy code" to achieve the same objective, but caution would be required to ensure that response time is not negatively impacted by this method.)

Application programming strategy will be carefully documented so that future maintenance/changes can be performed in a safe, secure manner.

Each application module used in the application software must be tested and documented prior to commissioning the program. Substitutes, such as use of counting for a timing function, are avoided. The application program subsystems are tested using simulation techniques.

After completing this phase, the subsystems are tied together in a logical method, proving the full system after each new subsystem is added.

Plans are developed for maintaining the application programming software integrity throughout the lifetime of the process. These are described in Chapters 4, 5, and 6. Therefore, it is particularly important that, throughout the design process, specific catalog numbers and revision levels of all aspects of the SIS hardware and software are documented. Any future changes in catalogue numbers and revision levels require detailed analysis to understand the impact of the change on the previous system operation and integrity.

Concurrent with the development and testing of the software, the UAS hardware is procured. This hardware is assembled utilizing mature, proven installation techniques, partitioned in a way that facilitates planned maintenance and operations, and documented "as-built," with provision for upgrading and reverification with each future change.

The SIS controller cabinets will be tested individually, then retested during each integration step. The installation will proceed through a planned, documented procedure that:

- Defines each step of a checkout procedure to assure that the installation is correct.
- Defines each step of a run-in procedure to assure the BPCS and SIS perform as designed and to detect/correct infant mortalities. (*Note:* No hazardous materials are used in this testing phase.)
- Outlines a planned, incremental, start-up procedure that will be supervised by the operating group (see Chapter 6).

An important key to a successful installation and testing phase is the involvement of plant operations and maintenance personnel to assure a clear understanding of all aspects of the process, the BPCS, and the SIS.

7.9 ADMINISTRATIVE PROCEDURES TO MAINTAIN INTEGRITY

In this example, all procedures that may be necessary to maintain the integrity of the control/protection systems are not detailed, but some of the key ones that must be implemented are outlined.

7.9.1 BPCS Related Procedures

It may be necessary to provide for conducting a Factory Acceptance Test (FAT) of the BPCS equipment by including this requirement in the purchase order for the system. The procedure outlined in Appendix H, can be used as a guide for this testing. The written test procedure is supplied to both the vendor of the BPCS equipment and the team performing the test prior to the time the FAT will be performed.

A procedure for validating the BPCS control logic is necessary. This should validate that implemented control algorithms perform as required including the sequence control of the batch reaction. The checklist format of this written procedure should allow for checking off each test as it is performed. The "green light" path through each module of code should be field verified, as well as each "failure and retry" path.

Standard Operating Procedures (SOPs) will be provided to the operators, describing the displays that should be used for performing each phase of the process operation, how control setpoints should be entered and changed, the documentation for special function control loops related to changing from heating to cooling of the reactor vessel, the limits of pressure and temperature ranges during the reaction, and what actions the operator should take when there is an indication of a problem with the control system.

Operator confirmation that he/she is ready to continue from one step to the next is required before automatic controls may proceed.

Emergency procedures are also provided which describe the operator's actions when safety critical alarm conditions occur (as well as other emergency conditions, such as bomb threats, emergency site shutdown and evacuation, etc.)

Another written procedure requires that the following steps be taken before any BPCS vendor updated system software is placed in service, to ensure that changes do not compromise control logic:

- Review basis for upgrade decision (old bugs corrected, functionality, etc.).
- Make and have immediately available back-up copies of current operating system and application software.
- Process operations are stopped during the validation period.
- All displays are validated as remaining correct.

- Control and sequencing logic functions are tested against design criteria (including switching from heating to cooling during startup).
- Limit controls on temperature and pressure remain in place.
- All changes in operation with new systems are communicated to operations personnel and their understanding of the changes is verified.

A formal system for making changes to the BPCS must be established, including procedures for requests, reviews, approvals and documentation. Master copies of all system configurations will be documented both in hard copy and soft media and placed in a secure location (plant's fireproof vault) for use in the event of a total system failure. A scheduled updating of these records should take place quarterly. Only the last three and the most recent versions will be maintained in the files.

7.9.2 SIS Related Procedures

Plant policy states that no SIS components may be bypassed or otherwise defeated during the time the reaction is proceeding to completion. No alarms pertaining to the SIS may be bypassed at any time during operation of the process.

Any calibration or maintenance work on SIS components must be done when the process is not in operation.

A written procedure states that, in the case of any abnormal events during operation of the SIS, the operator should initiate a manual shutdown of the reaction.

A policy states that no changes to SIS logic, initiating-variable setpoints, or final control elements are allowed without prior approval of the facility process safety management committee. All proposed changes shall receive an analysis using a formal method such as a HAZOP or FMEA.

A security procedure requires that access to the logic in the SIS be limited by a detached programming tool, password(s), and designated personnel who have appropriate knowledge of the systems' operations and functions. All maintenance activities related to the SIS shall be documented by noting the initiating problem observed, the identified cause, and the implemented solution. This documentation shall also identify the person(s) performing the work. Policy states that the same workstation shall not be used for programming or configuring both the BPCS and the SIS.

A procedure states that the SIS shall not be connected to any external system via a modem for the purpose of diagnostics.

A functional test shall be performed on the SIS prior to placing it in service for the first time and at intervals of no greater than the scheduled turn-arounds or annually, which ever is longer. The system test shall validate the following as a minimum:

- The operation and range of all input devices, including primary sensors and SIS input modules (including the field sensor and connecting wiring).
- The logic operation associated with each input device.
- The logic associated with combined inputs where appropriate.
- The trip initiating values (set points) of all inputs or the contact position of all switch inputs.
- The alarm functions that may be included.
- The operating sequence of the logic program.
- The function of all outputs to final control elements.
- The correct action of the final control elements.
- The first-out alarms.
- Any variable or output status indications that might be provided for operator monitoring (e.g., printed messages).
- The manual trip provided for bypassing the SIS logic program works to bring the system to its "fail-safe" condition.
- The software version in the processor is the correct one.
- The system action on loss of power (either instrument or utility) is correct.

7.9.3 Other Required Procedures

Training of operations, maintenance, and other support personnel on the function of both the BPCS and SIS shall be performed prior to placing the systems in operation and shall be updated at any time any changes are made to either system.

Documentation of the current control and safety logic implemented in both the BPCS and the SIS shall be maintained at all times. Any changes shall be documented at the time they are implemented. Hard copies of the documentation, fully describing the systems and their functions, shall be maintained in the control room for reference. The documentation of the logic performed by the SIS shall also describe the reasons for its actions.

An audit program shall require an examination of system documentation and any actions taken at the end of each operating year. A report shall be issued describing the results of the audit and any recommendations from the audit shall be flagged for follow-up (at quarterly intervals) until they are completed. The audit shall include:

- Review of all changes made since the last review and verification of correct documentation status.
- Review of all problems with equipment or logic associated with the SIS since the last review to ascertain if potential problems are developing that might degrade the system's function in the future.
- Functional check of the system operation during annual, or other turn-arounds.

- Verification that all official copies of documentation are in agreement.
- Review of operating personnel's understanding of the system's function and operation.
- Review of any proposed changes for compliance with the design intent of the system.

Emergency procedures are developed in conjunction with community local emergency planning committees. These procedures include periodic safety drills and also include requirements for periodic testing and maintenance of safety equipment.

7.10 APPROACH FOR AUTOMATION OF AN EXISTING FACILITY

The automation of an existing facility usually involves the upgrading of the existing BPCS and SIS or the complete upgrading of the BPCS or SIS technology (e.g. changing a pneumatic system to a PES-based system).

The first step in altering a control system for a chemical process that involves hazardous materials is to reestablish a Process Hazards Analysis (PHA) Team for analyzing the safety implications that the revised or new BPCS will have on the process. The analysis will be handled just like that for a new process, utilizing similar techniques to those already outlined in this chapter. Obviously, the experience gained in prior reviews and in operating the process has the potential for making this task easier. However, familiarity may lead to overlooking some key interactions. Thus, we make sure that the PHA team includes some participants who can ask the "what-if" questions without being influenced by knowledge of past performance.

If records are current, they provide a sound basis for evaluating the interactions between existing systems and the new systems involved in the upgrade. On an older plant, where records may be deficient or outdated, an essential first step is to bring the documentation on existing equipment and systems that will be retained up to date and to the level of detail described in this book.

A good rule of thumb is to plan for upgrading an existing system following the same procedures and methods that are necessary for the analysis and integrity verification of a totally new process system.

7.11 REFERENCES

7.1 CCPS/AIChE. *Guidelines for Hazard Evaluation Procedures, 2nd edition with worked examples.* New York: American Institute of Chemical Engineers, 1992.

7.2 CCPS/AIChE. *Guidelines for Chemical Process Quantitative Risk Analysis.* New York: American Institute of Chemical Engineers, 1989.

7.3 CCPS/AIChE. *Guidelines for Safe Storage and Handling of High Toxic Hazard Materials.* New York: American Institute of Chemical Engineers, 1988.

7.4 CCPS/AIChE. *Guidelines for Vapor Release Mitigation.* New York: American Institute of Chemical Engineers, 1988.

7.5 CCPS/AIChE. *Guidelines for the Technical Management of Chemical Process Safety.* New York: American Institute of Chemical Engineers, 1989.

7.6 Manufacturing Chemists Association. *Index of Case Histories of Accidents in the Chemical Industry.* Washington, D.C., 1951–1973.

7.7 Barkelew, C. H. Stability of chemical reactors. *Chemical Engineering Progress Symposium Series,* No. 25, Vol. 55 (1959).

8

THE PATH FORWARD . . . TO MORE AUTOMATED, SAFE CHEMICAL PLANTS

8.0 INTRODUCTION

Safe automation is achieved through a lengthy and structured work process. As discussed in this book, this safe automation process

- begins with the definition of the chemical process and its inherent hazards;
- requires the definition of a basic process control system (BPCS) for regulating the heat and material flows in the process;
- often includes specifying separate safety interlock systems (SIS) as part of the risk control measures defined by the process hazards analysis team;
- often makes use of programmable electronic system (PES) technology for both BPCS and SIS control functions; and
- ends with an operating plant where periodic safety reviews and sound management practices ensure continuing safety.

The importance and performance of each safety interlock in the typical chemical control system should be matched to the level of process risk assigned to the SIS for risk mitigation. A risk-based SIS design is achieved through the use of company-specific criteria that include detailed requirements for safety interlocks of differing integrity levels.

This book describes design criteria for three SIS integrity levels and a methodology for assigning the three-level design criteria to process hazards of varying likelihood and severity. In-plant procedures to manage and control changes to the BPCS and SIS, maintenance of control system documentation, and periodic tests to check performance of the installed SIS are essential to sustaining the level of protection afforded by safety systems designed according to these criteria.

These same fundamental design and operating procedures will apply to control systems through the 1990s. However, their use will become greater and of greater importance as the need for automation increases and as opportunities for applying new, more practical, effective and reliable automation systems develop through technological advances. The factors driving increased chemical plant automation include:

- governmental regulation and industry initiatives which seek better control of chemical processes to protect employees, the public and environment;
- management expectations and changing workplace factors, which necessitate better and sustained control of the process through automation; and
- new instrumentation and automation technology which create opportunities for more reliable and available Basic and Safety control systems.

Continued use of the procedures in these guidelines coupled with sound process safety management systems can provide safer, more efficient chemical and petrochemical operations into the twenty-first century.

8.1 GOVERNMENTAL REGULATIONS AND INDUSTRY INITIATIVES

Tenets of the environmental movements of the 1970s and 1980s have become part of mainstream America's beliefs. As a result of the trend in public attitudes toward greater environmental protection and increased safety to employees and plant neighbors, federal and state governments have enacted new regulations which address these concerns; and, industry groups have developed their own programs which are responsive to the public's concerns.

The federal law, SARA Title III enacted in 1986, aims at greater management control over a plant's operation through the requirement that public and local emergency response authorities be informed of the hazards posed by local plants to the safety of its employees, neighbors and the environment. Clean Air Act Amendments (CAAA) of 1990 seek better control of processes and reduction of chemical emissions through significantly increased monitoring and reporting of chemical releases to the environment. Limits on chemical releases were also established by this act.

As required by the 1990 CAAA, the Environmental Protection Agency is developing regulations that will require plants with amounts of hazardous chemicals above specified in-plant quantities to register and then develop and implement risk management plans to control and contain these chemicals. The EPA regulations, when developed, are expected to be similar to those contained in the Occupational Safety and Health Administration's 1992 rule on Process Safety Management of Highly Hazardous Materials (Ref. 8.2). The specific intent of the OSHA rule is to provide better protection for employees by requiring that plants with chemicals and hydrocarbons must develop programs to prevent and mitigate hazardous releases of the materials. California, New Jersey, Delaware and Nevada also have laws in place which seek to curtail hazardous emissions of chemicals. These rules and laws are the aftermath of several catastrophic chemicals and hydrocarbon accidents here and abroad, which caused multiple deaths and severe loss of property.

The U.S. chemical and petroleum industries have developed a number of initiatives in anticipation of and in response to these health and environmental concerns. Several years ago, the 200 member companies of the Chemical Manufacturing Association undertook the Community Action and Emergency Response (CAER) program to ensure better cooperation between plants and the local communities in a chemical emergency. More recently, CMA members adopted a Responsible Care program which defines the chemical plant management practices that members have pledged to follow when dealing with hazardous chemicals. In addition, the American Petroleum Institute has developed its Recommended Practice 750 which also addresses how its members should handle hazardous materials. Industry members have also participated in the activities of the AIChE/CCPS which, among other initiatives, has published a series of chemical process safety guideline books, including this one.

Although governmental regulations (which are written by separate agencies with different goals) differ in many respects from industry programs, they all focus on eliminating hazardous releases of chemicals. Provisions of these regulations and programs necessitate better control and design of processes, improved operating practices, and more effective management systems—all of which in some manner increase the need for automation and instrumentation in the foreseeable future. Consequently, work to reduce both chemical emissions and at-source waste products will have a significant impact on the number of future plant modifications and will increase the number of automation projects in the 1990s.

8.1.1 OSHA Rule 29 CFR 1910.119

The new OSHA rule, Process Safety Management (PSM) of Highly Hazardous Chemicals, has been developed for "preventing or minimizing the consequences of catastrophic releases of toxic, flammable or explosive chemicals." [Ref. 8.2]. The PSM regulation specifies a comprehensive safety management program that integrates technologies, procedures, and management practices. The rule addresses process hazard assessment, specification of risk control measures, evaluation of the consequence of failures of these controls, documentation of engineering controls, and scheduled maintenance to assure the on-going integrity of the protective equipment. Basic Process Control and Safety Interlock Systems must be considered in each of these segments of the PSM program. Chapters 2 through 6 of these CCPS *Guidelines* specifically address the place of process control systems in a chemical plant safety management program.

Excerpts from OSHA's PSM Rule that directly relate to chemical plant process control systems are included in the excerpts beginning on the next page. This rule became law in mid-1992 except for the provisions requiring process safety information and process hazard analysis. These two provisions will be phased in over a five-year period: 25% of the processes completed within two years, 50% after three years, 75% after four, and all by mid-1997.

EXCERPTS FROM OSHA'S 29 CFR 1910.119 RULE

Process Safety Management of Highly Hazardous Chemicals

(d) **Process safety information....** the employer shall complete a compilation of written process safety information before conducting any process hazard analysis required by the standard. The compilation of written process safety information is to enable the employer and the employees involved in operating the process to identify and understand the hazards posed by processes involving highly hazardous chemicals. The process safety information shall include information pertaining to the hazards of the highly hazardous chemicals used or produced by the process, information pertaining to the technology of the process, and information pertaining to the equipment in the process.

(1) Information pertaining to hazards of the highly hazardous chemicals in the process.—This information shall consist of at least the following:

  (i)   Toxicity information;
  (ii)   Permissible exposure limits;
  (iii) Physical data;
  (iv)   Reactivity data;
  (v)   Corrosivity data;
  (vi)   Thermal and chemical stability data; and
  (vii) Hazardous effects of inadvertent mixing of different materials that could foreseeably occur.

(2) Information pertaining to the technology of the process.

  (i)   Information concerning the technology of the process shall include at least the following:
    (A)   A block flow diagram or simplified process flow diagram;
    (B)   Process chemistry;
    (C)   Maximum intended inventory;
    (D)   Safe upper and lower limits for such items as temperatures, pressures, flows and/or compositions; and,
    (E)   An evaluation of the consequences of deviations, including those affecting the safety and health of employees.
  (ii)   Where the original technical information no longer exists, such information may be developed in conjunction with the process hazardous analysis in sufficient detail to support process analysis.

(3) Information pertaining to the equipment in the process.

  (i) Information pertaining to the equipment in the process shall include:
    (A)   Materials of construction;
    (B)   Piping and instrument diagrams (P&IDs);
    (C)   Electrical classification;
    (D)   Relief system design and design basis;
    (E)   Ventilation system design;
    (F)   Design codes and standards employed;
    (G)   Material and energy balances for processes built after May 26, 1992; and,
    (H)   Safety systems (e.g., interlocks, detectin and suppression systems).

(ii) The employer shall document that equipment complies with recognized and generally accepted good engineering practices.

(iii) For existing equipment designed and constructed in accordance with codes, standards, or practices that are no longer in general use, the employer shall determine and document that the equipment is designed, maintained, inspected, tested, and operating in a safe manner.

(e)   Process hazard analysis.

(l) The employer shall perform an initial process hazard analysis (hazard evaluation) on processes covered by this standard. The process hazard analysis shall be appropriate to the complexity of the process and shall identify, evaluate, and control the hazards involved in the process,. . .

(2) The employer shall use one or more of the following methodologies that are appropriate to determine and evaluate the hazards of the process being analyzed:

(i)   What-If;

(ii)   Checklist;

(iii) What-If/Checklist;

(iv)   Hazard and Operability Study (HAZOP);

(v)   Failure Mode and Effects Analysis (FMEA);

(vi)   Fault Tree Analysis; or,

(vii) An appropriate equivalent methodology.

(3) The hazard analysis shall address:

(i)   The hazards of the process;

(ii)   The identification of any previous incident which had a likely potential for catastrophic consequences in the workplace;

(iii) Engineering and administrative controls applicable to the hazards and their interrelationships such as appropriate application of detection methodologies to provide early warning of releases. (Acceptable detection methods might include process monitoring and control instrumentation with alarms, and detection hardware such as hydrocarbon sensors.);

(iv)   Consequences of failure of engineering and administrative controls; and,

(v)   Facility siting;

(vi)   Human factors; and

(vii) A qualitative evaluation of a range of the possible safety and health effects of failure of controls on employees in the workplace.

(4) The process hazard analysis shall be performed by a team with expertise in engineering and process operations, and the team shall include at least one employee who has experience and knowledge specific to the process being evaluated. Also, one member of the team must be knowledgeable in the specific process hazard analysis methodology being used.

(5) The employer shall establish a system to promptly address the team's findings and recommendations; assure that the recommendations are resolved in a timely manner and that the resolution is documented; document what actions are to be taken; complete actions as soon as possible; develop a written schedule of when these actions are to be completed; communicate the actions to operating, maintenance and other employees whose work assignments are in the process and who may be affected by the recommendations or actions.

(6) At least every five (5) years after the completion of the initial process hazard analysis, the process hazard analysis shall be updated and revalidated by a team meeting the requirements in paragraph (e)(4) of this section, to assure that the process hazard analysis is consistent with the current process.

(7) Employers shall retain process hazard analysis and updates or revalidations for each process covered by this section, as well as the documented resolution of recommendations described in paragraph (e)(5) of this section for the life of the process.

(f) Operating procedures.
(l) The employer shall develop and implement written operating procedures that provide clear instructions for safely conducting activities involved in each covered process consistent with the process safety information and shall address at least the following elements:
  (i)   Steps for each operating phase:
        (A)   Initial startup;
        (B)   Normal operations;
        (C)   Temporary operations;
        (D)   Emergency shutdown including the conditions under which emergency shutdown is required, and the assignment of shutdown responsibility to qualified operators to ensure that emergency shutdown is executed in a safe and timely manner.
        (E)   Emergency Operations;
        (F)   Normal shutdowns; and,
        (G)   Startup following a turnaround. or after an emergency shutdown.
  (ii) Operating limits:
        (A)   Consequences of deviation; and,
        (B)   Steps required to correct and/or avoid deviation.
  (iii) Safety and health considerations:
        (A)   Properties of, and hazards presented by, the chemicals used in the rocess;
        (B)   Precautions necessary to prevent exposure, including engineering controls, administrative controls, and personal protective equipment
        (C)   Control measures to be taken if physical contact or airborne exposure occurs;
        (D)   Quality control for raw materials and control of hazardous chemical inventory levels; and
        (E)   Any special or unique hazards.
  (iv) Safety systems and their functions. . . .

(i) Pre-startup safety review.
(1) The employer shall perform a pre-startup safety review for new facilities and for modified facilities when the modification is significant enough to require a change in the process safety information.
(2) The pre-startup safety review shall confirm that prior to the introduction of highly hazardous chemicals to a process:
  (i)   Construction and equipment is in accordance with design specifications;
  (ii) Safety, operating, maintenance, and emergency procedures are in place and are adequate;
  (iii) For new facilities, a process hazard analysis has been performed and recommendations have been resolved or implemented before startup and modified facilities meet the requirements contained in management of change paragraph (l); and,
  (iv) Training of each operating employee has been completed.

(j) Mechanical integrity

(l) Application. Paragraphs (j)(2) through (J)(6) of this section apply to the following process equipment:. . .

    (iv) Emergency shutdown systems; and,

    (v) Controls (including monitoring devices and sensors, alarms, and interlocks). . . .

(2) Written procedures.

    (i) The employer shall establish and implement written procedures to maintain the ongoing integrity of process equipment.

    (ii) The employer shall assure that each employee involved in maintaining the on-going integrity of the process equipment is trained in the procedures applicable to the employee's job tasks.

(3) Training for process maintenance activities. The employer shall train each employee involved in maintaining the on-going integrity of process equipment in an overview of that process and its hazards and in the procedures applicable to the employee's job tasks to assure that the employee can perform the job tasks in a safe manner.

(4) Inspection and testing.

    (i) Inspections and tests shall be performed on process equipment.

    (ii) Inspection and testing procedures shall follow recognized and generally accepted good engineering practices

    (iii) The frequency of inspections and tests shall be consistent with applicable manufacturer's recommendations and good engineering practices and more frequently if determined to be necessary by prior operating experience.

    (iv) The employer shall document each inspection and test that has been performed on process equipment. The documentation shall identify the date of the inspection or test; the name of the person who performed the inspection and test; the serial number or other identifier of the equipment on which the inspection or test was performed; a description of the inspection or test performed; and the results of the inspection or test.

(5) Equipment deficiencies—The employer shall correct deficiencies in equipment that are outside acceptable limits (defined by the process safety information in paragraph [d] of this section) before further use or in a safe and timely manner when necessary means are taken to assure safe operation.

(6) Quality assurance.

    (i) In the construction of new plants and equipment, the employer shall assure that equipment as it is fabricated is suitable for the process application for which they will be used.

    (ii) Appropriate checks and inspections shall be performed to assure that equipment is installed properly and consistent with design specifications and the manufacturer's instructions.

    (iii) The employer shall assure that maintenance materials, and spare parts and equipment are suitable for the process application for which they will be used. . . .

(l) Management of change.

(1) The employer shall establish and implement written procedures to manage changes (except for "replacements in kind") to process chemicals, technology, equipment, and procedures , and changes to facilities that affect a covered process.

(2) The procedures shall assure that the following considerations are addressed prior to any change:

(i)   The technical basis for the proposed change;
(ii) Impact of change on safety and health;
(iii) Modifications to operating procedures;
(iv) Necessary time period for the change; and,
(v)   Authorization requirements for the proposed change.
(3) Employees involved in operating a process and maintenance and contract employees whose job tasks will be affected by a change in the process shall be informed of, and trained in, the change prior to startup of the process or affected part of the process.
(4) If a change covered by this paragraph results in a change in the process safety information required by paragraph (d) of this section, such information shall be updated accordingly.
(5) If a change covered by this paragraph results in a change in the operating procedures or practices required by paragraph (f) of this section, such procedures or practices shall be updated accordingly.

8.1.2 Development of National and International PES Standards

The "safe automation work process" described in this Guidelines book complements the PSM rule. Furthermore, the content of these Guidelines will be harmonized with a parallel work product of ISA's S84 committee which is developing a national standard, "Programmable Electronic Systems for Use in Safety Applications." The S84 ISA standard is scheduled for issue in 1994. Furthermore the S84 standard is being developed with the expectation that this US document will become part of the IEC international standard on "Functional Safety of Programmable Electronic Systems" (SC 65A, WG10 specific standards sections; process control applications). These coordinated, worldwide standards development activities will provide additional guidelines for the use of PES technology in chemical plant safety systems during the mid-1990s.

8.2 MANAGEMENT EXPECTATIONS AND CHANGING WORKPLACE FACTORS

In recent years, the industrial sector has become intensely competitive, both in domestic and increasingly in international markets. Because of these competitive factors, management will continue to place high priority on cost containment and low cost operations. At the same time demands for zero defects in operations; higher and consistent product quality; and an emphasis on continuous improvement in manufacturing will continue and will increase. Customers demand products delivered to accommodate their just-in-time manufacturing; and, as discussed earlier, safety and environmental awareness are increasing pressures to reduce process emissions and wastes. Furthermore a company's level of risk acceptance at U.S. operations will establish worldwide operating standards. In each of these areas, increased automation will be a major factor in satisfying the requirements of the marketplace, the public, industry, and the government.

Now to look more closely at some of these areas: plant operations will continue to spiral toward "just-in-time" manufacturing techniques that require limited raw material storage at the plant site. In addition, business needs for rapid response to the market place will increase the value of flexible manufacturing facilities. "Flexible facilities" and "just-in-time" manufacturing can only be achieved by expanding the use of PES controls and smart instrumentation. Closely coupled process operations places greater importance on process control system communication and computing capacity, dynamic performance, and reliability. These changing manufacturing needs will require advanced software, programming languages, and database structures.

Research resulting in new process chemistries that are more environmentally acceptable will be standard practice during this period. Waste minimization and zero discharge process designs will be the goal as new facilities are constructed. New processes making use of biochemical reactions will increase in number, and most new processing facilities (those employing new and established processing technologies) will include many more sensors than traditionally have been used. Multivariable advanced process control strategies designed to ensure efficient, reliable, and safe plant operations will become the norm in the 1990s.

European countries have played a leading role in the development of international standards that regulate chemical plant safety and chemical product quality (e.g., the United Kingdom's HSE and the European Community's ISO 65 WG9, 10 and ISO 9000). The European approach to both safety and product quality in the control systems area includes an emphasis on disciplined design of the associated controls; detailed planning and documentation of maintenance procedures; and third party validation that design and maintenance is done according to plan. Chemical companies with US operations have begun to accept this approach to product quality assurance when obtaining ISO 9000 certification for a large number of products. ISO 9000 certification requires that the chemical facility meet substantial work practice documentation requirements and then subjects the site to periodic reviews by an independent auditor. This operating experience with "quality instruments" auditing may well lead to similar requirements in US plants for treatment of safety instrumentation by the late 1990s.

8.2.1 A Changing Workforce and Greater Automation

Several long term trends related to employment and staffing practices will amplify the need for increased automation in the remainder of this decade. As a nation, basic educational skills of young Americans, the ability to read, write and do mathematical calculations, have declined over the past 20 years. This is illustrated by the long term drop in average test scores of public school graduates on standard national tests. A recent survey of 4000 members of the National Association of Manufacturers revealed that "the average manufacturer rejects five out of every six candidates for a job, and that two-thirds of companies regularly reject applicants as unfit for the work environment. A third of the companies said that they regularly reject applicants because they cannot read or write adequately, and one fourth reject them because of inabilities with communications and basic mathematics [Reference 8.1]. Although successful activities by advocacy groups in the field of education have achieved major changes in public education standards, their effect will not be realized for several years. The result will be that a greater number of tasks in chemical processing facilities will be automated to ensure successful execution. The corollary is that future workers will have to be better and more frequently trained to understand and work with instrumentation, automation, and computers.

The number of chemical plant technical and site management staffs are being greatly reduced from the accepted staffing practices of the past. Flat organizations and use of self-directed work groups with limited technical supervision of daily activities are increasingly being used. Rotating operator and technical staff work assignments from process unit-to-process unit will continue in an effort to satisfy career development ambitions and to maintain staff flexibility. Improved training is essential if this rotation strategy is to continue but development of in-depth manufacturing skills will remain difficult. Continuing education and possibly expert systems or other artificial intelligence-based tools will become necessary in an effort to offset the lack of specific job related-experience and knowledge needed for safe process operation.

Use of engineering and maintenance contractors will be more common. Availability of support systems that allow use of these workers who have little historical knowledge of specific plant practices will be necessary. Documentation of quality design and maintenance practices and the availability of accurate documentation for existing facilities will become increasingly important to workplace safety when modifications to existing plants are required.

Increased automation of chemical plant operating tasks and of safety functions is an almost certain response to the reduced availability of highly skilled and experienced plant engineers, operators, mechanics and contractors. Furthermore, reliance on PES systems with powerful operator-support features—process alarm interpreters and other systems which make use of artificial intelligence technologies—as well as self-diagnostic features to aid with system maintenance will become even more necessary in the 1990s.

8.3 MEASUREMENT, CONTROL, AND COMMUNICATION TECHNOLOGIES FOR THE 1990s

8.3.1 PES Design Methodologies for Safety System Applications

Industrial design of process control systems (both BPCS and SIS) will continue to be based on risk-based, qualitative design criteria. These criteria which typically are part of a chemical company safety procedures will contain a similar methodology to that outlined in this CCPS Guidelines. Control system integrity is achieved through the use of macroscopic criteria such as requiring component and software diversity, separation of SIS from BPCS controls, use of read-only communication, etc. Movement to more refined, quantitative design methodologies will occur only when a proven method to specify software reliability is developed. Significant research on methods for validating the performance of computing modules and both system and application software is in progress. New test methodologies for programmable electronic systems will be realized during the period; and, when these technologies mature PES safety design practices will require modification (Ref. 8.28).

8.3.2 Process Sensors and Instrumentation

Significant changes are expected in the process sensor area. New applications of existing technologies (e.g., lasers and fiber optics, microcomputers, complex single processing algorithms, digital communication to field devices, etc.) and applications of new technologies (e.g., thin film and microchemical sensors, superconductivity, etc.) will contribute to major improvements in chemical process sensor performance.

Process sensors can be classified by their physical location relative to the process: *In situ* sensors are located in the process stream or vessel and are in physical contact with the process fluid. *At line* sensors are installed in the processing area (not in a plant laboratory) and may not have physical contact with the process fluid or may be in contact with a sample fluid extracted from the main process stream. Replacement of laboratory composition instruments with in situ or at line sensors is expected to be a major trend during this period. Innovative enhancements of existing process measurement technologies will be emphasized during the 1990s. Many of these "new" instruments will be the commercialization of established analytical technologies packaged for chemical plant use outside of a laboratory environment. These real-time instruments include

- mass spectrometers.
- nuclear magnetic resonance spectrometers.
- sonic acoustic and ultrasonic process phase and motion analyzers.
- finite element vibration analyzers.

Measurement system performance improvements expected in the 1990s include better measurement accuracy and instrument repeatability, wider range devices with improved reliability, and instruments with more predictable "fail-safe" characteristics. Fault-tolerant sensors which make use of embedded microcomputers, extensive self-diagnostics, read-only communications, etc., will make possible a supply of instruments with certified failure modes.

This section discusses several sensor technologies, in addition to these new applications of existing measurement technologies, which are expected to make process monitoring and control more pervasive in the future chemical plant.

8.3.2.1. *Fiber Optic Sensors*
Sensors, which are made from the fiber optic cable itself will be applied in many industrial settings. Process temperature, level and some compositions can all be measured using fiber optic sensors. Fiber optics used for process measurement is a simple technology which can readily be applied as a safe sensing technology for use in hazardous environments (Ref. 8.4).

8.3.2.2. *Full Spectrum Photometric Analyzers*
Several developments in the use of infrared, IR, and near-infrared light spectrum sensors for real-time chemical process composition detection are expected to make many more analytical measurements available in the control room for the operating staff of the 1990s. The concentrations of many chemical components found in chemical processes are the past process photometric analyzers have generally been limited to those applications where light was passed through the process stream and the absorbance of light at a given wavelength by a specific chemical component of interest was detected. The concentrations of many chemical components found in chemical processes are proportional to the degree of frequency-specific energy absorbance when a light wave passes through the medium. In the past process photometric analyzers have generally been limited to those applications where light was passed through the process stream and the absorbance of light at a given wavelength by a specific chemical component of interest was detected. A spectrometer produces individual light waves by using a light source and discrete interference filters, and up to six or more filters are used in some analyzers for multiple component measurements. This composition measurement technique works well when process stream components of interest have absorption bands which are not heavily overlapped (Ref. 8.29).

Recent advances in optics, lasers, and microprocessor electronics have resulted in rugged full spectrum analytical instruments that use absorbance characteristics of the full light spectrum to determine concentrations of a broad set of process stream components. The Fourier Transform Infrared

Spectrometer, FTIR, is now available from several manufacturers as a process analyzer. Although these devices are very complex in theory, they are simple in operation. The infrared spectrum of the process stream is analyzed to identify individual components in the process stream. Data analysis techniques used range from monitoring a few specific wavelengths to advanced statistical techniques (e.g., derivatives of the spectra, multiple linear regression, partial least squares, etc.) which can use hundreds of wavelengths to calculate stream composition. In essence, this single process analyzer has now become a powerful analytical tool equivalent to thousands of individual optical sensors each one tuned to a different wavelength of the IR spectrum. This on-line analyzer opens new avenues for improving process control and operating safety. Multiple components in the sample stream can be monitored continuously to provide the information necessary to keep the process on-aim and within established boundaries. In batch processes they can indicate precisely when endpoints are reached (Ref. 8.29).

Another advantage of this new generation of optical process analyzers results from the way the instrument interfaces with the process. Only a light wave contacts the process sample, the process sample is isolated from the analyzer. Attenuated Total Reflectance probes (probes that are prisms of optically transparent materials through which the light from the analyzer is guided) simplify the collection of composition indicating light signals. An ATR prism is immersed in the process stream allowing free flow of samples around the probe. The refractive indices of the prism and process fluid are such that light is totally reflected at the interface, but with a resultant standing wave that extends a small distance into the fluid. This evanescent wave is selectively attenuated by constituents of the process stream, so that the remaining light emitted from the end of the prism bears the imprint of stream compositions. The advantage is obvious. The multicomponent process analyzer interfaces to the process through a simple plug-in probe—there is no extractive sample system. Light can be guided from the spectrometer to the process via hollow light pipes (fiber optics for the IR region are still not readily available yet) and then interfaced to the process itself via an ATR probe (Ref. 8.29).

Full spectrum analyzers are also available in the Near Infrared region of the light spectrum. The NIR region differs from the mid-infrared in that the spectrum consists of harmonic overtones of the primary absorption bands of the IR region. These bands are heavily overlapped and usually require sophisticated signal analysis software to decipher. Advances in the field of chemometrics, a statistical procedure for extraction of information from infrared spectra, makes possible determination of a wide variety of unknown chemical compounds using a set of calibration factors determined from the spectra of known chemicals. Calibration factors are determined by scanning a part of the IR spectrum and correlating changes in the spectrum with changes in the composition of the known samples. A significant advantage of the NIR analyzer is that this region of the spectrum is compatible with fiber optics. Consequently NIR analyzers can be located remotely from the sample point and connected via intrinsically safe fiber optic media to fibre optic insertion probes which carry the IR sensing beam into the process stream rather than require extractive sampling (Ref. 8.29).

8.3.2.3. Thin Film Sensors
These sensors include electrically active films (often deposited on a silicon substrate) which are sensitive to a family of process properties, for example, temperature, humidity, and radiation. Changes in the process parameter of interest is transduced by resistive, capacitive, piezoelectric or photosensitive effects (Ref. 8.4).

8.3.2.4. Lasers
A LASER (Light Amplification by Stimulated Emission of Radiation) is an electromagnetic radiation source with emitted light which is characterized by the following properties:

- Laser light is almost monochromatic.
- Laser light is coherent as it emerges from the laser output mirror and remains so for a certain distance from the laser.
- The laser beam has very little divergence, typically 0.001 radian.
- Continuous operation lasers may have power outputs of 0.5 mW to 100 W or more. Pulse type lasers have power levels up to TeraWatt, but for only very short time pulses—microsecond or even nanoseconds in duration.

Industrial lasers are used for alignment, communication, cutting, drilling, heating, ranging vaporization, welding, etc. Increased use of laser-based measurement systems are expected in the 1990s.

Applications of Laser Systems
Laser-based environment surveillance. The presence of a foreign object or substance in a region may be detected by a reduction in laser beam intensity. An area is scanned with a fixed size laser beam; reflectors are used to structure the monitoring path. This technique is now being evaluated in industrial applications for perimeter monitoring. Interference problems have been encountered caused by the presence of rain, snow, dust, and humans.

Remote detection of gases in the atmosphere may be accomplished using the differential absorption lidar (DIAL) measurement technique. This established method employs two lasers (i.e., two light beams of differing wavelengths); one that is strongly absorbed by the gas to be measured; and one, that is not. The return signal for this technique is a result of the laser backscatter from either atmospheric aerosols or topographical targets. The return of the non absorbed wavelength is used as a reference or zero-level calibration for the system. The difference between the absorbed and non absorbed return signals is proportional to the concentration of gas along the optical path between the laser source and the back scattering medium. Only path-averaged concentration measurements are possible in the case of back scattering from topographical targets. The DIAL system produces quantitative concentration results (Ref. 8.5).

Another detection technique makes gases that are not visible to the human eye visible on a standard TV monitor. The backscatter/absorption gas imaging (BAGI) technique is inherently a qualitative, three-dimensional vapor detection scheme. The principle of operation is the production of a video image by back scattered laser radiation, where the laser wavelength is strongly absorbed by the gas of interest. Normally invisible gas is made visible on a TV monitor using this technology. The major components of the system are an infrared TV camera, an infrared carbon dioxide laser, and a TV monitor (Ref. 8.6).

The technique has been used to model and measure the dispersion of large, heavier-than-air gas clouds. The BAGI technique shows promise as a long-range detection system (>300 meters) capable of quickly and cost-effectively locating the sources of hazardous gas and monitoring the dispersion clouds even at very low concentration levels. A short-range (30 meters), man-portable version of the same system could be used to see the presence of vapor clouds and to identify the source of gas leaks in industrial plants and pipeline distribution systems. The technique has two basic constraints: (1) there must be a topographical background against which the gas is imaged, and (2) the system must operate in an atmospheric transmission window.

BAGI systems offer an alternative to conventional leak-detection methods used in chemical plants (dipping, soaping, pressure decay) and because of the real-time video aspect of this technology may become useful in automated, mechanical integrity testing systems. Since the CO laser is tunable, more than 80 different gases can be detected and identified from distances up to 30 meters. Monitoring of likely leak sources for fugitive chemical emissions is another expected application (Ref. 8.7).

Turbidity. Measurement of turbidity (which affects the propagation of light through a liquid that lacks clarity) is another laser application. A laser beam is split and passed through two samples to matched photo detectors. One sample is a carefully selected standard of allowed (acceptable) turbidity. The other is an in-line sample of the process liquid itself. If the in-line sample attenuates the light more than the standard, the signal conditioning system triggers an alarm or takes other appropriate action to reduce turbidity.

Distances and speed. Surveying instruments for measuring distance have been developed by measuring the time of flight of light pulses scattered off a distant object. Target velocity can be measured by an electronic computing system that records the changes in reflected-pulse travel time and computes velocity.

8.3.2.5. Microelectronics
Members of the SENSORS Editorial Advisory Board have identified

- silicon micromachining,
- microelectronics, and
- the addition of intelligence and self-diagnostic capabilities to the sensing element as the most significant, recent landmarks in sensing technology.

Systems-based approaches have largely replaced component-based solutions to sensing problems. Cost effectiveness and reliability are up (Ref. 8.8).

Board members anticipate continued growth in existing technologies over the next five years. The evolution of new materials and packaging that will allow sensors to detect a broader range of parameters and function in more extreme environments than those in which they operate today is expected to continue. Addition of intelligence (computing capability with memory) at the point of measurement will make possible redundancy and robustness of sensing systems not now possible. Biosensors promise bold innovations for both the health care and process industries. Continued progress in fiber optics and light-based sensing applications (for more than telecommunications) is expected.

Microchemical sensors

A microchemical sensor typically consists of multiple detectors integrated on a silicon chip which may utilize different deposited polymer coatings at each sensing site. Each sensor site detects different chemical species via ion transmission through the polymer coatings. The changing voltage signal created by the ion transmission across a resistor is amplified to the required level by "on-chip" integrated circuit electronics (Ref. 8.9).

These sensors are sold commercially (e.g., Microsensor, Inc.) and may make possible development of chemical composition sensors equivalent to a chromatograph-on-a-chip. Compositions could be monitored on every tray of a distillation column using this technology.

8.3.2.6. Superconductivity
Voltage supply and regulation components containing high-temperature superconductivity elements will be introduced into PES computing modules to further reduced random failures of process control systems. Furthermore spurious interruptions of chemical plant unit operations that in the past have been caused by grid voltage fluctuations and that often lead to equipment damage and to a rash of potentially hazardous process conditions will be reduced. Installation of very large (megawatt supply) and highly reliable electrical power systems for driving process equipment as well as control systems will be made possible by the application of superconductivity materials in power system UPSs.

8.3.3 New Control System Architectures

Industrial control system manufacturers will bring to market DCSs which make extensive use of the new 32-bit microprocessors, high-density VLSI memory chips, and low-cost reduced instruction set computer-based workstations during the 1990s. These DCSs will remove system performance constraints of the earlier generation PESs and will result in significant expansion of BPCS capabilities in supply of plant management information and in application of advanced process control strategies. These state of the art DCSs will

- provide imaginative, dynamic graphical displays for the Operator that make extensive use of statistical tools for analysis of plant status.
- provide special-purpose computing modules for execution of model-based data adjustment and plant optimization control programs.
- allow for bulk data collection and global plant status reporting.
- provide communication links to pass real-time plant measurements to multiple-site computer networks.
- allow for coupling of digital equipment from different manufacturers creating a single, distributed plant data base.
- maintain master records of DCS module configuration and instrument databases.

In these process control systems, "data highways" (local area networks) will function much as process utility systems do in other parts of the chemical plant. "Smart" instruments, HMI modules, data recorders, and special purpose control systems will be connectable to a BPCS highway without regard to manufacturer and without requiring the development of special communication protocols. *This architectural freedom and the emphasis on "open communication systems" will increase the importance (1) of separation of SIS and BPCS functions and (2) of the application of the process-safety related design practices outlined in Chapters 4 and 5 of this book.*

Maintaining functional integrity of the Safety Interlock System will become more of a challenge in the 1990s as the process control system is integrated into the global information system of the chemical company.

8.3.3.1. Control System Communication

PES control system performance characteristics depend on the communication of digital data. System characteristics which impact safe plant operations, namely,

- single control system data base in the plant.
- apparent real-time display of plant changes at the operator's workstation.
- increased functional modularity at the process controller level.

• redundancy available for all system modules without adding complexity for the user during configuration or later in DCS maintenance require a complex communication structure within the PES.

In addition to being a process control systems, these PESs will become increasingly complex information management systems in the 1990s.

Distributed control tasks and distributed control system modules will be efficiently integrated into a responsive system by the interconnection of single task-related communication systems, called Local Area Networks (LANs). These LANs are linked in an ordered fashion with other independent networks to provide information flow paths within the DCS. In the BPCSs of the 1990s, communication bridges will allow free flow of process data to site (and company) wide networks for use by all business and operating functions who have need for these data.

The need for data flow from LAN-to-LAN within the DCS is minimized since multiple microcomputers are grouped at the single task level to do the computational work required for process control. Extensive use is made of data compression techniques at each network Gateway so that new data is sent on to the next higher network only when operationally significant changes occur. Fast network speeds and special communication protocol—the combination of masterless point-to-point token-passing techniques with some broadcasting of preconfigured data—assure that module access time to the network and message transmission time is a constant value even though communication loading varies widely with events in the plant. This known communication frequency removes uncertainty for the operator when interpreting data at the Operator's Station and within control algorithms which must meet time-critical constraints.

8.3.3.2. Digital Field Bus

A major change will occur in the data-acquisition architecture of the PES process control systems by the late 1990s. The work of ISA's SP50 committee is pointing the way to a worldwide design standard for a digital Fieldbus which will result in PESs that have the capability of bringing process sensor data to the control center over a small number of digital data buses. The SP50 standard defines a standard communication protocol for this sensor bus and, perhaps of more importance, specifies standard software structures and functional requirements for field instrument algorithms. When this standard is fully accepted, smart field instruments (from pressure transmitters to valve positioners) may contain all of the control logic required for regulatory control loop applications. SP50 then will result in a further distribution of process control into smart modules distributed throughout the processing area.

The Fieldbus development is driven by cost reduction considerations (installation costs of control systems will be reduced significantly by this technology) and by the desire to increase user choices in selection of instrument suppliers. The SP50 fieldbus is an open communication system which will permit the connection of instruments from all manufacturers with compliant products. Since the SP50 standard is not complete, technical commentary on the use of this technology for process control cannot be offered; however, *CAUTION is advised!*

Early applications of Fieldbus will require close scrutiny during control system HAZOPs as *redundancy and fail-safe design options are not part of the first phase SP50 design specifications.* Use of Fieldbus within the SIS as now defined can not be recommended. Enhancements of Fieldbus communications integrity plus developments in bus media are expected, and these developments may some day allow use of selected fieldbus schemes in safety application.

8.3.3.3. Computing Architectures

Advances in fault-tolerant PES module design technology will essentially eliminate the concern for reliability of computing elements in the control systems of the 1990s.

The successful application of PES technology for chemical process control has been made possible by the coordinated use of multiple computer systems; i.e., by the use of parallel processing where many microprocessors work simultaneously with multiple data sets to solve a single problem instead of handling one piece of data at a time in a single central processing unit. This computing architecture will be extended and applied to other time-critical, computationally demanding tasks. These include the real-time simulation of chemical processes for model-based process control, operator training, and control room evaluation of alternative operating strategies.

8.3.4 Human/Machine Interfaces

VDU displays grouped to form a Workstation (and designed by ergonomic criteria) for the plant operator will continue as standard control room equipment into the 2000s. Supplementary display equipment ranging from wall-scale, interactive graphic interfaces to small head gear-mounted displays will also be part of the chemical plant control system by the late 1990s. Use of voice actuated commands for entering information into the control system will be used where hand-freedom is needed.

Operating information will be accessible at the Workstation through the selection of icons linked to multiple display windows on a single terminal (HMI technology refined for Apple's Macintosh personal computer). "Windowing" display technology will become standard in the industry, but limitations will be imposed on the number of windows that can be overlaid. Providing the operator with quick access to relevant process data during plant upsets remains a critical performance criteria for these Workstations. Measuring communication effectiveness of these human/machine interfaces during process upsets is a challenge and will continue to be an area of active research. A methodology to guide in the design and performance assessment of these multisegmented video displays is needed today as the business and social costs of chemical plant operating errors mount.

Innovative methods for maintaining operator alertness are needed when the processing conditions are normal in these darkened control rooms. Process simulations which permit operator training during quiet times may meet part of this need, but this clearly is a problem area where industrial psychologists and human factor professionals are expected to make a contribution.

8.3.5 Artificial Intelligence Applications

Artificial intelligence is a subfield of computer science that may be defined as the science of enabling computers to perceive, reason, understand, make judgments and learn (Ref. 8.10). The artificial intelligence umbrella covers three computer science application areas which are expected to aid in safe automation of chemical facilities:

- *Expert Systems*—Expert systems are computer applications which are repositories of task-related, in-depth human knowledge. These systems are particularly useful in situations where expertise is scarce; different parts of the expertise are distributed among many people; or the expertise is not reliably and continually available.
- *Artificial Neural Networks*—Neural networks are tools for modeling of work tasks. Neural network computer systems extract knowledge by learning from examples; that is, these systems learn by analysis of user supplied data sets. Development of a neural network tool requires neither the investment in definition of explicit physical relationships needed when engineering models are used for task simulation nor the knowledge extraction process required in the development of Expert Systems.
- *Natural Language Processing*—One major goal of artificial intelligence research has been to develop means to interact with machines in natural language. Natural language processing addresses the technology for developing effective natural language front ends to computer programs, computer-based speech understanding, text understanding and generation, and related applications.

Expert systems, artificial neural networks, robotics, vision systems, natural language processing, and smart sensors are complementary technologies. Research and development activity in each of these technologies is growing worldwide, and industrial applications are increasingly being reported. Each of these AI technologies appear to have the potential to make significant contributions to more efficient and secure chemical plant operations. Chemical facilities with integrated production units will emerge that make use of automated planning, scheduling, process control, and warehousing systems; and, elements of AI technologies will be part of each of these plant management computer systems.

The integrity of process control and safety system application software is expected to be advanced by programming development utilities and verification techniques that make use of artificial intelligence technologies. Smart control software development systems will greatly contribute to the acceptance of PES technology in risk prevention applications. In the far distant future, lights-out plant operation will be directed by intelligent robots with most of today's manual work done by automated robot carts and manufacturing machines.

8.3.5.1. Expert Systems

The key to the efficient building of expert systems is to use expert system development tools, or shells, with components that support all the requirements of the systems to be built. Shells contain the skeletal structure of an expert system without the rules or knowledge specific to each application.

DuPont was an early leader in the use of this technology and, by 1987, had developed and fielded over 300 individual expert systems (Ref. 8.11). In the manufacturing and product development areas expert systems were developed as aids for training, quality control, process control, equipment maintenance and troubleshooting, production scheduling, product selection and formulation, hotline support and on-line documentation .

Many successful chemical applications of Expert Systems have been reported:

- A rotating machinery maintenance and diagnostics system has been developed for use in interpreting machine temperature, vibration, relative component motion, and auxiliary system measurements (Ref. 8.12).
- A regulatory compliance expert system has been developed to help determine the legal consequences of proposed maintenance actions or changes in plant conditions (Ref. 8.12).
- Foxboro's Batch Reactor Consultant is used to design the control system for many different types of batch reactors (Ref. 8.13). The tool is essentially a menu-driven program for use on a personal computer wherein questions are posed serially after each one is answered. Expert systems offer the potential for achieving economies of scale in batch production rivaling those possible in mass production plants (Refs. 8.10, 8.14).

Expert systems have been proposed for process control applications. An expert controller reads several measured variables and processes this data according to the established rules of the system. The output of the expert controller may be presented as advice to an operator, or the controller may, itself, act and inform the operator that action has been taken. Expert Systems do require the existence of an expert; and when operating expertise is lost to a plant, the opportunity to apply this technology is also gone.

1. Expert Systems in Plant Controls. Expert systems are expected to be available to all chemical operating staff members by the late 1990s. These systems will place job-related instructions (knowledge) in the hands of workers who are less-trained and lower-salaried. The use of expert system shells makes practical the development of expert systems by factory floor personnel instead of requiring that expert systems applications be done by computer science practitioners.

Potential safety-related applications for expert systems in the chemical industry appear to be limited only by the imagination. Popular operator advisor applications include: alarm management, shutdown and start-up assistance, emergency shut-down assistance, management of spills and emissions, operator assistance, process monitoring, etc. Several successful, safety-related applications have been reported:

Safety Incident Prevention. Expert monitoring systems can alert operators to process deviations, provide a preliminary diagnosis of the probable cause, and recommend corrective action (Ref. 8.15). At one DuPont site, Sabine River Works, a process control expert system aids the central control room operator in monitoring processes which require seven computer screens and banks of instruments for process control (Ref. 8.16). Part of the expert system's function is to notify the operator before operating conditions exceed a known, safe operating envelope. The Norwegian "Senter for Industriforskning" (SI) has made its expertise in the use of chemical dispersants for oil spill cleanup widely available as an expert system to all who need it (Ref. 8.17).

Alarm Management. Alarm management systems which include expert system components are available from control system software suppliers as standard software packages for some process-specific applications (Ref. 8.10). This is a key application area for ES technology since alarm management is a challenge which is common to most chemical facilities. Typically a large amount of process data must be analyzed in a short period of time during a plant upset, and many operator have difficulty in focusing on the appropriate problem area when derivative alarms are being sounded.

Diagnosis and Fault Detection. DuPont, Foxboro, and the University of Delaware proved the capability of an expert system to detect and identify faults in an adipic acid reactor process. This application is well known as the FALCON (Fault Analysis Consultant) project (Ref. 8.11).

ABB Process Analytics provides an expert system-based troubleshooting and maintenance support system for gas chromatography (Ref. 8.10). The software package checks sensors and assesses chromatograms by applying a rule set to identify maintenance needs and to select corrective action suggestions. This is only one of many examples which could be cited of ways expert systems improve the effectiveness of on-line control system diagnostics. Reliance on these system to quickly and correctly diagnose PES module failures greatly reduces MTTR of modules in both the BPCS and SIS.

Process Control. DuPont reported significant savings from the use of a distillation column expert system at the Maitland, Ontario plant (Ref. 8.11). The system reads process data then asks the operator for additional plant operating conditions. From this it recommends adjustments to the column controller setpoints that will optimize its operation.

2. Expert System Challenges and Research. The present method of building expert systems requires that virtually each new system must be tailor-made, thus creating a knowledge engineer bottleneck. Current research is expected to result in expert systems which: (1) include utilities that facilitate identifying standard types of problems for which existing solutions can be adapted and, (2) include problem solvers capable of reasoning beyond the level of current systems.

Current systems are limited in their ability to communicate with the user. Expert systems of the future are expected to be built that include conversational dialogue between the plant user (operator, maintenance and engineering staff members) and the computer.

Access to plant data is a major limitation of present expert systems in many plants. Expert systems are one of the important safety-related applications which await the evolution of more computer-integrated manufacturing sites. Typically expert system applications require data which resides in distributed computing systems; and, access to this data requires a plant-wide and, for some applications, even a corporation-wide data base manager. Computer and communication technology exists to make this information readily available, but the investment requirements for management information networks, communication bridges, and data base management systems will restrain the broad deployment of real-time Expert Staff Advisor systems.

8.3.5.2. *Artificial Neural Networks*

Neural network computer systems demonstrate rule following behavior without requiring the explicit definition of rules. Computing elements that are analogous in function to the biological neuron make up these systems. Knowledge about a physical process is acquired through training rather than programming and is retained within the network by changes in node functions. A neural net recalls these learned states or cycles in response to the presentation of new cues. The power of neural computation comes from connecting the basic building block computing "neurons" into parallel networks (Refs. 8.18, 8.19, 8.20).

The objective of a neural network application is to develop a system for which a set of inputs will produce related set of outputs. To accomplish this, a neural network must be trained by sequentially applying input and output vectors and adjusting internal transmission weights. During training, the network tuning parameters gradually converge to values (i.e., learns) such that each input vector produces the desired output vector. The performance of the neural net is then limited to the domain of the information contained in the "teaching set" of data. Shifts in operating rates or process modifications can easily invalidate the learned model.

Neural network systems can perform many tasks that humans do well, such as, identifying faces, characters, and recognizing patterns, voices, and handwriting. They can mimic human thought and learn by example. Artificial neural networks are expected to become the preferred technique for a large class of pattern recognition tasks that conventional computers do poorly, if at all.

1. Applications of Neural Networks. The strength of neural networks is in monitoring and control of physical systems which change in unexpected or poorly understood ways. Applications reported for this emerging technology include:

Process diagnostics: (1) A neural net computer has been imbedded in an expert system as a preprocessor to analyze sensor data quality. (2) Fluorescence intensity spectra obtained from solutions of amino acids using an infrared spectrometer are input to a neural network, and the output of the neural net system estimates the concentration of fluorophores present (a measure of product quality) (Ref. 8.22).

Process control: (1) Workers at the University of Maryland have described in detail the use of neural nets in dynamic modeling and control of chemical process systems (Ref. 8.21). A back propagation algorithm is applied to model the dynamic response of pH in a reactor, and the model is then available for closed loop process control. In another application at the University of Maryland, a neural net system has been used to learn control configurations for distillation columns by example (Ref. 8.23). An effort to develop an expert system to perform this same design task found that knowledge extraction from the expert to be time consuming and difficult. The neural nets approach to the problem avoided these difficulties.

Neural nets and expert system technologies are often used together to solve problems in diagnosis and control (Ref. 8.24). Rule-based systems tend to be slow, and the tasks of collecting and codifying rules time consuming and difficult. When combining these two technologies, a neural net may be used to receive and analyze data; and its output , used as input to a rule-based system which selects among several discrete responses.

A 1989 conference sponsored by the Institute of Electrical and Electronic Engineers and the International Neural Network Society produced over 400 papers on a wide number of neural networks topics (Ref. 8.25). The conference included 64 papers on applications, of which 10 were presented by authors related to industries.

2. Neural Networks in the Future. Neural networks have limitations in their ability to learn and recall. Most troublesome is the long and uncertain training process. For complex problems days or weeks may be require to train the network, and it may not train at all. Neural network response to real-time process data can be unpredictable, and this AI technology does not include the ability to "explain" how they solve problems. *The use of nondeterministic computing modules in closed-loop safety systems is not expected to be acceptable in the foreseeable future.*

Neural computing in the future will likely be a complementary technology to, and not a replacement for, conventional control algorithms and other AI programs. There is promise of advances in this technology. For example, a VLSI chip can support billions of active cells; new methods of network node organization are being researched; and good designs have emerged for artificial retina and cochlea.

Potential applications for artificial neural networks include those human activities that require rapid responses and are governed by pattern recognition. Human tasks done rapidly and without conscious effort have often proven difficult for conventional computing methodologies to emulate. The ultimate challenge for experts in computer architecture is to exploit the two technologies of computers and neural nets, while presenting a single, flexible interface to the user. Questions regarding the reliability of neural network models must be resolved before this technology can be applied where human life or valuable assets are at stake.

8.3.5.3. Robotics

A robot is a flexible machine capable of controlling its own actions by utilizing stored programs for a variety of tasks. Basic task flexibility is achieved by the capability of reprogramming the robot. An industrial robot is a programmable, multifunction, and often multiaxis arm that is designed to move materials, parts, tools, or specialized devices through variable programmed motions for the performance of a variety of tasks. Most industrial robots have a single jointed arm with a gripper, and most of these robots are stationary. They are good production line machines, cost-efficient in hazardous and boring jobs. Mobile robots may walk or move along a preprogrammed path (AGV—Automatically Guided Vehicle), on-loading and off-loading at different stations (Ref. 8.26).

All developments are making more intelligent robots practical as researchers work on ways to add intelligence to these machines in the form of vision, tactile sensing, planning, and learning. Intelligent robots are not only able to perform a wider spectrum of manufacturing tasks, but these machines will begin to perform tasks outside the industrial assembly line environment. Thus, robots in fire fighting, undersea exploration, mining, batch chemical processing, and construction will appear in the 1990s.

1. Applications of Robot Systems. Industrial robots have found many chemical plant applications from performing process sample analysis procedures to materials handling operations.

2. Challenges. The complexity of tasks that robots can perform today is limited to some extent by deficiencies in sensors and mechanisms, notably the lack of dexterous end-effectors. Touch sensing evaluates objects on the basis of weight, shape, texture, vibration, direction, pressure, and temperature. Robots are expected to benefit from touch sensing because it provides information about areas they cannot see when their grippers are in the way. Touch will also allow for the more precise control needed in grasping delicate objects (Ref. 8.26).

Although some robots have mechanical and sensing capabilities that rival those of a human, without an equally sophisticated programming capability these machines are confined to simple repetitive tasks. The trend in programming robots is towards endowing robot systems with general purpose computer programming languages. Natural language processing technology developments are expected to reduce the time and effort required to program robots by simplifying the user interface when specifying task requirements (Ref. 8.27).

Artificial-intelligence-based methods for teaching robots are under development. One method involves entering the endpoints of a motion into a computer that automatically programs the full motion into the robot's memory. Speech recognition systems are expected to be incorporated into robots that will allow these machines to respond to user's vocal instructions. And eventually, AI technologies will enable robots to learn from experience and to automatically modify control programs used to perform new tasks.

8.3.6 Advanced Control Algorithms

Process control of chemical plants in the 1990s will move beyond the traditional application of multitudes of single input-single output feedback controllers to the widespread use of multivariable model-based predictive control algorithms. The introduction of powerful RISC technology computers into the DCS will make this computing-intensive technology generally available, and the competitive pressures to be a low-cost chemical producer will make the use of this technology necessary. Model-based predictive controllers drive the chemical process to maximize economic returns by simultaneously moving many manipulated variables without exceeding physical constraints. Although plants will be operated closer to safe boundaries by these smart controllers than has been the case in the past, this control technology has the capability to predict the future impact of present disturbances and to take appropriate control actions to avoid unsafe conditions.

Successful use of multivariable controllers places additional emphasis on the maintenance of the BPCS instruments and requires process control-trained staff for maintenance of the model-based controllers. The total effect of this move to advanced control applications is expected to be more stable plant operations and less operating risks created by plant upsets.

8.4 SUMMARY

In the 1990s, the path forward to increased safety of chemical plant operations includes an expanded reliance on chemical plant automation systems. Changes in community values; in the skills of the available work force; in process and plant designs; and, in process measurement, control, and communication technologies all influence the way in which automation systems will be applied for risk control.

The effective application of PES-based safety interlock systems requires a lengthy and complex work process. One methodology for definition of a risk-linked safety interlock system is outlined in these CCPS Guidelines. Risk mitigation provided by these instrumented safety systems continues over the life of the process only when in-plant administrative control of changes to the BPCS and SIS, maintenance of control system documentation, and periodic performance proving of the installed SIS is a daily priority of plant management.

8.5 REFERENCES

8.1   Cunniff, J., Lack of skilled workers boosts the jobless rate. *Charleston Daily Mail*, Charleston, WV, January, 1992.

8.2   29 Congressional Federal Register Part 1910, Section 119, "Process Safety Management of Highly Hazardous Chemicals"; Federal Register, Vol. 57, No. 36, February 24, 1992.

8.3   King, R. *Safety in the Process Industries*, London: Butterworth-Heinemann, 1990.

8.4   Rutledge, W. C. The future of sensors in control systems. *Control,* June 1989, pp. 104–105.

8.5   Murray, E. R., et al., Remote measurement of HCl, CH_4, and NO_2 using a single-ended chemical laser lidar system. *Applied Optics,* 15:3140–3148, 1978.

8.6   McRae, T., et al. "The Use of Gas Imaging as a Means of Locating Leaks and Tracking Gas Clouds." Paper presented at AIChE Petro Expo '86, New Orleans, LA, April 1986.

8.7   Bruno, R. P. Tracking gas leaks with active IR scanning. *Photonics Spectra*, February 1992, pp. 92–98.

8.8   Rosa, Dorothy (Ed.), Progress in sensing technologies: 1984–1994. *Sensors,* January 1989, pp. 9–21.

8.9   Rutledge, W. C. The future of sensors in control systems. *Control,* June 1989, pp. 104–105.

8.10 Bailey, D. J. Artificial intelligence in industry: Expert knowledge bases in control loops. *Control Engineering,* December 1986, pp. 45–48.

8.11 Baur, P. S. Development tools aid in fielding expert systems. *InTech,* April 1987, pp. 7–15.

8.12 Finn, E. A. Rules of thumb for implementing expert systems in engineering. *InTech,* April 1987, 33–37.

8.13 Blickley, G. J. Designing control systems with an expert system. *Control Engineering,* September 1987, 112–113.

8.14 Nisenfeld, A. E. and M. A. Turk. Batch reactor control: Could an expert advisor help? *InTech,* April 1986, 57–58.

8.15 Rajaram, N. S. Artificial intelligence: Its impact on the process industries. *InTech,* April 1986, 33–36.

8.16 Feilen, L. Expert systems fill voids at DuPont. *Control,* June 1989, pp. 43–44.

8.17 Waterbury, R. C. Expert systems extend expertise. *InTech,* December 1982, pp. 22–27.

8.18 Humelhart, D. E. and J. L. McClelland (Ed.), *Parallel Distributed Processing.* Vols. 1 and 2. Cambridge, MA: MIT Press, 1986.

8.19 Allman, W. F. *Apprentices of Wonder: Inside the Neural Network Revolution.* New York: Bantam Books, 1989.

8.20 Wasserman, P. D., *Neural Computing: Theory and Practice.* New York: Van Nostrand Reinhold, 1989.

8.21 Bhat, N. and T. J. McAvoy. "Use of Neural Nets For Dynamic Modeling and Control of Chemical Process Systems." Department of Chemical Engineering, University of Maryland, College Park, MD. Paper presented at the American Control Conference, Pittsburgh, PA, June 21–23, 1989.

8.22 McAvoy, T. J. et al. "Interpreting Biosensor Data via Backpropagation." Department of Chemical Engineering, University of Maryland, College Park, MD.

8.23 Birky, G. J. and T. J. McAvoy. "A Neural Net to Learn the Design of Distillation Controls." Department of Chemical Engineering, University of Maryland, College Park, MD. Paper presented at the IFAC DYCORD Symposium, The Netherlands, August 1989.

8.24 VerDuin, W. H. "Neural Nets for Diagnosis and Control." Ppaer presented at the 3rd Annual Expert Systems Conference, Detroit, MI, April 5, 1989.

8.25 *Proceedings of the International Joint Conference on Neural Networks.* Washington, DC, June 18–22, 1989.

8.26 Scown, S. J., *The Artificial Intelligence Experience: An Introduction.* Maynard, MA: Digital Equipment Corporation, 1985.

8.27 Grimson, W. E. L. and R. S. Patil (Eds.), *AI in the 1980s and Beyond: An MIT Survey.* Cambridge, MA: MIT Press, 1987.

8.28 Moon, I., G. J. Powers, et. al., Automatic verification of sequential control systems using temporal logic. *American institute of Chemical Engineering Journal,* 38(1):67–75, 1992.

8.29 Andrisani, J. M., "Modern Process Analyzer Technology—Key to Improved Process Control." Paper presented at the Chemical Manufacturers Association's Eight Process Control Users Forum, San Antonio, TX, April 1992.

8.30 Baker, D. G. *Monomode Fiber-Optic Design with Local-Area and Long-Haul Network Applications.* New York: Van Nostrand Reinhold, 1987.

APPENDIX A

SIS TECHNOLOGIES

A number of technologies can be selected for SISs. These include pneumatic and hydraulic, electrical logic technology, direct-wired systems, electromechanical devices, solid state relay, motor driven timers, solid state logic, fail-safe solid-state logic, PES technology and hybrid systems. Each of these technologies is discussed below.

A.1 PNEUMATIC/HYDRAULIC

Two fluid logic technologies found in SISs are pneumatic and hydraulic. Pneumatic and hydraulic systems can be used in hazardous areas because they are inherently safe (i.e., they are not capable of providing the energy to ignite gases, liquids, or dusts). Their major drawbacks for use in SIS applications are:

- Limited computing power.
- Difficult to interface to electronic systems.
- Long communication delay.
- Short transmission distance required to keep response times short.
- Difficulty in improving system reliability through redundancy.
- Unique maintenance requirements.

Where these drawbacks do not outweigh their inherently safe benefit, pneumatics and hydraulics continue to serve as acceptable SIS technology in processes controlled by pneumatic or hydraulic systems.

A.2 ELECTRICAL LOGIC TECHNOLOGY

Electrical SIS interlocks can be implemented by using direct-wired systems, electromechanical devices, solid-state relays, motor-driven timers, solid-state logic, or fail-safe solid-state logic. These technologies are discussed in this section.

365

A.2.1 Direct-Wired Systems

Direct-wired systems have the discrete sensor directly connected to the final element as shown in Figure A.1. Figure A.1a shows the direct-wired system in block diagram form; Figure A.1b shows the schematic (ladder) diagram form.

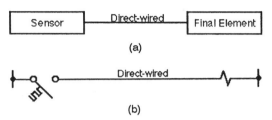

Figure A.1 Direct-wired logic system. (a) Block diagram format. (b) Schematic diagram format

Such systems may be connected to the Basic Process Control Systems (BPCSs) as shown in Figure A.2.

This scheme results in the sensor state and the BPCS output state being connected in series to the final element. The advantage of this approach is that the information is available in the BPCS network (note connection to BPCS input) without complicating or sacrificing safety integrity (note the direct-wiring approach that cannot be altered except through rewiring). The disadvantages of this approach are:

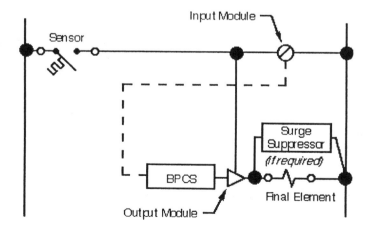

Figure A.2. Direct-wired logic interconnected with a BPCS (or SIS). Note: the BPCS output module performs an AND function in this application. The final element will not be energized unless the sensor is closed and the output module is turned ON by the BPCS. The sensor must be able to handle the load generated by the final element.

- It can only handle the simplest logic without the use of relays.
- Since the direct-wired interlock wiring is integrated into the BPCS wiring, it should be easily identifiable to avoid inadvertent wiring changes that could affect the SIS.

A.2.2 Electromechanical Devices

Electromechanical devices include relays and timers. Relays are often used where simple logic functions are adequate to provide the necessary safety logic and/or small applications where other technologies (i.e., FSSL, PES, etc.) are not justified. Extensive operating experience with relays and the mature technology make acceptance of this device as an SIS widespread.

Standards and guidelines for implementing electromechanical relays in SIS applications are available to users. Unsafe failure modes and fail-safe characteristics of relays can also be readily quantified, thus identifying the need for diagnostics. Electromechanical relays are often combined with other direct-wired, solid-state, or PES systems to form redundant SISs with diverse hardware and software.

Successful users of relays in safety applications have followed some simple guidelines. They include using a relay that:

- has a good, in plant track record,
- has the proper "fail-to-shelf" position characteristics when installed,
- is found reliable through life-cycle testing, and
- the producer has approved for safety applications.

The relay SIS has other attributes that should be considered:

- The on–off status can be readily obtained by checking contact position (e.g., open or closed).
- Its interconnected logic is very difficult to change (requires rewiring).
- It is simple and understood by plant personnel and can be easily supported.
- It is easily identified and secured as a critical control device.
- It has failure modes that can be isolated to prevent common mode failures.

Relay logic should not be considered inherently fail-safe, even if the relays are properly selected and applied so that the contacts should not weld and the spring should return the switching contacts to the de-energized position. Electromechanical relay logic systems should meet the following criteria to qualify as fail-safe (Ref. 5.1):

- Contacts open on coil deenergization or failure.
- The coil has gravity dropout or dual springs.

- Dual contacts of proper material are wired in series.
- Energy limiting load resistance is installed to prevent contacts from welding closed.
- Proper arc suppression of the contacts is provided for inductive loads.

Note: There are low energy loads (e.g., 50 V or below and/or 10 mA or below that require special contact materials or designs (e.g., hermetically sealed contacts). When utilizing these special contacts, specific failure mode analysis is needed for these contacts to ensure that a fail-safe electromechanical system is being designed.

Electromechanical relays for SIS interlocking may not be suitable for:

- Use in high duty-cycles resulting in frequent state changes.
- Use as timers (e.g., electropneumatic time delay relays [TDRs] have unsafe failure modes that should be identified and quantified) or latching relays.
- Executing complex SIS schemes involving math functions.

A.2.3 Solid-State Relays

Historically, solid-state relays were used in high duty-cycle application. However they have unsafe failure modes that can be identified and quantified. Appropriate design features should be added to handle these unsafe failure modes. Some additional applications of solid-state relays are described in the following paragraphs.

Solid-State Timers

Solid-state timers are used where the complexity doesnt warrant a PES. Solid-state timer technology can be categorized as either resistor–capacitor (RC) circuit or pulse counting. RC timing devices are not suitable for safety interlock applications because of poor repeatability and unsafe failure modes. Note that RC circuitry is often used in the time setting portion of pulse-counting timers; this does not preclude the use of these timers.

The pulse-counting timer, sometimes referred to as a digital timer, can use a number of methods to achieve pulse counting. These include:

- A line frequency (50 or 60 Hz).
- An electronic oscillator.
- A quartz crystal oscillator.

A user-approved safety crystal oscillator timer is recommended because of high repeatability and good reliability.

Input Signal Conditioners and Output Amplifiers (I/Os)
Input/output interface devices are special-purpose solid-state relays. They have unsafe failure modes that should be identified and quantified. Appropriate design features should be added to handle these unsafe failure modes before they can be approved for use in a SIS.

Input/output interfaces are required as the signal conditioners for solid-state logic systems or PESs. Input signal conditioners receive sensor signals at the strength required for suitable operation on the factory floor (e.g., 120 V, 48 V, 24 V, 4–20 mA). The purpose of the inputs and outputs in a solid-state SIS is to isolate the low energy logic system (typically low voltage DC) from the high energy field system (typical signal levels are 120 volt AC and 24 volt DC).

Low-energy signal levels are utilized in the logic system to achieve signal processing speed. High energy signal levels are used in the field devices to ensure a high signal-to-noise ratio over long transmission distances and to assure that contacts on discrete sensors used as input devices have sufficient power to provide appropriate contact-wetting. Output amplifiers receive the low energy signal from the solid state or PES logic solver and convert it to a signal strength suitable for driving the final element (e.g., load).

Current/Voltage Alarm Trips
Current/voltage alarm trips convert current and voltage (e.g., 4–20 mA or 0–10 V DC) analog inputs into discrete signal outputs. The trip value is field adjustable. These switches have unsafe failure modes; appropriate analysis and design features should be provided to ensure safe operation.

A.2.4 Motor-Driven Timers

Motor-driven timers have provided acceptable performance for key safety applications such as burner purge timing. Most motor-driven timers require a locking device or appropriate modification to eliminate tampering with critical settings. Motor-driven timers are also limited in timing resolution and the ability to handle high duty cycles.

A.2.5 Solid-State Logic

Solid-State logic refers to the transistor family of components like complimentary metal oxide semiconductor (MOS), resistor–transistor logic (RTL), transistor–transistor logic (TTL), and high noise immunity logic (HNIL). These components are assembled in stand alone modules, plug-in board modules, or in highly integrated, high-density chips. They differ from typical computer-type equipment in that they have no central processing unit (CPU). They perform according to the logic obtained by the direct-wiring techniques of interconnecting the various logic components such as ANDs, ORs, and NOTs. These systems have limitations in fail-safe requirements (e.g., indeterminate failure modes) that should be recognized.

Solid-state logic, such as TTL, has generally been integrated with direct-wiring and relay schemes to provide safety interlocking. Solid-state logic, like PES, is not recommended for SISs unless the solid-state logic was provided with features allowing its use in SIS interlocking. PESs are sometimes used as a diagnostic tool to make solid-state logic systems fail-safe.

A.2.6 Fail-Safe Solid-State Logic

Fail-safe solid-state logic (FSSL) is a mature technology marketed mostly in Europe. Typically, FSSL generates a pulse with a specified amplitude and period. Generation of a pulse train is recognized as a logic "true" or "one," whereas all other signals (e.g., grounds and continuous "on" or "off") are recognized as a logic "false" or "zero."

Some of the limitations of FSSL were its cumbersome fault-tolerant I/O, the space required, the heat generated, and the lack of communication capability. These limitations are being addressed by the use of semiconductors and the implementation of PESs for communication. When compared with PES technology, FSSL limitations may include limited communication, advanced control, and new technology integration capability.

Note that solid state logic can be considered fail-safe if it meets the requirements noted in Section A.2.5.

A.3 PES TECHNOLOGY

The PES logic solver can be a programmable controller (PLC), a distributed control system controller (DCSC) or an application-specific stand alone micro-computer. They usually have good diagnostics and can provide information well beyond conventional hardware systems. PESs should be analyzed to ensure that critical safety data are not being put into a part of the system that is not suitable for handling those data. *As of the date of this book, personal computers should not be used as SIS logic solvers.*

PESs have many failure modes. The use of software and hardware in infinite combinations results in many difficult to recognize failure modes, many of which can be classified as unsafe (Appendix B).

Some of the techniques that can be used to minimize the inherently unsafe characteristics of PES, are:

- Use of watchdog timers, both internal and external (see Appendix C)
- Use of pulsed outputs to detect output module failures (see Section 5.1.2.7)

- Use of redundancy, fault tolerant (e.g., 2-out-of-3), and similar configurations (see Section 5.4)

PESs (and other solid-state logic devices) require extensive diagnostics to provide for safe operation. This aspect of PESs is discussed in more detail in Section 5.1.2.7. Select PES technology for SIS interlocking when:

- Electronic or electrical signals are used.
- Logic requirements are complex, large numbers of I/O are required, or the logic includes computational functions.
- Extensive signal sharing with the BPCS is required.
- Electromechanical relay scheme deficiencies make their application unacceptable.
- Frequent changes to SIS logic is required.

Programmable logic controllers (PLCs) are the primary PES technology used for SIS interlocking. However, single-station digital controllers (SSDCs), distributed control system controllers (DCSCs), and single-station sequencing controllers (SSSCs) may be used for some applications. SSDCs and SSSCs refer to a family of microcomputer-based instruments that resemble the single-loop analog pneumatic and electronic controllers, but with increased capabilities. The primary function of an SSDC is PID control; it may have some sequencing capabilities. The primary function of an SSSC is sequencing; it may have some PID control capabilities.

A.4 HYBRID SYSTEMS

Hybrid systems refer to SISs using two or more technologies to provide the safety interlocking. Examples are a combination of an electromechanical system with a PES or a combination of a pneumatic system with a PES. They are commonly found in:
- Non-PES electrical systems when timing or analog I/O is required.
- SISs with two layers of safety protection, one using relays for their simplicity and the other using PESs for their computing power.

It may be necessary to read the information out of the SIS. Although direct-wired techniques to accomplish this are possible, they require many interconnections, so multiplexing techniques are typically utilized to read information from the SIS on a human/machine interface (HMI). Microprocessor systems are often embedded in solid-state logic systems to provide communications to the solid-state logic and to provide diagnostics for the solid-state logic.

The selection of the appropriate SIS technology is discussed in more detail in Section 5.1.2.1.

APPENDIX B

SEPARATION

Separation can be accomplished by a number of methods. These include no connectivity, direct-wired connectivity, read/write communication, read only communications, configurable read only communication, and memory mapping. Figures B.1 thru B.6 are provided to illustrate these schemes. Application considerations are discussed with each figure to assist in selecting the correct separation scheme for an application.

No interconnectivity (see Figure B.1) is frequently used in processes where there is no need for information transfer between the BPCS and SIS.

However, most PES-based BPCSs require some degree of interconnectivity to the SIS.

Direct-wired connectivity, shown in Figure B.2, is the traditional method of communicating between a BPCS and SIS that do not have digital communication capability (e.g., electromechanical relays and non-microprocessor based electronic systems). Direct-wiring is the communication method used when the SIS is relay-based. Direct-wiring is an alternate communication method sometimes used between PES-based BPCSs and PES-based SISs. This is used when minimal information interchange is required (because of the impracticality of using this approach for large amounts of information interchange) or serial communications do not offer satisfactory speed and security.

Figure B.3 uses serial communications and should only be used when the SIS application program cannot be altered by the read/write communications (e.g., FSSL, or ROM) and the serial communication is point-to-point (i.e., no network), User-Approved Safety.

Figure B.1. Separation—no interconnectivity.

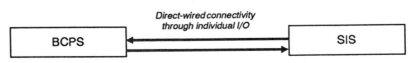

Figure B.2 Separation—direct-wired connectivity.

Figure B.3. Separation—serial communication (read/write).

Figure B.4 is the scheme recommended for communication between PES-based BPCSs and SISs. The communication is point-to-point, user-approved safety, serial and read-only. "Read only" means the SIS program cannot be altered from the BPCS during normal process operation. This ensures that unintentional changes to the SIS program do not occur. "Read only" does allow commands (e.g., stop, run) to be transmitted from the BPCS to the SIS during normal operation.

This is accomplished with a User Approved Safety gateway ("G" in Figure B.4) that communicates to the BPCS. This same gateway also communicates to the SIS. The BPCS can read/write to the gateway (similar to Figure B.3), but it cannot write directly to the SIS. This gateway prevents direct BPCS/SIS information interchange via its hardware architecture and/or application program and provides the overall read-only communications between the BPCS and SIS. Therefore, the overall communication between the BPCS and the SIS functions effectively as read only. Because of the read-only communication conducted at the BPCS, the SIS application program cannot be altered on-line, and only with appropriate security off-line, from the BPCS. The read-only communication is used to transport the status of the SIS and its program to the BPCS controller and HMI; the BPCS may use this information

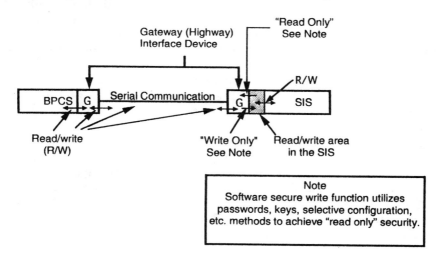

Figure B.4 Separation—serial communication (read only).

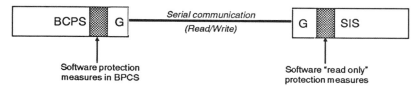

Figure B5. Separation—"soft" read only.

for diagnostic purposes and for operator information interface functions (e.g., alarms, diagnostics).

There is a need for communication between logic systems in safety applications. The most common need is between the BPCS and SIS. In some applications, there may only be a need for the BPCS to be able to read data from the SIS (e.g., the status of safety interlocks within the SIS so they can be displayed on a HMI in the BPCS). Figure B.4 addresses this need. The main requirement is that the BPCS be able to accomplish this "read" without corrupting the memory of the SIS.

A method that may be acceptable is called "Soft Read Only" (Figure B.5). Here the BPCS and SIS offer software security (e.g., passwords, key locks) to protect against inadvertent changes to the SIS. Consider this method for low-integrity SISs and for higher integrity SISs where supplemented by hardware "write" protect feature. PHA team approval is required.

The PHA team will review the risk reduction strategy to understand if the SIS is the only IPL protecting against the hazard. If other IPL(s) exist understand their separation, diversity, independence, capability to be audited, risk reduction capabilities, and if they have any common mode faults with the BPCS and the SIS. If these non-instrumented IPLs can provide suitable risk reduction and do not have common mode faults with other IPL(s), then the PHA team may accept "soft read only" separation techniques and minimal hardware diversity.

This method is acceptable for IL1 and for IL2 with PHA approval. Hardware "write" protect is required for IL3. The hardware "write" protect prevents programming of the SIS with the "hardware" write protect set in a defined physical position.

Figure B.6 illustrates an infrequently used separation concept found in some packaged safety systems. This approach integrates the BPCS and SIS logic into a single SIS. The BPCS program is partitioned from the SIS program to minimize the potential for inadvertent changes to the SIS program while working on the BPCS program. The BPCS portion should adhere to all the rules that are applied to the SIS. *Note: This does not provide the same degree of separation as the previous techniques, and may not be acceptable for the highest integrity level systems.* PHA approval is required.

Figure B.6. Separation—SIS with embedded BPCS functions.

APPENDIX C

TYPICAL WATCHDOG TIMER CIRCUITS

C.1 INTERNAL WATCHDOG TIMERS

Internal watchdog timers include software, hardware, and communication diagnostic subsystems provided by the manufacturer, within the PES.

PESs for critical applications should provide diagnostics for all elements of the PES. A WDT system within a PES may provide user selectable options ranging from the shutdown of an input or output card to total shutdown of the system. Internal WDT diagnostics check items the PES manufacturer considers most important. The limitations of an internal WDT include:

- It may fail for the same reason the PES failed and be unable to perform its monitoring functions.
- It may not provide the user with PES availability status.
- It may fail to monitor the complete application.
- It is not application program specific.

C.2 EXTERNAL WATCHDOG TIMERS

The limitations inherent in internal WDTs require the addition of external WDTs for PESs performing safety interlocks. The use of external WDTs in no way eliminates the need for internal WDTs for safety interlocks.

The external WDT frequently used today is an electronic timing device. In its most basic form, the external WDT is continuously pulsed by application logic located in the PES user program. The concept generally employed is to program several groups of instructions (that are widely separated in key memory locations) to generate a square wave with a desired period. Two other features typically included in the external WDT include

- input(s) (connected to output with direct connection program) to detect input system faults,
- appropriate output (s) (with no applications program) that detects output system faults and Figure C.1 shows the principle.

The width of the pulses in this square wave can be varied by changing the application program in the square wave generator. Figure C.2 shows a sample

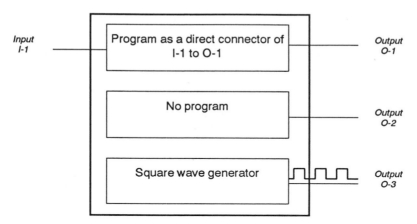

Figure C.1 Application program for use with the external WDT.

external WDT circuit for a Modicon 984-480 PLC. The dual timer circuit is used to generate the square wave that is used to generate output O-3. This output is used as the input to the external WDT. The lengths of the pulses in the square wave can be changed by varying the settings of the two timers. Figure C.3 is a timing diagram that shows the output of the square wave generator and the output of the WDT circuit of Figure C.1.

This square wave drives an output (0–3 in Figure C.4) on and off in the correct timing sequence that keeps the WDT output (1CR in Figure C.4) energized. Note that the external WDT has built in adjustable on-delay and off-delay timer functions. The ON-delay and OFF-delay timer settings of the external WDT are set so that neither delay should time out. If the WDT times out, 1CR drops out , and the system may be shut down and/or alarmed (see the auxiliary schematics in Figure C.5).

Design features to improve WDT monitoring capability include:

- PES square wave generation for the WDT utilizing the same instruction set used by the safety interlocking program.
- Dedicated PES inputs, typically one per SIS subsystem, monitor the state of the PES input system buses to detect abnormal operation. The following is an example of how one safety input subsystem is monitored.
  —Dedicate input I-1 (see Figures C.1 and C.2) to detect abnormal input bus operation.
  —Program I-1 (see Figure C.1) so output O-1 will energize (or be pulsing) when I-1 is on (or pulsing).
  —Locate I-1 on the same input board as safety interlock inputs.
  —Note that input I-1 is connected to neutral on the "common" side and disconnected on the "switched" side (see Figure C.4).

- Note that outputs O-1 and O-2 are connected to relay 2CR. 2CR has four contacts:
  —2CR-1 that seals in around O-1 and O-2 whenever either output is energized (or pulsing), causing C to stay energized.
  —2CR-2 that deenergizes the load device (see Figure C.5).
  —2CR-3 that activates an alarm (see Figure C.5).
- Distribution of the WDT program across various memory locations of the PES, if possible, that will best monitor total memory functionality. The program may include unique passive or active diagnostics.

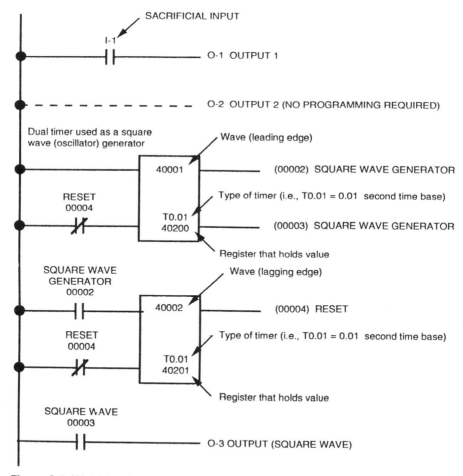

Figure C.2 Watchdog timer application program (square wave generator). Typical program for a Modicon 984-480 programmable controller. (Adapeted with permission of AEG Modicon.)

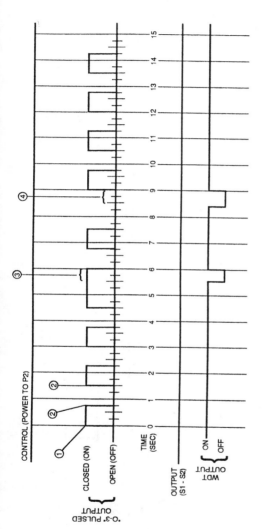

Notes:

① Closing the control circuit energizes the output.

② Opening and reclosing the control circuit before the set time interval (assume set at 1 second) is complete keeps the output (0-3) energized. The output (0-3) remains energized as long as the monitored pulsing continues to provide at least 1 transition per set time interval.

③ If the monitored control stays on longer than the preset time ③, the output (R1) de-energizes.

④ If the monitored control stays off longer than the preset time ④, the output (R1) de-energizes.

Note: Refer to Figure C-4 Outputs O-1 and O-2, reset, and test functions are not included in the above timing diagram.

Figure C.3 WDT timing diagram.

379

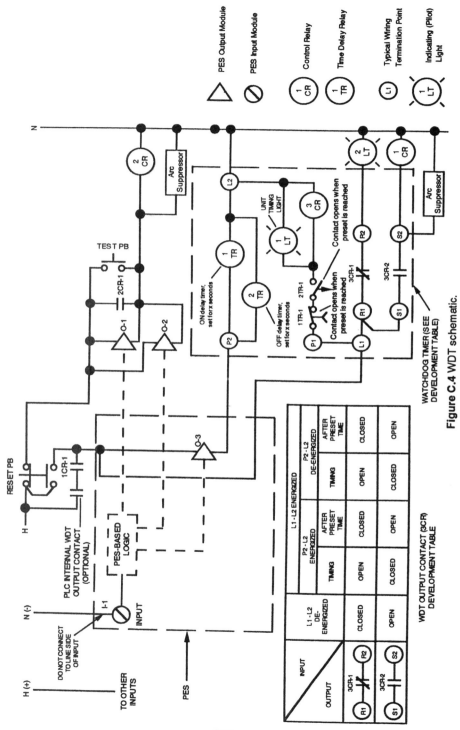

Figure C.4 WDT schematic.

	PES Output Module
	PES Input Module
① CR	Control Relay
① TR	Time Delay Relay
① L1	Typical Wiring Termination Point
① LT	Indicating (Pilot) Light

WDT OUTPUT CONTACT (3CR) DEVELOPMENT TABLE

INPUT OUTPUT	L1-L2 DE-ENERGIZED	L1-L2 ENERGIZED			
		P2-L2 ENERGIZED		P2-L2 DE-ENERGIZED	
		TIMING	AFTER PRESET TIME	TIMING	AFTER PRESET TIME
3CR-1 (R2/R1)	CLOSED	OPEN	CLOSED	OPEN	CLOSED
3CR-2 (S2/S1)	OPEN	CLOSED	OPEN	CLOSED	OPEN

380

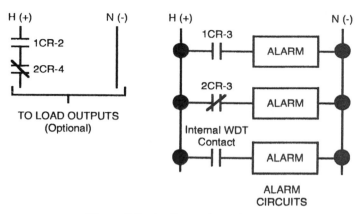

Figure C.5   Auxiliary schematics

- Full system check of PES communications. To accomplish this assign the PES I/O for the WDT (I-1, O-1, O-2, O-3, etc. in Figure C.4) to remote I/O locations. This will inherently check full communication operation of the PES I/O links.
- The possible need for test and/or reset buttons. A test button may be desirable. A reset button will be required if the WDT is interlocked down at start-up or upon shutdown (see Figure C.4).
- Dedicated PES outputs, typically one per safety subsystem, monitor the state of the PES output system buses to detect abnormal operation. The following is an example of how one safety output subsystem is monitored.
  —Dedicate output O-2 (see Figure C.4) to detect abnormal output bus operation.
  —Locate O-2 on the same output board as the safety subsystem it is monitoring.
  —Note that output O-2 is connected in parallel to output O-1 (see Figure C.4).
  —Note that O-2 has no application program. It will always be OFF unless there is a malfunction in the output bus operation that inadvertently turns (or pulses) the output ON. If this happens, the input to the external WDT will be constantly in the energized (ON) state (via contact C-2), and the WDT will trip.
  —Note that outputs O-1 and O-2, may be combined into one output. Sufficient output addresses allowing monitoring of PES output subsystem communications determine whether one, two, three, or more outputs will be used for the overall WDT.
- Dedicating output O-3 as the pulsing output and connecting O-3 to the WDT and in parallel to contact C-2 (see Figure C.4) .

- A surge suppressor to dampen the inductive interaction to the electronics from 1CR and 2CR (see Figure C.4). Review the application for additional power line conditioning requirements such as:
  —Undervoltage protection.
  —Electrical noise suppression.
  —Lightning protection, etc.

Relays 1CR and 2CR (see Figure C.4) should be user-approved safety relays.

It is preferable to connect the internal WDT output contact, if process shutdown is required, parallel to the reset button of the external WDT as shown on Figure C.4. A separate alarm for internal WDT operation is desirable, accordingly an additional relay may be required (see Figure C.4).

APPENDIX D

COMMUNICATIONS

This appendix addresses the relationship of information transmitted over communications links and the suitability of various communications techniques for use in SISs. SISs will use various serial and/or parallel communications at the board-level; over the back-plane; to peripherals; and to various other components in the system (e.g., I/O, HMIs, back-up controllers, engineer's workstations, and programming panels).

Safety interlock systems should utilize *User-Approved Safety* communications; the BPCS generally uses *User-Approved* communicators. Figure D.1 shows local I/O with bus connectivity (see Figure D.1a) and ribbon-cable connectivity (see Figure D.1b). Vendor installation requirements should be followed to realize proper PES operation.

Figure D.2 shows remote I/O as used for both SISs and BPCSs. The serial communications that are designed for a specific SIS configuration have better diagnostics and security (see Figure D.2a) than do the general purpose serial

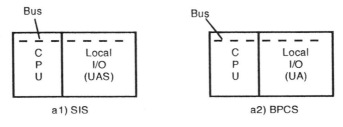

a) Bus Connectivity

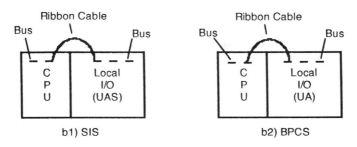

b) Ribbon-Cable Connectivity

Figure D.1 Local I/O.

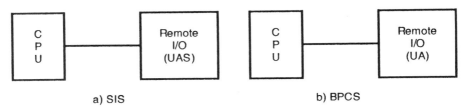

Figure D.2. Remote I/O

communications or parallel communications that are used in many BPCSs (see Figure D.2b).

Figure D.3 shows communications between two SISs or two BPCSs. The direct-wired or dedicated, serial communications used in Figure D.3a makes this communications suitable as User-Approved Safety. The general-purpose serial communications or parallel communications used in Figure D.3b may not be as secure and may not have the necessary features to allow it to be classified as User Approved Safety.

There are some limitations on what data can be communicated outside of the confines of the SIS (e.g., within the BPCS). Separation is required between the BPCS and SIS (See Section 5.1.2.4).

D.1 COMMON COMMUNICATIONS TECHNIQUES

There are a variety of communications techniques in common use. This section discusses some of these techniques for several types of SIS technology.

Electromechanical Devices to PESs
Relays (and other electromechanical devices) are typically wired directly to the PES I/O (see Figure D.4).

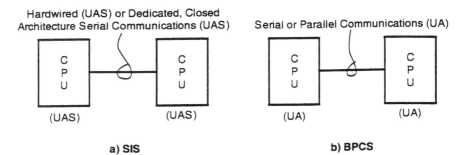

Figure D.3 SIS–SIS or BCPS–BCPS communications

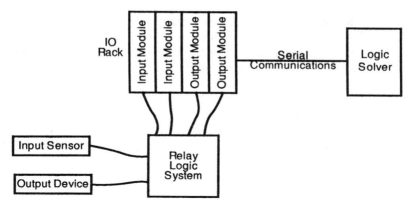

Figure D.4. Electromchanical logic to PES communications.

Solid-State Logic to PESs
The solid state-logic interface emulates the electromechanical interface to PESs. The difference is that the connections between the solid-state logic system and the PES I/O can be low level signals (less than 24 V) if desired. The advantage of the solid-state logic interface to the PES is that it can be done at the low energy levels so that the input interfaces themselves are less complex and usually smaller in size. The disadvantage of using low-level signals is that they are more sensitive to noise pickup.

Pneumatic and Hydraulic to PESs
Pneumatic and hydraulic systems also need to interface to the human/machine interface. These are typically used for local controls with limited monitoring. These signals must be converted into electrical signals so that they can be treated in the same manner as electromechanical and solid-state logic. The configuration of this system would be similar to that of Figure D-4 except a converter would be needed to convert the pneumatic or hydraulic signals to electrical signals before they could be connected to the PES I/O.

D.2 I/O COMMUNICATIONS

The I/O in a PES SIS offer some different mounting options such as:

- At the controller (i.e., local I/O).
- Remote from the controller (i.e., remote I/O).
- In a daisy-chain format (e.g., with the newer technology distributed I/O).
- With a separate I/O controller (e.g., in the I/O typically provided with a distributed control system).

A positive aspect of local I/O is that there is indication of I/O status with the CPU and the I/O in one location. The advantage of remote I/O is to reduce the installation complexity and allow the location of the input/output line conditioners adjacent to the sensors and final elements. The disadvantage of this system is that the controller is at some distance from the I/O and its status may only be available through some programming panel located out in the field providing diagnostics information. Because local I/O are located next to the controller and typically housed in the same cabinet as the controller, there is a perception among producers that communication does not have to be as secure as a remote I/O device. The result is that often local I/O is done on parallel communications or a less secure communication technique. If the cabinet is arranged properly and if the equipment is mounted properly, this should not be a problem. If the user or vendor makes some mounting mistakes resulting in I/O communication losing EMI/RFI field/INS protection, experience has shown that there have been some noise problems with this type of I/O communication.

Many use remote I/O even though the I/O is located locally. Remote I/O (Appendix F, Table F.5) is ideally suited for serial communications. An inherently higher security is often built into remote channel communications than into local channel communications, because remote channels are more prone to communication problems. Unfortunately, remote channel communications diagnostics may slow down processor response time. This is not important unless the application requires high speed response. Usually, remote I/O is used on safety applications even though the I/O is located adjacent to the processor.

The newer technology distributed I/O (e.g., fieldbus) has not had the proven field experience needed to recommend it for SIS applications, except for one layer of a multiple protection layer SIS application. The star-type configuration that is typical of user approved safety controllers DCS has the necessary communications security and proven field experience to allow it to be used for SIS applications.

APPENDIX E

SENSOR FAIL-SAFE CONSIDERATIONS

The actual principle behind the sensing device should be investigated. An example is provided to show how two common sensors are analyzed and, where safety is a consideration, applied to achieve optimum safety.

Figure E.1 shows two common ultrasonic point level switches. In the level switch in Figure E.1a, the diaphragm (face of the probe) vibrates whenever energy is supplied to the drive coil in the probe. If the probe is covered with a gas, the diaphragm is free to move, and this energy is transferred to a pickup coil. When the diaphragm motion is damped by a noncompressible liquid, the energy transferred to the pickup coil decreases and the output relay drops out (switch trips). This type of level switch is more fail-safe when it is used as a high level switch where it is normally exposed to a compressible gas, and the output relay is energized. The output relay will then drop out if the probe sees liquid, if the face of the probe is coated with solids, or if power is lost.

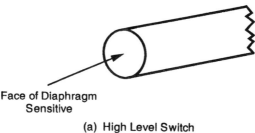

Face of Diaphragm
Sensitive

(a) High Level Switch

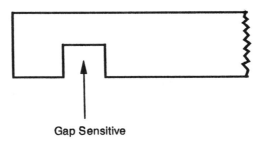

Gap Sensitive

(b) Low Level Switch

Figure E.1 Two types of ultrasonic point level switches

Figure E.1b shows another ultrasonic level switch that is more fail-safe when it is used as a low level switch. This device has a transmitter transducer on one side of the gap and a receiver transducer on the other side of the gap. The control unit generates an electrical signal that is converted to an ultrasonic signal at the transmitter side of the gap. If the gap is filled with liquid, the signal is transmitted across the gap and is received by the receiver. This signal is amplified in the control unit and energizes the output relay. When liquid falls below the sensor gap, the signal is attenuated and the output relay drops out. This type of level switch should be used in a low level application where the gap is normally filled with liquid, and the output relay is energized. The output relay will then drop if the probe sees a compressible gas or vapor, if the faces of the probe are coated with solids, or if power is lost.

APPENDIX F

SIS EQUIPMENT SELECTION CONSIDERATIONS

Guidelines are presented to aid in the application of five aspects of the SIS equipment in relation to Figure F.1: (1) separation, (2) diversity, (3) software, (4) diagnostics, and (5) communications. Note that the PHA team's evaluation of the risk control requirements is utilized to determine the extent that each of these five aspects are required in the SIS. From this PHA guidance, proper SIS equipment selection can be made.

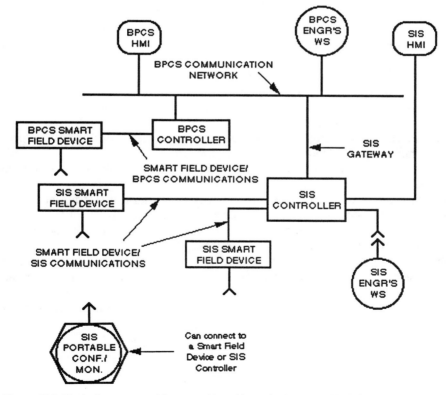

Figure F.1. Typical system architecture. *Note:* Many devices used in SISs are not "smart" (e.g., an electromechanical relay). "Smart" devices are shown in Figure F.1 so that application guidance can be provided. Generally, the same guidance provided applies to both "smart" and conventional field devices.

Figure F.1 shows a typical system that will be used to illustrate how some of the concepts discussed in Chapter 5 can be applied. This figure assumes that only one SIS controller (logic solver) is needed. That is not true in many applications. Even when only one SIS controller is used, there may be a need for multiple field devices.

Additional definitionS for the devices shown in Figure F.1 follow:

BPCS COMMUNICATION NETWORK—the data highway that allows communications within the BPCS

BPCS CONTROLLER—the controller that is part of the BPCS and that is used for normal process control [e.g., could be a distributed control system controller (DCSC), a programmable controller (PLC), or a single-station digital controller (SSDC)].

BPCS ENGR'S WS—the engineer's workstation that is used for application programming and troubleshooting for the BPCS.

BPCS HMI—BPCS human/machine interface used for operator display and action.

BPCS SMART FIELD DEVICE—a programmable electronic device incorporating one or more microprocessors (e.g., a smart transmitter) that is used as an input to the BPCS.

BPCS SMART FIELD DEVICE/BPCS COMMUNICATIONS—the serial communications link between a smart field device and the BPCS.

SIS CONTROLLER—a logic solver (e.g., a PLC, electromechanical relays, etc., as discussed in Appendix A).

SIS ENGR'S WS—the engineer's workstation that is used for application programming and troubleshooting for the SIS.

SIS GATEWAY—a safety gateway that is used to connect the BPCS to the SIS (see Appendix B, Figure B.4, page 373).

SIS HMI—SIS human/machine interface device used for operator display.

SIS PORTABLE CONF./MON.—A portable SIS configurator/monitor that can be used to configure a smart field device or SIS and that can also be used to monitor the operation of these devices.

SIS SMART FIELD DEVICE—a programmable electronic device incorporating one or more microprocessors (e.g., a smart transmitter) that is used as an input to the SIS.

SIS SMART FIELD DEVICE/SIS COMMUNICATIONS—the communications link between a smart field device and the SIS.

F.1 SEPARATION APPLICATION GUIDELINES

Separation (Appendix B) must be considered for hardware, system software, and application programming. When a SIS as shown in Figure F.1 is implemented, the following separation is recommended:

- Hardware
  - BPCS controller from SIS controller
  - BPCS smart field device from SIS smart field device
  - Smart field device–BPCS communications from smart field device–SIS communications
- System software
  - BPCS controller from SIS controller
  - BPCS smart field device from SIS smart field device
- Application programming
  - BPCS controller from SIS controller
  - BPCS smart field device from SIS smart field device

When more than one SIS controller is required, separation is also recommended (in addition to the above recommendations) between the two SIS controllers, between the smart field devices for the two SIS controllers, and between the communications links to the smart field devices for the two SIS controllers.

F.2 DIVERSITY APPLICATION GUIDELINES

Diversity (Section 5.1.2.5) should be considered for hardware, system software, and application programming. Hardware and software diversity may be achieved by the use of:

- different technologies (see Appendix B), or
- two different manufacturers' or vendors' products, or
- two different products from the same manufacturer, and
- different application programming teams.

When a SIS as shown in Figure F.1 is used, diversity should be considered when selecting the following:

- Hardware
  - BPCS controller and SIS controller
  - BPCS smart field device and SIS smart field device
  - Smart field device–BPCS communications and smart field device–SIS communications
- System software
  - BPCS controller and SIS controller
  - BPCS smart field device and SIS smart field device
- Application programming
  - BPCS controller and SIS controller
  - BPCS smart field device and SIS smart field device

Note that the two SIS smart field devices in Figure F.1 are independent sensors measuring the same variable. Diversity may be needed when selecting the two SIS smart field devices in Figure F.1 to achieve the necessary level of measurement integrity.

When redundant paths are required, as encountered in IL2 and always in IL3, Table F.1 is used. In these cases each SIS controller may have multiple sensors measuring the same variable for a single SIS controller.

F.3 SOFTWARE APPLICATION MATRIX

Software functionality found in a typical process control system (see Figure F.1) includes:
- BPCS COMMUNICATION NETWORK software
- BPCS CONTROLLER application program and software
- BPCS ENGINEER'S WORKSTATION software

Table F.1 Diversity Application Matrix for a Single SIS with Redundant Paths

BPCS/SIS Component	IntegrityLevel 1	Integrity Level 2	Integrity Level 3
Hardware			
BPCS controller and SIS controller	O	O	O(1)
SIS controller and SIS controller	NA	R(1),(2)	R(1)
SIS smart field device and SIS smart field device	O	O	R(1)
System Software			
BPCS controller and SIS controller	O	O	O(1)
SIS controller to SIS controller	NA	R(1),(2)	R(1)
SIS smart field device and SIS redundant smart field device	O	O	R(1)
Application Programming			
BPCS controller and SIS controller	O	O	R(1)
SIS controller and SIS controller	NA	R(1),(2)	R(1)
SIS smart field device and SIS redundant smart field device	O	O	R(1)

NA = not applicable   O = Optional   R = Recommended
(1) Diversity can cause serious problems when reliability is sacrificed to achieve diversity—this should not be done. Utilize diversity only when reliable components (e.g., sensors, final control elements, logic solvers etc.) are available for each of the diverse systems, subsystems, etc.
(2) Applies when 2 SIS controllers needed to meet IL2.

- BPCS HMI console software (application programming may not be required in the HMI)
- BPCS SMART FIELD DEVICE application program and system software
- BPCS SMART FIELD DEVICE/BPCS COMMUNICATIONS software
- SIS CONTROLLER application program and system software
- SIS ENGINEER'S WORKSTATION software
- SIS GATEWAY application program and system software
- SIS HMI (used for operator display) software
- SIS PORTABLE CONFIGURATOR/MONITOR software
- SIS SMART FIELD DEVICE application program and system software
- SIS SMART FIELD DEVICE/SIS COMMUNICATIONS software

Table F.2 is a software matrix that discusses system software and application programming for the SIS shown in Figure F.1.

Some examples of the use of the matrix are given below:

- The matrix shows that it may not be acceptable to use the BPCS HMI as the only HMI for the SIS, regardless of the integrity level. The exception is when the PHA team has determined that the process can be shut down safely upon loss of the BPCS HMI, and the plant operating procedures state that the process must be immediately shut down upon loss of the BPCS HMI. Loss of the BPCS HMI must not mask critical SIS status

Table F.2 Software Application Matrix

SIS Software Functions	Integrity Level 1	Integrity Level 2	Integrity Level 3
SIS Logic Solver Application Software Installation			
BPCS controller (User Approved) only	NR	NR	NR
BPCS controller (User-Approved Safety) only	O	O(1)	O(1)
SIS controller only	R	R	R
SIS controller w/mirroring in BPCS	O	O	O
Maintenance and Design Software			
BPCS engineer's workstation	O	O(1)	O(1)
SIS engineer's workstation	R	R	R
Portable SIS configurator/monitor	O	O	O
Programming Utility Software			
Third party vendor supplied	O(1)	O(1)	O(1)
SIS hardware vendor supplied	R	R	R

NR = not recommended   O = Optional   R = Recommended
(1) Implementation of this software function should not proceed without PHA approval.

information needed by the operating staff during a BPCS "loss of view" emergency. It is acceptable to use the SIS HMI alone while recovering from loss of the BPCS HMI, but the use of both the BPCS HMI and the SIS HMI is typical. The PHA team should define the acceptability of either approach.

- Using the SIS controller and also mirroring this logic in a BPCS controller is discussed in Section 5.1.2.7; it adds more software complexity to the system, but it does provide a method of providing additional diagnostic coverage for the SIS.
- A User-Approved (UA) BPCS controller, as the only SIS controller, should not be used for any of the three integrity levels. The use of a User-Approved Safety (UAS) BPCS controller is acceptable for integrity level 1 and may be acceptable for IL2 and as one of the controllers for IL3; to proceed in this manner there must be diversity in other IPLs, and PHA team approval is required.
- The SIS engineer's console is recommended for maintenance and coding of the SIS application program. The portable SIS configurator/monitor is optional.
- Caution should be exercised when using third party software for developing and maintaining the application program. Software that has been approved by the manufacturer of the hardware is recommended because of their knowledge of the equipment, the ease of maintaining version-control of the software, and good user experience with these products.

F.4 DIAGNOSTICS APPLICATION MATRIX

Active diagnostics are recommended for Integrity Level 3 interlocks. Although passive diagnostics are suitable for Integrity Level 2 interlocks, this is a minimum requirement; active diagnostics are preferred. Passive diagnostics are acceptable for Integrity Level 1 interlocks. Active diagnostics can improve the integrity of the SIS. Table F.3 provides diagnostic application guidance for SISs.

F.5 COMMUNICATIONS APPLICATION GUIDELINES

A communications matrix for selecting the appropriate communication techniques for the example shown in Figure F.1 is shown in Table F.4. For example, use of the BPCS HMI for displaying SIS information is typical. If the BPCS HMI fails, there may be procedures in place to restore the BPCS to normal operation. An SIS HMI should be provided to allow appropriate monitoring during this procedural phase. The recommended communications technique

Table F.3 Diagnostics Application Matrix for a Single SIS with Redundant Paths

SIS Device	Integrity Level 1	Integrity Level 2	Integrity Level 3
Inputs (sensor through input module)			
BPCS controller (User-Approved Safety) as an SIS logic solver	R_P	$R_P(1)$	$R_A(1)$
Diverse SIS controllers	R_P	R_P	R_A
SIS smart field devices	R_P	R_P	R_A
Outputs (output module through final control element)			
BPCS controller (User-Approved Safety) as an SIS logic solver	R_P	$R_P(1)$	$R_A(1)$
Diverse SIS controllers	R_P	R_P	R_A
SIS smart field devices	R_P	R_P	R_A
Internal PES Diagnostics (includes internal watchdog timer)			
BPCS controller (User-Approved Safety) as an SIS logic solver	R_P	$R_P(1)$	$R_A(1)$
Diverse SIS controllers	R_P	R_P	R_A
SIS smart field devices	R_P	R_P	R_A
External Watchdog Timer			
BPCS controller (User-Approved Safety) as an SIS logic solver	R_P	$R_P(1)$	$R_A(1)$
Diverse SIS controllers	R_P	R_P	R_A
SIS smart field devices	R_P	R_P	R_A
Communications Diagnostics			
SIS gateways	R_P	R_A	R_A
SIS smart field devices/ SIS communications	R_P	R_P	R_A

R_A = Active Diagnostics Recommended
R_P = Passive Diagnostics Recommended
(1) The use of the BPCS controller, even if it is User-Approved Safety, requires PHA team approval as one of the SIS logic solvers for Integrity Level 2 and 3 interlocks.

between the SIS smart field device and the SIS controller is User-Approved Safety analog. User-Approved Safety digital transmission is optional as long as a hard-wired (such as a removable jumper) write protect scheme inhibits programming during normal operation. Only User-Approved Safety communications link should be used for any safety integrity level interlocks.

Table F.5 is an I/O application matrix that will assist in the selection of the proper type of I/O communications for an SIS.

Table F.4 Communications Matrix			
SIS Function	Integrity Level 1	Integrity Level 2	Integrity Level 3
Display of SIS logic functions using the BPCS HMI via the SIS gateway and the BPCS communications network	O(1)	O(1)	O(1)
Diagnostics of the SIS with the BPCS engineer's workstation via the SIS gateway and the BPCS communications network	O	O	O
Display of the SIS from the BPCS HMI via the SIS safety gateway and the BPCS "read" only communications network	O(2)	O(2)	O(2)
Configuration of the SIS controller from the engineer's workstation via the SIS safety gateway	O	O(3)	O(3)
SIS Smart Field Device to SIS Controller:			
User-approved safety, analog communications link	R(4)	R(4)	R(4)
User-approved safety, digital communications link, no remote programming	O(4)	O(4)	O(4)

O = Optional   R = Recommended
(1) Redundant display paths required.
(2) With User-Approved Safety SIS gateway and BPCS communications network.
(3) Acceptable if an additional independent protection layer is provided.
(4) Communication links that are at "user-approved" status only, are not recommended.
(5) Requires PHA team approval.

Table F.5 I/O Communication Application Matrix

Type of I/O Communications	Block Diagram Example	Integrity Level 1	Integrity Level 2	Integrity Level 3
Local I/O (Backplane Connected)	PES | I/O	R	R	R
Local I/O (Ribbon cable connected)	PES I/O	O	O	O
Remote I/O (Direct Communications)	PES — I/O	O	O	R(1)
Remote I/O (Daisy-chained racks)	PES / R R R / I/O I/O I/O	O	O	O
Distributed I/O	PES / I/O I/O I/O	O*	O*	O*
Remote I/O (with separate I/O controller)	I/O / PES — I/O / I/O Controller	R**	R**	R**

O = Optional   R = Recommended

*Care should be taken because of the lack of proven field experience.

**This SIS controller should be User-Approved Safety and should be separate from the BPCS.

Note: All communications are User-Approved Safety.

(1) Remote I/O communications has UAS approved security and error checking, hence the "recommended" guidance.

APPENDIX G

POTENTIAL PES FAILURE MODES

The table that begins on the facing page provides a list of some potential failure modes for a PES with one processor, one I/O scanner, one rack adapter board, and one AC and one DC I/O module (each of the I/O modules has 16 channels).

This list is submitted to reflect the depth of analysis that a PES vendor should provide before the vendor's PES is certified for critical interlock applications (e.g., User-Approved Safety). Typically the vendor provides a quantitative analysis (events versus time) of control system fail-to-danger modes. Vendors may utilize FMEA of the equipment, Markov models, field data, or a combination of these methods to obtain these numbers (see Chapter 3).

The use of redundant, fault tolerant (e.g., 2-out-of-3), and similar configurations require additional considerations (e.g., communication failure modes between controllers providing incorrect information, 2-out-of-3 board vote to on failure, etc.). The vendor should provide the list of dangerous failure modes for the equipment.

Module	Failure Mode	Effect
Discrete Input	Erroneous output bit No real input signal present—one channel	False indication to processor of an input true condition
Discrete Input	Loss of strobe out	One input module left unattended (16 inputs)
Discrete Output	False output with no input present—one channel	False output on one channel
Discrete Output	No output Data present one channel	Failure to set one output to on state (1)
Discrete Output	Loss of output reset capability—16 channels	Inability to reset 16 outputs of one module
Discrete Output	Loss of fuse group—Loss on all 16 outputs	Loss of blown fuse fault detection (2)
Memory	Failure to properly decode address resulting in memory loss. (Not checked by parity. Read/write locations do not correspond.) (3)	Erroneous data transfer to and from memory Effective loss of memory Random outputs on all outputs.
Processor	Loss of program counter failed diagnostic Loss of decision logic test diagnostic Loss of parity check capability Loss of clock fail check	Unattended I/O racks in event of counter failure Inability to detect major fault (clock fail). Racks left unattended Inability to detect major fault (clock fail). Racks left unattended Inability to detect major fault (clock fail). Racks left unattended
Processor	Loss of I/O fault present line (I/O or input parity)	Loss of fault detection capability and ability to reset all outputs
Processor	Loss of comparator capability	Loss of timer, counter, functions
Processor	Loss of 1 to 4 bits of data or instruction word to memory. Possible word is assigned parity, even if incorrect	Erroneous program execution
Processor	System software bug that affects counter, preset values, math, and timer functions	Loss of counter, math, preset values, and timer capabilities
I/O Scanner	Loss of data in control line	Discrete input modules left unattended

Module	Failure Mode	Effect
I/O Scanner	Loss of enable drivers signal	Parity check on address data to adapter module–detected failure (1)
I/O Scanner	Loss of data out control line	Discrete output modules left unattended
I/O Scanner	Loss of read / write line	Modules left unattended
I/O Scanner	Loss of internal watchdog timer	I/O racks left unattended
I/O Scanner	Loss of I/O communications capability	All modules left unattended
I/O Scanner	Loss of scan control/time base—stalled scanner	I/O racks left unattended
I/O Scanner	I/O address counter stopped or malfunction and/or loss of reset slot address and loss of alternate input/output line	I/O racks left unattended
I/O Scanner	Loss of outputs off (fault diagnostic)	Inability to shut down system in event of a major fault. (2)
I/O Scanner	Loss of minor fault (fault diagnostic)	Inability to detect blown fuse in 8 or 16 discrete output channels (2)
I/O Adapter	Loss of data in requested to modules	All input modules left unattended
I/O Adapter	Incorrect data bit to module	Erroneous output on one channel discrete
I/O Adapter	Loss of rack present to cable	Checked by processor diagnostics (2)
I/O Adapter	Loss of data strobe	Checked by processor diagnostics (2)
I/O Adapter	Loss of output reset line	Inability to reset all outputs in one rack
I/O Adapter	Loss of data parity check and parity generator at rack	Parity checked at processor (2)
I/O Adapter	Loss of "all modules not addressed" check	Possible unattended module—discrete 16 inputs or outputs
I/O Adapter	Loss of fault indication from modules	Inability to detect blown fuse on 8 or 16 channels (2)
Line Filter Receiver	Output fails to high state	Various—for reference only
Line Filter Receiver	Output fails to low state	Various—for reference only
Typical High Level Driver	Output fails high	Various—for reference only

NOTES TO TABLE:

(1) Typically a problem in Energize-to-Trip systems only.
(2) This is a diagnostic problem that may result in serious integrity degradation of the SIS, if the diagnostics area with the problem is part of the SIS fault tolerant design.
(3) The information requires interpretation by microcomputer users in some cases. An example is the reference to parity. Software security (e.g., checksum, CRC checks, etc.) may be substituted for parity as required.

APPENDIX H

FACTORY ACCEPTANCE TEST GUIDELINES

Before a BPCS/SIS is shipped from its staging site to the final plant site, a factory acceptance test (FAT) may be performed. Use of guidelines such as the following will ensure a successful FAT. The user may select alternate method(s) to accomplish similar objectives.

H.1 PURPOSE OF THE FAT

The purpose of the FAT is to test both the software and hardware functionality of the BPCS/SIS system as an integrated unit. The goal is to identify and resolve any problems in the system before it arrives at the plant site. There may be some areas where certain problems cannot be corrected until the equipment is received at the plant, but these should be kept to a minimum.

The FAT should include testing of all hardware components and software in the system. This may be accomplished by having the testing performed by the vendor or the user, or a combination of the two. The latter approach is the most desirable.

H.2 PARTICIPANTS IN THE FAT

Participants of the FAT represent both the vendor and the user. The combined team should consist of the following personnel:

- The user engineer responsible for the system specification.
- The user and/or vendor personnel responsible for software programming/configuration.
- The user and/or vendor personnel responsible for hardware integration.
- The user and/or vendor support personnel.

One of the team members, generally the user engineer, should be selected as leader of the FAT. The number of participants depends on the technologies involved and the size, and complexity of the system to be tested.

H.3 PREPARATION FOR THE FAT

As a part of the purchase order, the user team members should specify the test requirements by including a list of items that the test will encompass. This should include, at a minimum:

- A general description of the test procedure based on the system specification.
- A definition of responsibilities for each task during the FAT.
- A schedule showing the daily activities during the FAT.
- A list of the documents that are required at the FAT site during the test.
- A list of the test equipment and instruments required to conduct the FAT.
- A detailed procedure to be followed during the FAT.

The FAT team leader should present the list of requirements to the vendor through written correspondence prior to the scheduled FAT. The vendor may wish to modify some items, but this should only be done with the agreement of all team members. The final document may be referred to as the FAT Preparation Document.

H.4 GENERAL PROCEDURE

The general testing procedure should outline the details to be addressed in the test. These should include the following considerations, as a minimum:

- A statement of the location and dates of the FAT.
- A description of the general approach.
- A description of the format of the FAT punch-list.
- Specification of the revision levels of the hardware and software to be tested.
- Specification of the exact configuration of equipment being tested.
- Personnel safety issues that may apply during the test.

Documentation of procedures and results is key to accomplishing the test's intended purpose. One method of ensuring that all findings are documented is to maintain a punch-list for logging any errors found during the test. Each item on the punch-list should be numbered serially, dated when found, described in an understandable format, dated when corrected, and dated when rechecked.

H.5 RESPONSIBILITIES

The general assignment of responsibilities should depend on the application. Table H.1 indicates several scenarios of distribution. The FAT Preparation Document should identify the general responsibility assignments. In all instances, the user should approve the test(s).

Table H.1 Possible Testing Scenarios					
Case No.	Hardware Integration*	Software Programming/ Configuration*	Responsible for Conducting Test	Hardware Corrections	Software Correction
1	S	S	S	S	S
2	S	S	U	S	S
3	S	S	S/U	S	S
4	S	U	S/U	S	U
5	S	U	U	S	U
6	U	U	U	U	U

S = responsibility assigned vendor   U = responsibility assigned user
* = determined system specification

Case 1: The vendor is responsible for integrating the hardware and programming (or configuring) the software. The test is run by the vendor, and problems are corrected by the vendor. The user team members witness the test.

Case 2: Responsibilities are the same as Case 1 except the user team members conduct the test, and the vendor corrects any problems found.

Case 3: Responsibilities are the same as Case 2 except that the vendor team members assist the customer in conducting the test.

Case 4: The user is responsible for the software program/configuration. The test is conducted by both user and vendor. The user is responsible for correcting problems found in software programs/configurations.

Case 5: The user is responsible for software programming/configuration and conducting the test. The vendor is responsible for correcting any hardware problems.

Case 6: This is an example of a BPCS/SIS for which the user purchased off-the-shelf hardware, integrated the system in-house, and performed the programming/configuration. The user is responsible for conducting the test and resolving all problems encountered during the test.

If possible, the FAT team should be separated into two groups: one group conducting the test; the second group correcting the problems identified by the first. The benefit of this method is that the FAT may continue nearly uninterrupted for one shift while problems are corrected on the following shift. This will require some overlapping of the hours of work of the two teams to ensure information is transferred in a timely and accurate manner.

Each individual team member should be assigned specific tasks during the FAT. The FAT Preparation Document should define these task assignments as clearly as possible.

H.6 FAT SCHEDULE

The FAT schedule should show a daily list of activities, identifying each item in the system to be checked on a particular day. The schedule should include enough time to review the punch-list.

The schedule should allow enough slack time to permit one day each week, usually on the weekend, when testing is discontinued. This day should be used to reduce the punch-list to a manageable size. It is the responsibility of the FAT team leader to utilize the slack time for such purpose.

H.7 DOCUMENTS

The purchase order for the system should list the documents required for the test. An example of such a list follows:

- System Functional Requirements Specification
- System I/O List
- Process Control Description
- Control Logic Specification Sheets
- Configuration Worksheets
- Logic Flow Diagrams
- Instrument Specifications
- Process Flow Sheets
- Graphic Design Drawings
- Program Listings (Documented)
- System Self-documentation Printouts
- System Arrangement Drawings
- Termination Lists
- Vendor Manuals
- Punch-list Form

Responsibility of supply should be assigned to either the user or the vendor for each document on the list. The FAT should not only check the functionality of the system, but should also check the accuracy of the documentation. At the end of the FAT, all documentation should accurately describe the system. Any exceptions should be included on the punch-list.

H.8 TEST EQUIPMENT

The purchase order should list the test equipment required to perform the FAT. The list should be submitted separately to the vendor for confirmation.

Both parties should agree upon responsibility of supply for each item on the test equipment list.

The test equipment list should include all items required to perform 100% of the test. The list should also include equipment required for troubleshooting. The user team leader should verify the availability of required tools such as screwdrivers, pliers, wrenches, special tools, jumpers, soldering iron, etc.

An example of a typical test equipment list follows:

Device	Usage
Digital Multi-meter	Monitor system outputs and troubleshoot wiring
DC mA Source	Input current signals (instrument simulation)
Two-wire Transmitter Simulator (instrument simulation)	Input current signals
DC mV Source	Thermocouple or other low-voltage input simulation
Pulse Generator	Input pulse signals
Pulse Counter	Monitor pulse signals
Resistance Box	RTD simulation
Variable DC Voltage Source (instrument simulation)	Input voltage signals
Breakout Box	Troubleshoot data communication links
Discrete Device Simulation Panel (with discrete switch devices, lights, and power supply if needed)	Test sequence logic involving (valve, switches, and contactI/O simulation)

H.9 DETAILED TEST PROCEDURE

The detailed test procedure should ensure that all aspects of the system are checked against system documentation. It should include, at a minimum, the following:

- A description of a typical loop test for each type of I/O in the system using the proper test equipment. Inputs should be simulated at 0%, 50%, and 100% signal input. Outputs should be monitored at 0%, 50%, and 100% of output level.
- A description of a typical method for testing each classification of software programs in the system (i.e., interlocks, special calculation blocks, batch control programs, etc.).
- A description of what checks are to be made on graphic displays. These should include graphic layout, color specifications, text, touch areas, paging functions, point addressing, and any system-specific items.
- A description of the method to ensure proper point (DCS tag) configuration. Included in this check should be point range, signal conditioning, alarm settings, and point type.
- Provision for a method for checking all other aspects of the system (i.e., visual checks, trends, logs, system failures, etc.).

H.10 EQUIPMENT ACCEPTANCE

The equipment may be released for shipping at the conclusion of the FAT. An official document should be signed by both the user team leader and by the vendor.

The acceptance document should state whether open items on the punchlist are to be resolved in the field or prior to shipping.

The user should have the right to return to the factory to back-check any items agreed to be resolved at the factory. For convenience, the user may choose to waive this right.

H.11 FAT PREPARATION DOCUMENT

The FAT Preparation Document should include the items described in Sections H.4 through H.9 and should be submitted for review and acceptance to all team members prior to the issuance of the purchase order for the system. This document should then become a part of the purchase order.

INDEX

409